ESSAYS IN THEORETICAL PHYSICS

IN HONOUR OF DIRK TER HAAR

Related Pergamon Titles of Interest

Books

GINZBURG
Physics and Astrophysics (A Selection of Key Problems)

LANDAU & LIFSHITZ
Course of Theoretical Physics (in 10 volumes)

TER HAAR
Elements of Hamiltonian Mechanics, 2nd edition
Collected Papers of P. L. Kapitza, volumes 1 to 4*
Lectures on Selected Topics in Statistical Mechanics
Men of Physics: L D Landau, Volume 2
Problems in Undergraduate Physics

Volume 1: Mechanics
Volume 2: Electricity and Magnetism

Journals

Vistas in Astronomy

Full details of all Pergamon publications/free specimen copy of any Pergamon journal available on request from your nearest Pergamon office.

*Volumes 1 to 3 available through MIMC, a subsidiary of Pergamon Press Inc. Volume 4 is in preparation.

DIRK TER HAAR

ESSAYS IN THEORETICAL PHYSICS

IN HONOUR OF DIRK TER HAAR

Edited by

W. E. PARRY

Fellow of Oriel College, Oxford

PERGAMON PRESS

OXFORD · NEW YORK · TORONTO · SYDNEY · PARIS · FRANKFURT

U.K.	Pergamon Press Ltd., Headington Hill Hall, Oxford OX3 0BW, England
U.S.A.	Pergamon Press Inc., Maxwell House, Fairview Park, Elmsford, New York 10523, U.S.A.
CANADA	Pergamon Press Canada Ltd., Suite 104, 150 Consumers Road, Willowdale, Ontario M2J 1P9, Canada
AUSTRALIA	Pergamon Press (Aust.) Pty. Ltd., P.O. Box 544, Potts Point, N.S.W. 2011, Australia
FRANCE	Pergamon Press SARL, 24 rue des Ecoles, 75240 Paris, Cedex 05, France
FEDERAL REPUBLIC OF GERMANY	Pergamon Press GmbH, Hammerweg 6, D-6242 Kronberg-Taunus, Federal Republic of Germany

First edition 1984

British Library Cataloguing in Publication Data

Essays in theoretical physics.
1. Mathematical physics — Addresses, essays, lectures
I. Parry, W. E. II. Haar, Dirk ter
530.1'5 QC20.5

Library of Congress Cataloging in Publication Data

Main entry under title:
Essays in theoretical physics.
Includes index.
1. Statistical mechanics—Addresses, essays, lectures.
2. Astrophysics—Addresses, essays, lectures.
3. Haar, D. ter. I. Parry, W. E. (William Edward)
QC174.82.E87 1984 530.1'3 83-22006
ISBN 0-08-026523-5

Printed in Great Britain by A. Wheaton & Co. Ltd., Exeter

Dedication

Dirk ter Haar was born on 22nd April, 1919. He was a student of Leiden University. He has been a research fellow at Bohr's Institute in Copenhagen, a professor at Purdue, and a lecturer at St. Andrew's. Since 1956 he has been at Oxford where he is a fellow of Magdalen College and University Reader in Theoretical Physics. He has been an editor of Physics Letters since 1952.

He has published many papers, chiefly in astrophysics and the many-body problem, and has written books on Statistical Mechanics and Hamiltonian Mechanics. He has edited the collected papers of Landau and Kapitza, and is the general editor of Pergamon's Natural Philosophy Series.

In 1947 he married Christine Janet Lound; they have two sons and one daughter.

The remarkable breadth of Dirk's interest in physics is to some extent reflected in this volume, which is a product of the affection of his pupils and colleagues, and of the high regard in which they hold his own research, his inspiration to others in their research, and his work for the scientific community.

Contents

List of Contributors

J. BINNEY	Oxford University, United Kingdom, and Institute for Advanced Study, Princeton, USA.
B.G.S. DOMAN	Liverpool University, United Kingdom.
T. FUJIWARA	Dept. of Mathematical Science, U.O.P. 591 Sakai, Japan.
V.L. GINZBURG	P.N. Lebedev Physical Institute of USSR Academy of Sciences, USSR.
A.C. HEWSON	Imperial College of Science and Technology, London, United Kingdom.
F.K. LAMB	University of Illinois, at Urbana-Champaign, USA .
A.J. LEGGETT	Cornell University, USA.
C.A.S. LIMA	Universidade Estadual de Campinas, Brazil.
M.E. LINES	Bell Laboratories, Murray Hill, N.J. USA.
L.C.M. MIRANDA	Centro Tecnico Aerospacial, Brazil.
W.E. PARRY (Editor)	Oriel College, Oxford, United Kingdom.
C.J. PETHICK	NORDITA, Copenhagen, Denmark and University of Illinois at Urbana-Champaign, USA.
A.J.R. PRENTICE	Monash University, Australia.
S. RAMADURAI	Tata Institute of Fundamental Research, Bombay, India.
S. STENHOLM	University of Helsinki, Finland.
R.A. TAHIR-KHELI	Temple University, Philadelphia, USA.
N.G. VAN KAMPEN	University of Utrecht, The Netherlands.
H. WERGELAND	Hurdal, Norway.

Stellar Dynamics with Non-classical Integrals

James Binney
Institute for Advanced Study, Princeton N.J.
and
Department of Theoretical Physics, Keble Road, Oxford OX1 3NP

ABSTRACT

Effective integrals of motion that do not derive from a manifest Lagrangian symmetry appear to play a vital role in structuring galaxies. Recent work suggests that these effective integrals may sometimes by understood in terms of potentials that lead to separable Hamilton-Jacobi equations. However, studies of the action integrals of numerically integrated orbits in given rotating potentials suggest that many effective integrals of astronomical interest cannot be understood in this way. A hybrid approach to stellar dynamics in which the angle-action representations of orbits in model potentials are obtained by numerical integration and are then analyzed by the techniques of Hamiltonian perturbation theory may lead to a more satisfactory understanding of the relationship between potentials and the orbits that they support.

1. Introduction

From Newton through Laplace, Gauss and Jacobi to Poincaré it may be safely asserted that dynamics was driven by astronomy. In recent years dynamical problems have attracted the attention of scientists from a wider range of backgrounds and astrophysicists can no longer claim to be in undisputed possession of the frontier of dynamical theory. However, several important branches of astrophysics still turn on problems in dynamics and the interest of astrophysicists in developing this theory is as keen as ever. In this article I review some recent work in one such branch of astrophysics — the dynamics of large groupings of point masses, or 'stellar dynamics' as it is usually rather confusingly called. My aim in this article has been to tell physicists what astrophysicists have been up to recently, and to describe two new developments which promise to bring mathematical concepts that were developed by Jacobi and others in connection with the dynamics of the solar system, into contact with the empirical-numerical approach to stellar dynamics that has grown up since electronic computers became widely available. I have tried to make the mathematical development in this article as self-contained as possible, but I have assumed a familiarity with the basic ideas of Hamiltonian mechanics at the level of Berry's excellent review article (Berry, 1978).

This article is organized as follows: § 2 introduces the mean-field approximation that underlies all work in stellar dynamics. §3 provides a very brief summary of the classical 'third integral' problem in galactic dynamics. §4 reviews recent work on the nature of orbits in triaxial gravitational potentials, with particular emphasis on the classification of orbits into families and the natural explanation which these families find in the context of certain separable potentials. §5 introduces the spectral approach to stellar dynamics and describes its application to studies of the action integrals of orbits that have been obtained by numerical integration of the equations of motion. These studies throw an interesting light on how widely separable potentials may be applicable to problems in stellar dynamics. §6 sums up and indicates some possible directions of future work.

2. The Mean-Field Approximation

The central idea of all stellar dynamics is the use of a mean-field approximation for the gravitational field in which individual stars move: we calculate the orbits of individual stars as they move through the potential of an idealized model of the system under study. In this model mass is smoothly distributed rather than concentrated into individual stars. Clearly an orbit that is calculated using this model potential will not be exactly the same as the true orbit of the star. The relaxation time $t_{\mathbf{rel}}$ is a measure of the length of time during which orbits in the model potential remain a reasonable approximation to the true orbits. Orbits in our model potential are entirely untrustworthy after a time $t \geqslant t_{\mathbf{rel}}$ and are a good approximation to the truth for $t \ll t_{\mathbf{rel}}$.

The true orbit deviates from the model orbit for two main reasons:

1) Stars sometimes pass close to each other and experience short-lived bursts of relatively intense acceleration toward one another. No such bursts

of acceleration occur along a model orbit.

2) Since there are only a finite number of stars in even a large volume of the stellar system, the number of stars in such a volume will fluctuate more-or-less randomly about the mean value on which the mean-field model is based. Large-scale density fluctuations of this nature generate large-scale fluctuations in the gravitational potential of the system and these deflect the true orbit from the path calculated within the context of the mean-field model.

These processes that cause true orbits to deviate from model orbits are generally referred to as 'encounters'. Process (1) above is that of 'close encounters', while process (2) is rather misleadingly described as that of 'distant encounters'. Since Jeans' pioneering investigations (Jeans, 1913) many authors (Chandrasekhar 1941, Cohen, Spitzer and Routly 1950) have calculated the relaxation time $t_{\mathbf{rel}}$ to which encounters lead. The relative importance of close and distant encounters still generates a surprising amount of controversy. Fortunately the order of magnitude of $t_{\mathbf{rel}}$ is not controversial and one may safely adopt Spitzer's (Spitzer and Hart 1958) estimate

$$t_{\mathbf{rel}} = \frac{(2/3)^{3/2}\, v^3}{2\pi G^2 \rho m \log_e (0.4N)} \,. \qquad (1)$$

Here G is Newton's gravitational constant, ρ is the mass density of (identical) stars of mass m, and v is the rms speed of the N stars in the system. If the system has a characteristic radius r, one has $v^2 \simeq GNm/r \simeq 4\pi G\rho r^2/3$ by the virial theorem, so

$$t_{\mathbf{rel}} = \frac{(2/3)^{5/2}\, rN}{\log_e (0.4\,N)\, v} \simeq \frac{0.36\,N}{\log_e (N)}\, t_{\mathbf{cross}} \qquad (2)$$

where $t_{\mathbf{cross}} \equiv r/v$ is the characteristic dynamical or 'crossing' time of the system.

In some smaller stellar systems, for example globular clusters (which have $N \simeq 10^5$ and $t_{\mathbf{cross}} \simeq 10^6\, y$), $t_{\mathbf{rel}}$ is less than the age of the Universe, $t_{\mathbf{Hubble}} \simeq 10^{10}\, y$, and one cannot ignore the effects of encounters. However, whenever $N \geqslant 100$, equation (2) yields $t_{\mathbf{rel}} \gg t_{\mathbf{cross}}$, so the mean-field approximation gives a description of the dynamics of the system that is valid over many dynamical times. Thus it is expedient to treat encounters among cluster stars as perturbations within a mean-field model; these perturbations cause the mean-field model to evolve on the long time-scale $t_{\mathbf{rel}}$.

A typical giant galaxy like our own has $N \geqslant 10^{11}$ and $t_{\mathbf{cross}} \geqslant 10^8\, y$, so $t_{\mathbf{rel}} \simeq 10^{18}\, y \gg t_{\mathbf{Hubble}}$. Hence the mean-field approximation should provide an entirely sound basis for understanding the structure of galaxies.

The first step towards constructing a self-consistent mean field model of a stellar system is to understand the structure of orbits in the mean-field potential. Once these orbits have been conveniently catalogued, we can try to populate them in such a way that the net density due to all the stars in the system is at each point equal to the known density of the system.

It should be emphasized that we are interested in the structure of orbits over only a few hundred to some thousands of characteristic dynamical times. This greatly simplifies the task of cataloguing stellar orbits as it implies that we do not have to worry about subtle effects that might become apparent if we were to integrate the equations of motion for millions or even thousands of millions of orbital times. This situation contrasts with that obtaining in celestial mechanics and plasma physics. In those fields one does have to worry about the long run. In stellar dynamics we can safely argue that in the long run we are all dead.

3. Orbits in Axisymmetric Potentials

What sort of potentials do galaxies have ? While the detailed distribution of gravitating matter in even our own Galaxy is still highly controvertial. it is certain that few if any galaxies have spherically symmetric potentials. Therefore the highest symmetry that we need consider is that of axisymmetry. Studies of motion in flattened axisymmetric potentials became popular in the mid 1960's, when electronic computers became widely available, and we now have a pretty clear picture of the types of orbits that these potentials support.

If (R,ϕ,z) is a system of suitably oriented cylindrical polar coordinates, motion in an axisymmetric potential, $\Phi(R,z)$ can be reduced to the two-dimensional problem of motion in the 'comoving meridional plane' $(R\ ,\ z)$ by replacing $\dot{\phi}$ in the R-equation of motion by $\dot{\phi}=L_z/R^2$ where L_z is the star's component of angular momentum about the symmetry axis of the potential. The equations governing the two-dimensional motion are

$$\ddot{R}=-\partial\Phi_{\mathbf{eff}}/\partial R\ ,\quad \ddot{z}=-\partial\Phi_{\mathbf{eff}}/\partial z$$

where

$$\Phi_{\mathbf{eff}}=\Phi+L_z/2R^2$$

is the sum of the gravitational and 'centrifugal' energies. $\Phi_{\mathbf{eff}}$ has a minimum at $z=0$ and $R=R_c$ where R_c is the radius of the circular orbit with angular momentum L_z. Expanding $\Phi_{\mathbf{eff}}$ in a Taylor series

$$\Phi_{\mathbf{eff}}(R,z)=\Phi_0+\tfrac{1}{2}\kappa^2(R-R_c)^2+\tfrac{1}{2}\omega_z^2\,z^2+\ \ldots$$

about this minimum, one can regard the star's motion in the (R,z) plane as that of an harmonic oscillator that is perturbed by the cubic and higher-order terms in the excursions $R-R_c$ and z from the point of equilibrium.

If the deviation of $\Phi_{\mathbf{eff}}$ from harmonicity is not very great, we would expect the motion to admit two integrals

$$E_{\mathbf{R}} \equiv \frac{1}{2}(\dot{R}^2 + \kappa^2 R^2) + \ldots \qquad (3a)$$

$$E_{\mathbf{z}} = \frac{1}{2}(\dot{z}^2 + \omega_{\mathbf{z}}^2 z^2) + \ldots \qquad (3b)$$

that are in lowest order κ times the radial action for motion in the plane $z = 0$, and the 'z-energy' that Oort and Lindblad had already used in the 1930's to interpret observations of the motions through our Galaxy of stars that pass near the Sun. It is often convenient to regard $E_{\mathbf{R}}$ as dependent on $E_{\mathbf{z}}$, $L_{\mathbf{z}}$ and the total energy E; $E_{\mathbf{R}} = E - E_{\mathbf{z}} - \Phi_0$. $E_{\mathbf{z}}$ is then called the 'third integral'.

In 1960 Contopoulos showed that one can use standard Hamiltonian perturbation theory to obtain higher-order terms in the expression (3b) for $E_{\mathbf{z}}$ (Contopoulos 1960, 1963). As Henon and Heiles (1964) subsequently demonstrated by writing down a special polynomial approximation to $\Phi_{\mathbf{eff}}$, this approach to the problem of motion in an axisymmetric potential can be misleading; the Henon and Heiles potential exhibits a wealth of stochastic orbits which are not constrained by a third integral. However, numerous numerical studies (Ollongren 1965, Martinet and Mayer 1975) have shown that most orbits in realistic axisymmetric potentials are in fact constrained by third integrals of the type discussed by Contopoulos.

4. Orbits in Triaxial Potentials

In recent years evidence has been accumulating that the majority of galaxies do not have axisymmetric potentials, but may have potentials that are of more-or-less ellipsoidal form [see Binney (1982) for a review]. Consequently it is important to understand the sorts of orbit that are supported by 'triaxial' potentials of this type.

The figure of a triaxial galaxy is likely to rotate with respect to inertial space, so it is necessary to allow the possibility that the frame of reference in which the potential is stationary (assuming such a frame exists) rotates with some angular speed ω. However, it is helpful to consider first the important special case $\omega = 0$.

4.1 *Schwarzschild's Potential*

Schwarzschild (1979) presented the first systematic investigation of orbits in the potential of a body that might resemble a triaxial elliptical galaxy. The particular potential studied by Schwarzschild is that generated by a body whose isodensity surfaces are roughly elliptical with axis lengths in the ratio 1:0.625:0.5. If we orient the body with its longest axis along the x_1 axis, and its shortest axis along the x_3 axis, Schwarzschild's distribution is

$$\rho(x) = \rho_0\left[1 + \sum_{i=1}^{3}\left(x_i/a_i\right)^2\right]^{-3/2}. \tag{4}$$

If light were emitted from each volume of this body in proportion to the mass in that volume, then in projection the body would closely resemble an elliptical galaxy.

Over the timescales of astronomical interest (a few hundred orbital times) Schwarzschild found by direct numerical integration of the equations of motion that the great majority of orbits are confined to three-dimensional regions of phase space. This result suggests that in addition to the classical Hamiltonian integral $H(\boldsymbol{x},\boldsymbol{p})$, two further isolating integrals (or possibly quasi-integrals) exist at most points in phase space. Since these additional integrals do not arise via Noether's theorem from a manifest symmetry of the Lagrangian for motion in the given potential, Schwarzschild called them 'non-classical integrals'.

Superficially Schwarzschild's demonstration that most orbits in a particular ellipsoidal potential admit three effective isolating integrals may appear no more than a minor extension of the earlier work on motion in axysymmetric potentials. However, if we exclude a small and physically unreal region near the centre of Schwarzschild's model, the potential of the model cannot be regarded as that of a perturbed harmonic oscillator. The potential generated by the spherical density distribution $\rho = \rho_0\,[1 + (r/a)^2]^{-3/2}$ is

$$\Phi(r) = -4\pi G\rho_0\, a^3\, \frac{1}{r}\, \log_e\left[\frac{r}{a} + \left(\frac{r^2}{a^2} + 1\right)^{\frac{1}{2}}\right] \tag{5}$$

and if we expand this in powers of (r/a) around the point of equilibrium at $r = 0$, we obtain a series which diverges beyond $r = a$. Many of Schwarzschild's orbits reach $r = 20a$. Thus Schwarzschild's result cannot be easily understood in terms of perturbations of a three-dimensional harmonic oscillator.

Of course orbits at very large binding energies do not move to $r \gg a$, and one can attempt to understand these orbits in terms of the motions of a perturbed harmonic oscillator. De Zeeuw and Merritt (1983) have taken this approach with a certain measure of success, but recently de Zeeuw (1983) and de Zeeuw and Lynden-Bell (1983) have shown that there is a much better way of understanding Schwarzschild's numerical results in analytic terms. The approach taken by de Zeeuw and Lynden-Bell is to go back to the classical work of Stäckel (1893) on separable potentials. This development is sufficiently interesting to merit a digression into the nature of motion in Stäckel's potentials.

4.2 *Stäckel Potentials*

Stäckel's potentials are most conveniently expressed in terms of ellipsoidal coordinates. An ellipsoidal coordinate system is defined by choosing three numbers $\alpha_1 < \alpha_2 < \alpha_3$: the ellipsoidal coordinates of the point $\boldsymbol{x}$ are then the roots $q_1(\boldsymbol{x}) > q_2(\boldsymbol{x}) > q_3(\boldsymbol{x})$ of the cubic equation for q

$$\sum_{i=1}^{3} \frac{x_i^2}{q+\alpha_i} = 1 \quad (\alpha_1 < \alpha_2 < \alpha_3 \text{ arbitrary constants}). \tag{6a}$$

The ranges of variation of the q_i are given by

$$-\alpha_3 \leqslant q_3 \leqslant -\alpha_2 \leqslant q_2 \leqslant -\alpha_1 \leqslant q_1. \tag{6b}$$

The coordinate system provided by the q_i proves to be orthogonal. That is

$$\sum_{i=1}^{3} dx_i^2 = ds^2 = \sum_{i=1}^{3} g_i\, dq_i^2 \tag{7a}$$

where $g_i(q_1, q_2, q_3)$. A straightforward but tedius calculation shows that

$$1/g_i = -\frac{2}{\Delta} \sum_{j,k=1}^{3} \varepsilon_{ijk}(q_j - q_k) \prod_{\ell=1}^{3} (q_i + \alpha_\ell). \tag{7b}$$

where

$$\Delta = \tfrac{1}{6} \sum_{ijk} \varepsilon_{ijk}(q_i - q_j)(q_j - q_k)(q_k - q_i) = -\tfrac{1}{2} \sum_{ijk} \varepsilon_{ijk}\, q_i^2 (q_j - q_k). \tag{7c}$$

Stäckel considered potentials of the form

$$\Phi(\boldsymbol{x}) = \frac{V_1(q_1)}{g_1} + \frac{V_2(q_2)}{g_2} + \frac{V_3(q_3)}{g_3} \tag{8}$$

where the V_i are three arbitrary functions of one variable. All such potentials lead to a separable Hamilton-Jacobi equation as the following analysis shows.

If S is Hamilton's principal function, the Hamiltonian for motion in a potential Φ may be written

$$\begin{aligned} E &= \frac{1}{2m} (\nabla S \cdot \nabla S) + \Phi \\ &= \frac{1}{2m} \sum_{ij} g^{ij} \frac{\partial S}{\partial q_i} \frac{\partial S}{\partial g_j} + \Phi = \frac{1}{2m} \sum_{i} \frac{1}{g_i} \left(\frac{\partial S}{\partial q_i}\right)^2 + \Phi \end{aligned} \tag{9}$$

where $g^{ij} = (1/g_i)\,\delta_{ij}$ is the inverse of the covariant metric tensor $g_{ij} = g_i\,\delta_{ij}$. With (8) the Hamilton-Jacobi equation becomes

$$\sum_i \frac{1}{g_i} \left[\frac{1}{2m} \left(\frac{\partial S}{\partial q_i} \right)^2 + V_i(q_i) \right] = E\,. \tag{10}$$

When one substitutes from equations (7) for the g_i, this becomes

$$\sum_{ijk} \varepsilon_{ijk}(q_i - q_k)\left\{q_i^2 E - 4\prod_\ell (q_i + \alpha_\ell)\left[\frac{1}{2m}\left(\frac{\partial S}{\partial q_i}\right)^2 + V_i\right]\right\} = 0 . \tag{11}$$

If S is of the form $S = \Sigma S_i(q_i)$, we obtain a solution to this equation with the required two arbitrary constants A and B in addition to E, by requiring that the S_i satisfy the three separated equations

$$q_i^2 E - 4\prod_\ell (q_i + \alpha_\ell)\left[\frac{1}{2m}\left(\frac{\partial S}{\partial q_i}\right)^2 + V_i\right] = -(Aq_i + B). \tag{12}$$

Since the momentum p_i conjugate to q_i is $p_i = \partial S/\partial q_i = \partial S_i/\partial q_i$, we have finally

$$\frac{1}{2m} p_i^2 = \frac{Eq_i^2 + Aq_i + B - V_i^*(q_i)}{4\prod_\ell (q_i + \alpha_\ell)} \tag{13}$$

where V_i^* is defined by

$$V_i^* = 4\prod_\ell (q_i + \alpha_\ell) V_i(q_i). \tag{14a}$$

This last definition implies that

$$\Phi = \frac{-1}{2\Delta}\sum_{ijk} \varepsilon_{ijk}(q_j - q_k)\, V_i^*(q_i). \tag{14b}$$

As is well known (e.g. Berry 1978), all orbits in an integrable potential are confined to tori in phase space. In the present case, each triplet (E, A, B) defines such a torus through the three restrictions (13) on the six phase-space coordinates $(\boldsymbol{q}, \boldsymbol{p})$. In general the phase point associated with a given orbit winds around its torus in such a way as to pass close to every point of the torus. Consequently if we wish to know whether the orbit passes close to any given point $\boldsymbol{q}(0)$ of coordinate space, we have to ask only whether the space of all momentum vectors $(\boldsymbol{q}(0), \boldsymbol{p})$ associated with $\boldsymbol{q}(0)$ has any point in common with the torus. In other words, we have to ask whether the equations (13) all have real roots for the $p_i[q_i(0)]$ $(i = 1, 3)$. Correspondingly, the edge of the region in coordinate space that the orbit visits coincides with the points q_i (edge) at which $p_i[q_i(\text{edge})] = 0$ and thus the roots cease to be real. Hence all orbits in Stäckel's separable potentials lie between the coordinate surfaces on which

$$Eq_i^2 + Aq_i + B - V_i^*(q_i) = 0 . \tag{15}$$

In the light of this result, it is clearly important to investigate the shape of the surfaces $q_i = \text{constant}$.

Figure 1 shows some surfaces q_i = constant. The important features of these surfaces may be summarized as follows:

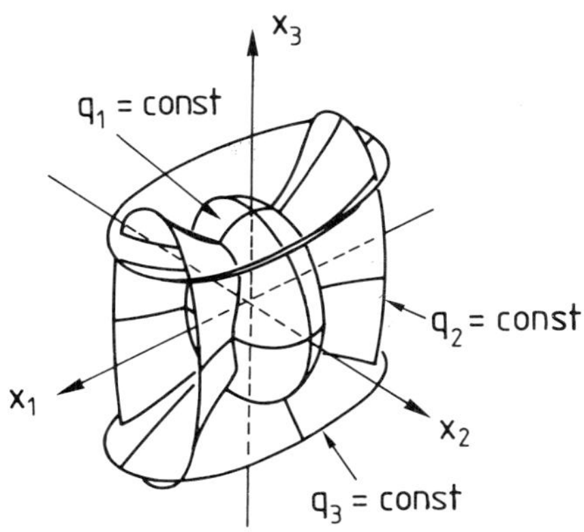

Fig. 1 Surfaces of constant q_1 , q_2 **and** q_3.

1) Surfaces on which constant = $q_1 > -\alpha_1$ are ellipsoids. When $q_1 \gg -\alpha_1$ these ellipsoids tend to spheres of radius $q_1^{\frac{1}{2}}$, and as $q_1 \to -\alpha_1$ they degenerate into an elliptical disk in the (x_2, x_3) plane.

2) Surfaces on which constant = $q_2 > -\alpha_2$ are flaring elliptical tubes around the x_1 axis. When $q_2 \to -\alpha_1$ these tubes open out to cover the portion of the (x_2, x_3) plane that lies outside the elliptical disk $q_1 = -\alpha_1$, and when $q_2 = -\alpha_2$ they degenerate into the portion of the (x_1, x_3) plane around the x_1 axis that is bounded by the two branches of the hyperbola

$$x_1^2/(\alpha_2 - \alpha_1) - x_3^2/(\alpha_3 - \alpha_1) = 1 .$$

3) The equations constant = $q_3 > -\alpha_3$ define a hyperboloids of two sheets. These hyperboloids surround the x_3 axis; as $q_3 \to -\alpha_2$ they degenerate into the parts of the (x_1, x_3) plane that lie outside the surface $q_2 = -\alpha_2$. When $q_3 \to -\alpha_3$ they coincide with the whole (x_1, x_2) plane.

If one were able to choose the functions V_i^* freely, one could arrange that for some values (E, A, B) of the constants, the equations $p_i(q_i) = 0$ had roots at any desired points $q_i(\min)$ and $q_i(\max)$, and hence construct a potential in which there is an orbit which is confined between any given set of coordinate surfaces. However, if the potential (14b) is to resemble the potential of a galaxy, the V_i^* cannot be chosen independently of one another; if the underlying density distribution is to be smooth and non-singular on the curves $q_1 = -\alpha_1 = q_2$ and $q_2 = -\alpha_2 = q_3$, the V_i^* and their first three derivatives must coincide there. The most straightforward way of ensuring satisfaction of these conditions is to set $V_i^*(q) = V_j^*(q) \equiv V^*(q)$ for all i, j and q. Furthermore, the general form of V^* is restricted by the requirement that far from the centre the potential should tend to zero roughly as $\Phi \sim -1/r \sim -q_1^{-\frac{1}{2}}$. These restrictions on the V_i^* limit the sets of coordinate surfaces between which orbits can be confined.

De Zeeuw (1983) has shown that the potential generated by the mass distribution

$$\rho(x) = \rho_0 \left[1 + \sum_i \left(\frac{x_i}{a_i}\right)^2\right]^{-2} \qquad (a_1 > a_2 > a_3) \tag{16}$$

is of Stäckel's form with $\alpha_1 = 0$, $\alpha_2 = a_1^2 - a_2^2$, $\alpha_3 = a_1^2 - a_3^2$ and

$$V^*(q) = -2\pi G \rho_0 a_1 a_2 a_3 \alpha_3^{3/2} \left[f(t) - t^2 f(1)\right] \tag{17a}$$

where

$$t^2(q) \equiv \frac{q}{\alpha_3} + 1$$

$$f(t) \equiv t^4 \int_0^{(\alpha_3/a_3^2)^{\frac{1}{2}}} \left\{(1+v^2)\left[1 + \left(1 - \left(\frac{\alpha_2}{\alpha_3}\right)\right) v^2\right]\right\}^{\frac{1}{2}} \frac{dv}{1 + t^2 v^2}. \tag{17b}$$

De Zeeuw shows that only three types of bound orbit occur in this potential:

i) *Short-axial loops*. q_1 oscillates between $q_1(\min) > -\alpha_1$ and $q_1(\max)$, while q_2 circulates through its full range $(-\alpha_2 \leqslant q_2 \leqslant -\alpha_1)$ and q_3 oscillates in the restricted range $-\alpha_3 \leqslant q_3 \leqslant q_3(\max)$. Stars of these orbits rotate in a fixed sense around the x_3 axis.

ii) *Box orbits*. q_1 oscillates in the range $-\alpha_1 \leqslant q_1 \leqslant q_1(\max)$ while both q_2 and q_3 oscillate in the restricted ranges $-\alpha_3 \leqslant q_3 \leqslant q_3(\max)$ and $-\alpha_2 \leqslant q_2 \leqslant q_2(\max)$ and $-\alpha_2 \leqslant q_2 \leqslant q_2(\max)$. Stars on these orbits have no fixed sense of rotation about any axis. They lie within boxes whose six curved sides fall into three pairs of opposite sides on which $q_i = q_i(\max)$ $(i = 1, 3)$.

iii) *Long-axial loops*. q_1 oscillates in the range $-q_1(\min) \leqslant q_1 \leqslant q_1(\max)$ while q_2 oscillates in the range $q_2(\min) \leqslant q_2 \leqslant q_2(\max)$ and q_3 circulates through its full range $-\alpha_3 \leqslant q_3 \leqslant -\alpha_2$. Stars on these orbits rotate in a fixed sense about the x_1 axis, while oscillating in distance from and along this axis.

Figure 2 shows how these different possibilities arise when the constant B in equation (13) is varied. At sufficiently negative values of E, no curves $Eq^2 + Aq - V^*(q)$ have the form shown in Figure 2, and all orbits are box orbits.

The density distribution (16) whose Stäckel potential admits these orbits differs in some important respects from that (5) studied by Schwarzschild, but the classification of orbits in de Zeeuw's Stäckel potential is exactly that

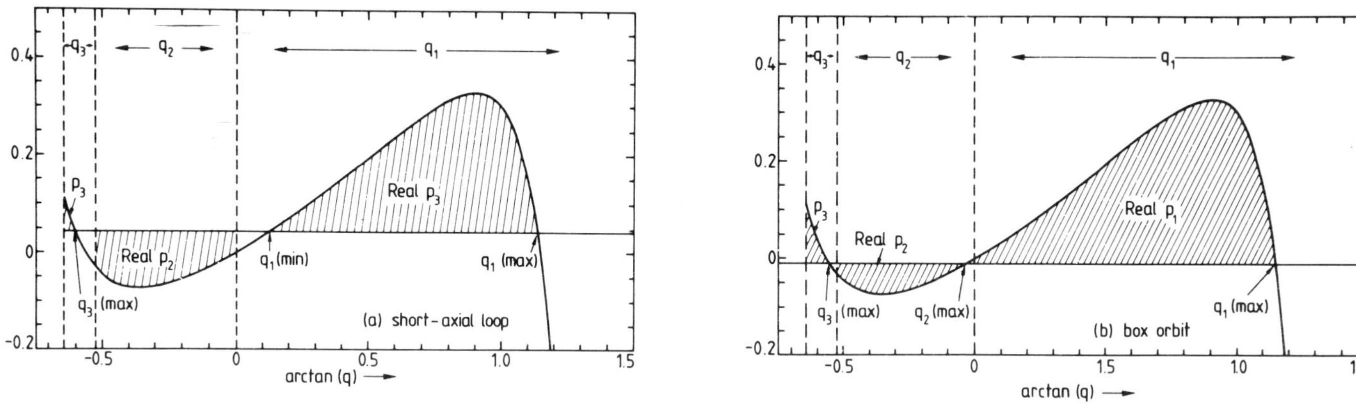

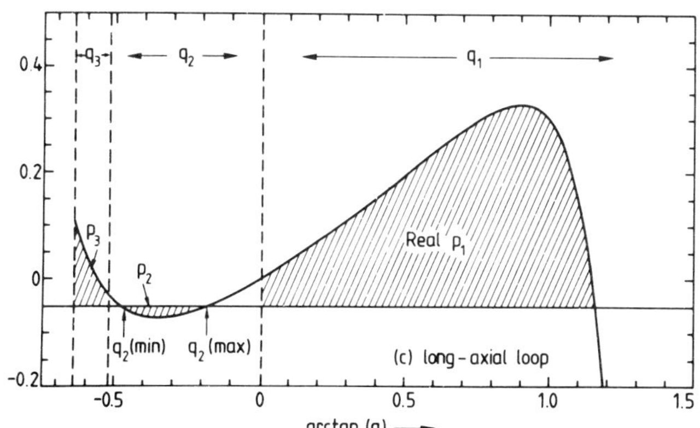

Fig. 2 The curves in each panel are plots of $EQ^2 + Aq - V^*(q)$ for $E = 0.46\,\Phi(0)$ and $A = 0.9\,E$, where $V^*(q)$ is given by (17a) for $a_1 : a_2 : a_3 = 1 : 0.625 : 0.5$. Vertical lines mark the allowed ranges of the q_i. The horizontal lines correspond to possible values of B. In (a) a negative value of B generates a short-axial loop orbit. In (b) a larger value of B generates a box orbit, and in (c) a long-axial loop orbit is generated by a still smaller value of B. The portions of the diagrams that correspond to real p_1 are shaded.

which Schwarzschild derives from his numerical work. This strongly suggests that the potentials of non-rotating triaxial galaxies are close enough to Stäckel potentials to be understood as perturbed Stäckel potentials. We shall see at the end of the next section, however, that the situation is much less favourable if the figure of the potential rotates with a finite angular velocity $\omega \neq 0$.

5. Action Integrals

Much of our current understanding of galactic dynamics derives from n-body models. In these numerical models one allows from a few hundred to some tens of thousands of mass points to move in an approximation to the mean gravitational field that would be generated by a larger ensemble of mass points.

N-body models have enabled us to study the way stellar systems that are initially far from virial equilibrium relax towards quasi-steady states. The models have shown that after a few dynamical times most systems reach approximate virial equilibrium and that the rate of subsequent evolution is slow. In the process of attaining virial equilibrium, the system appears to forget the details of the configuration from which it started to relax, but certain large-scale features of the initial state — for example whether it was spherical or elongated — are reflected in the virialized equilibrium to which the system finally settles. This capacity of collisionless stellar systems to remember how they were born is tantalizing because it suggests that stellar dynamics may enable us to probe the obscure epoch in the history of the Universe between the end of the first three minutes, when helium and deuterium production were coming to an end, and the end of the first few billion years when light from the most distant quasars set out towards us.

However, I believe that we have hitherto not exploited the full potential of n-body models because we have been too exclusively concerned with the coordinate-space structure of the models; the configuration of a galaxy of N stars must be specified by the positions of each star in six-dimensional phase space. We can no more hope to know such a model fully from its projection onto three-dimensional coordinate space than Plato's troglodytes could hope to know the things of the outside world from the shadows that they cast on the cave wall. What is needed is a simple way of gauging the full phase-space structure of n-body models. Fortunately the Hamiltonian reformulation of Newton's equations of motion provides the tool for this job.

5.1 *Evaluating Actions of Numerically Integrated Orbits*

Straightforward numerical integration of the equations of motion of stars in realistic triaxial potentials shows that the majority of orbits move through three-dimensional subsets M of the full six-dimensional phase space. Suppose we now assume that:

1) The subsets M can be characterized as the intersections of the level surfaces $H(\boldsymbol{x},\boldsymbol{p})$ = constant, $I_1(\boldsymbol{x},\boldsymbol{p})$ = constant, $I_2(\boldsymbol{x},\boldsymbol{p})$ = constant of the Hamiltonian and some two other isolating integrals I_1 and I_2.

2) The Poisson bracket $[I_1, I_2] = 0$.

Then we can deduce (e.g., Berry 1978) that the subsets M have a very special structure. In particular, it is possible to construct a special coordinate system $(\theta_1, \theta_2, \theta_3, J_1, J_2, J_3)$, the *angle-action* coordinate system, for the portion of phase space that is occupied by the M. In this system the J_n label the sets M and the θ_n tell one where one is on the given M. Since each M proves to be a three-torus, increasing any θ_n by a certain number, which may be chosen to be 2π, brings one back to the same place. Consequently the Cartesian coordinates $(\boldsymbol{x},\boldsymbol{p})$ of a point on M can be expanded as a Fourier series in the θ_n;

$$\boldsymbol{x}(\theta_1, \theta_2, \theta_3) = \sum_{\ell mn} \boldsymbol{X}_{\ell mn}\, e^{i(\ell\theta_1+m\theta_2+n\theta_3)} \tag{18}$$

where ℓ, m, n take on integer values and for each triple (ℓmn), $\boldsymbol{X}_{\ell mn}$ is a complex vector.

Hamilton's equation for the motion of θ_n is

$$\dot{\theta}_n = \frac{\partial H}{\partial J_n} \equiv \omega_n . \tag{19}$$

Since H is constant on each torus M, it may be expressed as a function of the J_n only. Hence we have that $\omega_n = \omega_n(J_1, J_2, J_3)$ is constant along the orbit and (19) immediately integrates to

$$\theta_n(t) = \omega_n t + \theta_n(0) \tag{20}$$

Substituting into (18) we find

$$\boldsymbol{x}(t) = \sum_{\ell mn} \boldsymbol{X}_{\ell mn}\, e^{i(\ell\omega_1+m\omega_2+n\omega_3)t} \tag{21}$$

where we have absorbed the $\theta_n(0)$ into redefined constants $\boldsymbol{X}_{\ell mn}$.

From equation (21) it follows that if we obtain $\boldsymbol{x}(t)$ for an orbit that is constrained by three isolating integrals by numerically integrating the equations of motion, we will find that the Fourier transform of the time evolution of any coordinate $x_i(t)$ contains only discrete lines. The frequencies at which these lines occur in the computed spectrum may be expressed as integer linear combinations of the three fundamental frequencies ω_n;

$$\omega_{\mathrm{line}} = \ell\omega_1 + m\omega_2 + n\omega_3$$

The complex amplitude associated with ω_{line} is clearly to be identified with $X_{\ell mn}$. Hence we may use the Fourier decomposition of the $x_i(t)$ and equation (18) to reconstruct $\boldsymbol{x}$ as function $\boldsymbol{x}(\theta_1, \theta_2, \theta_3)$ of the angle variables.

The actions associated with any orbit may be found by exploiting the fact that the angle-action coordinates are canonical. We start by noting that

$$J_n = \frac{1}{2\pi}\int_0^{2\pi} J_n \, d\theta_n . \tag{23a}$$

We now convert this line integral into a surface integral in the (θ_n, J_n) plane;

$$J_n = \frac{1}{2\pi}\int_0^{2\pi} d\theta_n \int_0^{J_n} dJ_n' . \tag{23b}$$

Poincaré proved that one can evaluate surface integrals of this kind using any canonical coordinate system. Hence

$$J_n = \frac{1}{2\pi}\int\int \sum_i dx_i \, dp_i \tag{23c}$$

where the integral is still over a portion of the (θ_n, J_n) plane. The final step is to convert this surface integral back into a line integral around the boundary of the surface in question;

$$J_n = \frac{1}{2\pi}\int_0^{2\pi} \boldsymbol{p}(\theta_i) \cdot \frac{\partial x(\theta_i)}{\partial \theta_i} \, d\theta_n . \tag{24}$$

If the coordinates $\boldsymbol{x}$ are inertial, we have $\boldsymbol{p} = \dot{\boldsymbol{x}}$ and the integral over θ_n may be evaluated explicitly. A straightforward calculation leads to the result

$$J_1 = \sum_{\mathrm{lines}} \ell\omega_{\mathrm{line}} \left| X_{\mathrm{line}} \right|^2 . \tag{25}$$

In practice one finds that in order to obtain satisfactory actions by substituting into equation (25) the amplitudes of lines in the Fourier spectra of numerically integrated orbits, one needs to integrate each orbit for about twenty orbital times. The potential of an n-body model should be made strictly time-independent before these integrations commence.

Equation (25) and its generalization to rotating frames of reference (Binney and Spergel, 1983) provide a means of determining the action of any orbit that is restricted by three effective integrals.

Actions enjoy two notable advantages over other phase-space coordinates:

1) The volume $\int d\theta_1 \, d\theta_2 \, d\theta_3 \int dJ_1 \, dJ_2 \, dJ_3$ in six-dimensional phase space that is occupied by any set of adjacent tori is simply $(2\pi)^3$ times the volume $\int dJ_1 \, dJ_2 \, dJ_3$ of three-dimensional action space within which the actions of

these tori fall. Thus the volume that a set of orbits occupies in action space is a fair measure of the *a priori* probability of these orbits.

2) When the potential is deformed on a time-scale that is long compared to the characteristic orbital time-scales, the actions of individual orbits may remain unchanged. When the actions are adiabatically invariant in this way, we may use this property to study the slow evolution of stellar systems without explicitly following the full time evolution.

It will be very interesting to calculate the actions of orbits in typical n-body models and to discover whether these attractive features of action integrals will enable us to develop a better feel than we have at the moment for the connection between the structure of a relaxed n-body model and the way in which the model was formed. Work along these lines is now in progress at a number of institutions. As a preliminary to these investigations it is instructive to study the actions of orbits in simple model potentials.

5.2 *Actions and separability*

In Oxford we have recently used the formalism described in the last sub-section to investigate the actions of orbits in a variety of simple barred potentials. To date we have studied only orbits that lie in the equatorial plane of the potential. Our main conclusions can be summarized as follows (Binney and Spergel, 1983):

1) When the figure of the potential is fixed with respect to inertial space, the actions of the two principal families of orbits in this plane — the short-axial loop orbits and the box orbits — fit together to form a single continuum. When the potential is slowly distorted in such a way as to cause an orbit to switch from a loop orbit to a box orbit, or vice versa, the actions of the orbit do not change.

2) The actions of orbits in a potential whose figure rotates steadily with respect to inertial space do not fit together into a single continuum. When a slow distortion of the potential causes an orbit to switch from a loop orbit to a box orbit, or vice versa, its actions suffer a sudden change. Thus the actions of orbits in a rotating potential can change discontinuously no matter how slowly the potential is deformed.

These numerical experiments show that the phase-space structure associated with orbits in a rotating bar differs fundamentally from the corresponding phase-space structure in the non-rotating case. An investigation of the actions of orbits in separable potentials suggests the likely significance of this result.

Consider as a specific example orbits in de Zeeuw's Stäckel potential. Through the separation of the Hamilton-Jacobi equation, each p_i becomes a known function of E, q_i and the integrals A and B. Thus the action of the orbit labelled by (E,A,B) is given by the integral

$$J_i(E,A,B) = \frac{1}{2\pi}\oint p_i\, dq_i = \frac{1}{\pi}\int_{q_i(\min)}^{q_i(\max)} \left[\frac{Eq_i^2 + Aq_i + B - V^*(q_i)}{4\prod_\ell (q_i+\alpha_\ell)}\right]^{\frac{1}{2}} dq_i \qquad (26)$$

Here the integral is over all values of q_i for which $p_i^2 \geqslant 0$. When the numbers (E,A,B) pass from a set for which some coordinate q_i circulates through its full range, to an adjacent set (E',A',B') for which this coordinate oscillates in the restricted range $-\alpha_i \leqslant q_i \leqslant q_i(\max)$, the actions J_i change continuously. Hence one expects the actions of orbits in a separable potential to form a single continuum that embraces all orbital families. Furthermore, the actions of orbits should not change when the parameters α_i are slowly changed so as to deform the potential into another separable potential.

Since numerical experiments show that the actions of orbits in barred potentials show these qualitative features only when the bar does not rotate, the implication is that no separable potential will serve as a satisfactory model of orbits in rotating bars in the same way that de Zeeuw's Stäckel potential provides an analytically soluble model of the orbital structure of inertial bars.

6. Discussion

Orbits in triaxial gravitational potentials are constrained by effective integrals of motion that do not derive from any manifest symmetry of the Lagrangian. Since a large proportion of galaxies have roughly triaxial potentials, these effective integrals are likely to play a crucial role in structuring galaxies.

Our understanding of galactic structure has advanced considerably in the last few years as a result of numerical experiments. The first indications that triaxiality might be a natural consequence of the way in which galaxies formed came from n-body experiments (Aarseth and Binney 1978. Hohl and Zang 1979). These showed that triaxial configurations were long-lived and probably maintained by non-classical effective integrals. Soon after these n-body experiments, Schwarzschild (1979, 1982) showed how one could construct self-consistent triaxial models by studying the orbits of individual stars in a given potential.

Schwarzschild's approach contributed enormously to our understanding of the fundamental processes at work in the n-body models, and by implication in the galaxies themselves. In particular, Schwarzschild's work and the subsequent analysis of n-body models by Wilkinson and James (1982), suggested that the majority of orbits in a non-rotating triaxial potential belong to one of three major orbital families (short-axial loops, box orbits and long-axial loops).

Two very recent developments have put this picture of major orbital families in a new light. De Zeeuw (1983) has shown that Schwarzschild's families are a natural feature of the orbital structure associated with a separable potential of the type introduced a century ago by Stäckel (1893). These potentials have hitherto found comparatively little application, possibly because until de Zeeuw's work it was widely believed that no potential of this class is generated by a physically realistic density distribution. The discovery that such an entirely reasonable density distribution as that given by equation (16) generates a fully separable potential is likely to prove a major step forward; there seems every reason to hope that a wide range of astrophysically interesting potentials can be treated as perturbations of one of the basic Stäckel forms.

A likely limitation of this approach emerges from a recent study (Binney and Spergel 1983) of the action integrals of orbits in simple barred potentials . In this study Fourier transforms of the time-developments of the coordinates along each orbit were used to calculate the action integrals of the orbits. When the figure of the potential is fixed in inertial space, the disposition of the orbits in three-dimensional action space qualitatively resembles that expected from a separable Stäckel potential. However, when the potential rotates, the structure of action space is more complex. This suggests that a rapidly rotating potential cannot be treated as a perturbed form of any separable potential.

Although this work suggests that we are unlikely to be able to extend Stäckel potentials to the case of appreciable figure rotation, it is possible that the spectral approach to stellar dynamics will enable us to use the methods of perturbation theory even in the absence of a suitable analytic system to perturb. This possibility arises because spectral analysis yields the angle-action representation of any quasi-periodic orbit. By perturbing these quasi-periodic orbits we may be able to obtain useful insights into such fundamental questions as the density of stochastic orbits in rotating potentials.

A more elementary application of spectral methods arises in the analysis of the phase-space structures of n-body models. The distribution functions of these models can be most easily expressed as functions of three appropriately chosen action integrals. Possibly analyses of these distribution functions will yield valuable insights into the connection between the structure of a self-consistent stellar system and the manner of its formation. Such a step would be a step towards an adequate understanding of the all-important process of galaxy formation.

Acknowledgements

It is a pleasure to thank T. de Zeeuw, O. Gerhard, D. Lynden-Bell, M. Schwarzschild and D. Spergel for many stimulating discussions. I am especially indebted to de Zeeuw for communicating his results to me in advance of publication and to P. Hut for many constructive comments on an early draft of this article.

References

Aarseth, S.J. and Binney, J.J., 1978. Mon. Not. Roy. Astr. Soc., **185**, 227.
Berry, M.V., 1978. *Topics in Non-Linear Mechanics*, (S. Jorna, ed.) p.16 (New York: A.I.P.)
Binney, J.J. 1982. Ann. Rev. Astron. Astrophys., **20**, 399.
Binney, J.J. and Spergel, D., 1983. Mon. Not. R. Astr. Soc. (in press).
Chandrasekhar, S., 1941. Astrophys. J., **94**, 511
Cohen, R.S., Spitzer, L. and Routly, P.McR., 1950. Phys. Rev., **80**, 230.
Contopoulos, G., 1960. Zeitschrift Astrophys., **49**, 273.
Contopoulos, G., 1963. Astrophys. J., **138**, 1297.
de Zeeuw, T., 1983. (in preparation)
de Zeeuw, T. and Lynden-Bell, D. 1983. (in preparation).
de Zeeuw, T. and Merritt, D., 1983. Astrophys. J., **267**, 571.
Henon, M. and Heiles, C., Astron.J., **69**, 73.
Hohl, F. and Zang, T.A., 1979. Astron.J., **84**, 585.
Jeans, J.H., 1913. Mon. Not.Roy. Astr. Soc., **74**, 109.
Ollongren, A., 1965. Ann. Rev. Astron. Astrophys. **3**, 113.
Martinet, L. and Mayer, F., 1975. Astron. Astrophys., **44**, 45.
Schwarzschild, M., 1979. Astrophys. J., **232**, 236.
Schwarzschild, M., 1982. Astrophys. J., **263**, 599.
Spitzer, L. and Hart, R., 1958. Astrophys. J., **127**, 544.
Stäckel, P., 1983. Math. Ann., **42**, 537.
Wilkinson, A. and James, R.A., 1982. Mon. Not. Roy. Astr. Soc., **199**, 171.

A Mean-field Treatment of Charge Density Waves in a Strong Magnetic Field

B. G. S. Doman

Department of Applied Mathematics and Theoretical Physics, Liverpool University

1. Introduction

The organic charge transfer compounds like TTF — TCNQ which was one of the earliest to be discovered, have excited a lot of interest for their very high electrical conductivity in one direction, and their low conductivity in perpendicular directions. A practical interest in these materials arose from the possible realisation of high temperature superconductivity by the Fröhlich mechanism (Fröhlich 1954). In all cases discovered until most recently the tendency to form charge density waves (CDWs) pinned to the crystal lattice first predicted by Peierls (1956) has dominated. An energy gap has been formed at low temperatures and the material has become an insulator with a periodic CDW and associated lattice distortion with period some rational multiple of the crystal period — the commensurate case — the most commonly considered case being where the multiple is two. In the other mainly considered case — the incommensurate case — the ratio between the period of the CDW and the crystal period is an irrational number. Here in practice, the CDWs are found to be pinned by imperfections in the crystal. If the conduction band of the material is exactly half full, the Peierls mechanism leads to a commensurate CDW with period twice that of the crystal's. If the conduction band is made slightly more (or less) than half full, e.g. by doping, we find at low enough temperatures an incommensurate CDW which takes on the appearance of a succession of regions with a commensurate CDW which are successively out of phase with each other, separated from each other by domain walls — kinks — or antikinks, giving rise to charged quasi-particles — solitons. The transport properties of these solitons give rise to a number of very interesting effects.

Although we shall be concentrating on the one dimensional properties of these materials, it is important to realise that the effects we are discussing are essentially three dimensional. In a truly one dimensional system with the inevitable impurities and imperfections, all the electrons will be localised by the random potential and electrical conductivity in a macroscopic crystal will be zero. In the crystals we are considering, the possibility of hopping from one chain to another giving rise to non-zero perpendicular conductivity, and the three dimensional nature of the phonon spectrum with coupling of the lattice displacement in adjacent chains dominate the localisation effects of the random potential. We shall therefore, in this review, neglect impurity scattering effects and treat the interaction with the phonons in a molecular field approximation. This simple approximation is the only approximation which has lead to a significant increase in our understanding of CDW phenomena. However, it suffers from a number of well known defects, notably that it overestimates the tendency for a system to become ordered, and in particular, it predicts the possibility of an ordered state for a one dimensional system in the absence of three dimensional effects.

2. The Fröhlich Hamiltonian and the Mean-Field Approximation

The one dimensional Fröhlich Hamiltonian is

$$\mathcal{H} = \sum_{k\sigma} (\varepsilon_{\mathbf{k}} - h\sigma)\, c^+_{\mathbf{k},\sigma} c_{\mathbf{k},\sigma} + \frac{1}{\sqrt{N}} \sum_{kq\sigma} g_{\mathbf{q}}\, c^+_{\mathbf{k+q},\sigma} c_{\mathbf{k},\sigma} (b_{\mathbf{q}} + b^+_{-\mathbf{q}}) + \sum_{\mathbf{q}} \hbar\omega_{\mathbf{q}}\, b^+_{\mathbf{q}} b_{\mathbf{q}} \tag{2.1}$$

where $c_{\mathbf{k}\sigma}$ and $b_{\mathbf{q}}$ are electron and phonon destruction operators, $\varepsilon_{\mathbf{k}\sigma} = -W\cos ka$ is the electron energy in the tight-binding approximation with band width $2W$ and $h\sigma$ is the energy in a magnetic field. $\omega_{\mathbf{q}}$ is the phonon energy, N the number of lattice sites, and $g_{\mathbf{q}}$ the coupling constants for electrons and optical phonons.

In the mean-field approximation, we replace the phonon operators $b_{\mathbf{q}}$ by their averages, introducing the gap parameters

$$\Delta_{\mathbf{q}} = g_{\mathbf{q}} \left(\langle b_{\mathbf{q}} \rangle + \langle b^+_{-\mathbf{q}} \rangle \right) \Big/ \sqrt{N} \, . \tag{2.2}$$

This leads to the mean field Hamiltonian for the electrons

$$\mathcal{H}_{\mathrm{MFA}} = \sum_{\mathbf{k}\sigma} (\varepsilon_{\mathbf{h}} - h\sigma)\, c^+_{\mathbf{k},\sigma} c_{\mathbf{k},\sigma} + \sum_{\mathbf{kq}\sigma} \Delta_{\mathbf{q}}\, c^+_{\mathbf{k+q},\sigma} c_{\mathbf{k},\sigma} \tag{2.3}$$

If we suppose that all the qs are multiples of $Q = 2\pi/na$, we can diagonalise the Hamiltonian by the transformation (Kotani and Harada 1980)

$$a^+_{\lambda,\mathbf{k},\sigma} = \sum_{\mathbf{m}=0}^{\mathbf{n}-1} \langle k + mQ | \lambda, k \rangle\, c^+_{\mathbf{k+mQ},\sigma} \, . \tag{2.4}$$

The energy eigenvalues satisfy the secular equation

$$(\varepsilon_{\mathbf{k+mQ}} - E_{\lambda,\mathbf{k}})\langle k+mQ|\lambda,k\rangle + \sum_{\ell=1}^{\mathbf{n}-1} \Delta_{\ell\mathbf{Q}}\langle k+(m+\ell)Q|\lambda,k\rangle = 0 . \tag{2.5}$$

In (2.4) and (2.5), the band splits up into n sub-bands labelled by λ with $-\frac{\pi}{na} \leqslant k \leqslant \frac{\pi}{na}$ and where necessary $\langle q|\lambda k\rangle$ is continued for $|q| > \frac{\pi}{a}$ periodically, with period $\frac{2\pi}{a}$. We have also used $\Delta_{\mathbf{n}-\ell} = \Delta_{\ell}^{*}$.

The free energy per lattice site is given by

$$F = -\frac{k_{\mathbf{B}}T}{N} \sum_{\lambda,\mathbf{k},\sigma} \ln\left\{1+\exp\left[-\beta(E_{\lambda,\mathbf{k}}+h\sigma-\mu)\right]\right\} + \sum_{\ell=1}^{\mathbf{n}-1} |\Delta_{\ell\mathbf{Q}}|^{2}/\lambda_{\ell\mathbf{Q}} \tag{2.6}$$

where $\lambda_{\ell\mathbf{Q}} = 4|g_{\mathbf{q}}|^{2}/\hbar\omega_{\ell\mathbf{Q}}$ and μ is the fermi energy. The gap parameters are determined by minimising F, i.e. using $\partial F/\partial\Delta_{\ell\mathbf{Q}} = 0$. This gives

$$\Delta_{\ell\mathbf{Q}} = -\frac{\lambda_{\ell\mathbf{Q}}}{2N} \sum_{\lambda,\mathbf{k},\sigma} \sum_{\mathbf{m}=0}^{\mathbf{n}-1} f(E_{\lambda\mathbf{k}}+h\sigma-\mu)\langle\lambda,k|k+(\ell+m)Q\rangle\langle k+mQ|\lambda,k\rangle . \tag{2.7}$$

3. The Commensurate Phase

We consider the commensurate phase where the wavelength of the lattice distortion is twice the lattice spacing of the crystal, i.e. 2a. In this case there is only one gap parameter Δ which can be taken to be real. The energies $E_{\mathbf{k}}$ are given in this case by

$$E_{\mathbf{k}} = \operatorname{sgn}(\varepsilon_{\mathbf{k}})\sqrt{\varepsilon_{\mathbf{k}}^{2}+\Delta^{2}} . \tag{3.1}$$

The free energy becomes (Mertsching and Fischbeck 1980; Doman 1981)

$$f_{\mathbf{e}} = \pi W(F+\mu) + 2W^{2} \approx \Delta^{2}[\ln(\Delta/\Delta_{0}) - \tfrac{1}{2}]$$
$$- k_{\mathbf{B}}T \sum_{\sigma,\pm} \int_{0}^{\infty} d\varepsilon \ln\left\{1+\exp\left[-\beta(\sqrt{\varepsilon^{2}+\Delta^{2}} \pm \mu + h\sigma)\right]\right\} \tag{3.2}$$

where $\Delta_{0} = 4W\exp(-\pi W/\lambda_{\mathbf{q}})$ is the gap parameter at absolute zero for $\mu = h = 0$ and $\beta = 1/k_{\mathbf{B}}T$. Differentiating with respect to Δ gives the gap equation

$$\frac{1}{2\Delta}\frac{\partial f_{\mathbf{e}}}{\partial\Delta} \approx \ln\left(\frac{\Delta}{\Delta_{0}}\right) + \frac{1}{2}\int_{0}^{\infty} \frac{d\varepsilon}{\sqrt{\varepsilon^{2}+\Delta^{2}}} \sum_{\sigma,\pm} f(\sqrt{\varepsilon^{2}+\Delta^{2}} \pm\mu - h\sigma) = 0, \tag{3.3}$$

provided that $\mu \ll W$ and $h \ll W$.

The number of electrons can be obtained by differentiating $f_{\mathbf{e}}$ with respect to μ

$$n-1 = \pi W \int_{0}^{\infty} \sum_{\sigma} [f(\sqrt{\varepsilon^{2}+\Delta^{2}} - h\sigma - \mu) - f(\sqrt{\varepsilon^{2}+\Delta^{2}} - h\sigma + \mu)]\, d\varepsilon \tag{3.4}$$

and the net magnetisation by differentiating $f_{\mathbf{e}}$ with respect to h

$$m = \pi W \int_{0}^{\infty} \sum_{\pm} [f(\sqrt{\varepsilon^{2}+\Delta^{2}} - h\sigma \pm\mu) - f(\sqrt{\varepsilon^{2}+\Delta^{2}} + h\sigma \pm\mu)]\, d\varepsilon . \tag{3.5}$$

The spin density $S(z)$ is given by

$$S(z) = \frac{1}{N} \sum_{\mathbf{q},\mathbf{k},\sigma} \sigma_{\mathbf{z}}\, c^{+}_{\mathbf{k+q},\sigma} c_{\mathbf{k}\sigma}\, e^{\mathbf{iqz}} = \frac{1}{N} \sum_{\mathbf{q}} (\rho_{\mathbf{q}\uparrow} - \rho_{\mathbf{q}\downarrow})\, e^{\mathbf{iqz}} \tag{3.6}$$

where

$$\rho_{\mathbf{q},\sigma} = \sum_{\mathbf{k}} \left\langle c^{+}_{\mathbf{k+q},\sigma} c_{\mathbf{k}\sigma} \right\rangle$$

can be obtained from (2.4) and (2.5). We find

$$\rho_{\frac{\pi}{\alpha},\sigma} = \frac{\Delta}{2} \sum_{\mathbf{k}} \left\{ f(\sqrt{\varepsilon_{\mathbf{k}}^2 + \Delta^2} - \mu - h\sigma) + f(\sqrt{\varepsilon_{\mathbf{k}}^2 + \Delta^2} + \mu + h\sigma) - 1 \right\} \Big/ \sqrt{\varepsilon_{\mathbf{k}}^2 + \Delta^2} \tag{3.7}$$

and

$$\rho_{\frac{\pi}{\alpha}\uparrow} - \rho_{\frac{\pi}{\alpha}\downarrow} = \frac{N\Delta}{\pi W} \int_0^\infty \frac{d\varepsilon}{\sqrt{\varepsilon^2 + \Delta^2}} \Big\{ f(\sqrt{\varepsilon^2 + \Delta^2} - \mu - h\sigma) + f(\sqrt{\varepsilon^2 + \Delta^2} + \mu + h\sigma)$$
$$- f(\sqrt{\varepsilon^2 + \Delta^2} + \mu - h\sigma) - f(\sqrt{\varepsilon^2 + \Delta^2} - \mu + h\sigma) \Big\}. \tag{3.8}$$

A number of consequences follow from these equations. Firstly non-zero μ and/or h tends to inhibit the formation of a CDW. If h and μ are both non-zero, then the CDW will be accompanied by a net magnetisation and a spin density wave, with the same wavelength as the CDW. For very low temperatures and small μ and h such that $|\mu| + |h| \ll \Delta$ the system will be insensitive to variations in μ and h. In particular, for $\mu = 0$, the system will have no magnetic moment for $|h| < h_c \sim \Delta$, and no spin density wave for any h. And finally from (3.4), for very low temperatures and $n > 1$, μ will adjust to a value such that $\mu + h > \Delta$.

4. The Incommensurate Phase

In this section we shall consider the phase boundary between the normal phase and that with an incommensurate CDW. (Doman 1981b). Starting from (2.4) to (2.7), we shall suppose that the number of sub-bands n is large but finite. We shall leave consideration of what happens when the wave vector of the incommensurate CDW approaches π/a, the commensurate CDW, until later on.

Let us now consider the situation with T just below the transition temperature. The gap parameters Δ_ℓ will be very small, much smaller than the energy width of the sub-bands $\sim\varepsilon_{\mathbf{Q}/2}$. A careful examination of equations (2.7) and (2.5) shows that under these circumstances the equations for Δ_ℓ decouple to first order in Δ_ℓ.

Thus only one of the Δ_ℓ and $\Delta_{\mathbf{n}-\ell}$ will have any significance, the others will from (2.7) be of order Δ_ℓ^2 or higher and thus negligible. At a slightly lower temperature $\Delta_{\ell\pm1}$ will satisfy the first-order approximation to (2.7). $\Delta_{\ell\pm1}$ will then be $0(\Delta_\ell)$. In this way the number of Δ_ℓ of significance increases as T falls.

It will be more convenient now to work in the extended sub-zone scheme with $-\frac{\pi}{a}<k<\frac{\pi}{a}$. The sub-band index will however be retained to distinguish the true wave functions $|\lambda,k\rangle$ from the free-electron $|k\rangle$ for all $\Delta_\ell = 0$. The solution of (2.5) for given k will be that which reduces to $\varepsilon_{\mathbf{k}}$ as every $\Delta_\ell \to 0$. It is easily shown that to first order for all k, except on a set of measure zero,

$$\langle k,\lambda|k+mQ\rangle\langle k|k,\lambda\rangle = \Delta_{\mathbf{m}}\,\mathrm{sgn}\,(\varepsilon_{\mathbf{k}}-\varepsilon_{\mathbf{k+mQ}})/[(\varepsilon_{\mathbf{k}}-\varepsilon_{\mathbf{k+mQ}})^2 + 4|\Delta_{\mathbf{m}}|^2]\,. \quad \mathbf{(4.1)}$$

Functions of k will be presumed to be continued periodically outside $-\frac{\pi}{a}<k<\frac{\pi}{a}$ where necessary.

If we ignore contributions to the gap equation (2.7) which are first order in $\Delta_{\mathbf{m}}$ over a region of k which is itself of first order, we find that provided all $\Delta_{\mathbf{m}}$ are very small or zero:

$$1 = -\frac{\lambda_{\mathbf{m}}a}{4\pi}\int_{-\pi/a}^{\pi/a} dk \sum_{\sigma} f\,(E_{\lambda,k} + h\sigma - \mu)$$
$$\times\left(\frac{\mathrm{sgn}\,(\varepsilon_{\mathbf{k}}-\varepsilon_{\mathbf{k+mQ}})}{[(\varepsilon_{\mathbf{k}}-\varepsilon_{\mathbf{k+mQ}})^2 + 4|\Delta_{\mathbf{m}}|^2]^{\frac{1}{2}}} + \frac{\mathrm{sgn}\,(\varepsilon_{\mathbf{k}}-\varepsilon_{\mathbf{k-mQ}})}{[(\varepsilon_{\mathbf{k}}-\varepsilon_{\mathbf{k-mQ}})^2 + 4|\Delta_{\mathbf{m}}|^2]^{\frac{1}{2}}}\right). \quad (4.2)$$

When only one $\Delta_{\mathbf{m}}$ is different from zero, the energy $E_{\mathbf{k},\lambda}$ for $0 \leqslant k \leqslant \frac{\pi}{2a}$, $m < \frac{n}{2}$ or $p < \frac{\pi}{a}$, where p is the wavevector for the incommensurate charge-density wave is

$$E_{\mathbf{k},\lambda} = \tfrac{1}{2}\left\{\varepsilon_{\mathbf{k}} + \varepsilon_{\mathbf{k-p}} + \mathrm{sgn}\,(\varepsilon_{\mathbf{k}} - \varepsilon_{\mathbf{k-p}})\,[(\varepsilon_{\mathbf{k}}-\varepsilon_{\mathbf{k-p}})^2 + 4|\Delta_{\mathbf{m}}|^2]^{\frac{1}{2}}\right\}. \quad (4.3)$$

For $\frac{\pi}{2a} < k \leqslant \frac{\pi}{a}$ we have the same formula but with $-p$ replaced by p while for negative k we have $E_{\mathbf{k},\lambda} = E_{-\mathbf{k},\lambda}$.

For a band which is almost half-filled, with the Fermi energy μ small, the value of p for which we get the first gap-parameter $\Delta_{\mathbf{m}}$ will be close to $\frac{\pi}{a}$. If the zero-temperature gap-parameter for the half-filled band $\Delta_0 = 4W\exp(-\lambda/\pi W) \ll W$, the gap equation determining $\Delta_{\mathbf{m}}$ very close to the transition temperature, and thus $T_{\mathbf{c}}$ and p for given non-zero μ and k takes the form

$$\ell n\left(\frac{\Delta_{\mathbf{m}}}{\Delta_0}\right) + \frac{1}{4}\int_0^\infty \frac{d\varepsilon}{\surd(\varepsilon^2 + |\Delta_{\mathbf{m}}|^2)} \sum f\left(\surd(\varepsilon^2 + \Delta^2) \pm K \pm \mu \pm h\right) = 0\,, \quad (4.4)$$

where $K = W(\pi - pa)/2$, and the summation is carried out over the eight different combinations of + and − signs. In this equation we have neglected terms independent of $\Delta_{\mathbf{m}}$ which cancel on replacing f by a step function. We have also assumed Δ_0, μ and k to be very small compared with W.

For a band which is not almost half-filled. with $\mu \pm h$ not nearly zero, we find

$$2\,\ell n\left(\frac{\Delta_m}{\Delta_0^-}\right) + \ell n\left[\tan\left(\frac{\theta_+}{2}+\frac{pa}{4}\right)\right] + \ell n\left[\tan\left(\frac{\theta_-}{2}+\frac{pa}{4}\right)\right]$$
$$+\sum_\sigma \int_0^\infty \frac{d\varepsilon}{\sqrt{(\varepsilon^2+|\Delta_m|^2)}}\Big\{f(\sqrt{(\varepsilon^2+|\Delta_m|^2)}-W_+-\mu+h\sigma)$$
$$+f(\sqrt{(\varepsilon^2+|\Delta_m|^2)}+W_++\mu+h\sigma)\Big\} = 0\,, \qquad (4.5)$$

where $\Delta_0^- = 4W\exp(-2\pi W_-/\lambda)$, $W_- = W\sin(pa/2)$, $W_+ = W\cos(pa/2)$ and $W\cos\theta_\pm = \mu \pm h$.

In deriving (4.5) we have noted that in the integral of (4.2) with $k > 0$ the first term in the brackets is small and independent of Δ_m everywhere that f is non-negligible. For this integral we have approximated f by a step function. For $\mu \pm h > 0$ it is easy to see that the equation (4.5) remains true.

For $h = 0$, we have a charge-density wave with a wavelength determined by $W\cos(pa/2) = -\mu$. As h increases from zero its only effect first of all is to decrease the transition temperature with no effect on the wavelength of the charge-density wave. Above a certain critical value of h, the wavelength of the charge-density wave is shifted nearer to π/a, and $W_+ + \mu$ is no longer zero. In fact we see that in (4.5), $W_+ + \mu$ varies with h in the same way as K in (4.4) varies with μ for $h = 0$, (or h for $\mu = 0$). The variation of T_c will also be the same as in (4.4). The difference in the effective gap-parameter incorporating the $\ell n\tan$ terms in (4.5) being the only significant difference. The function of the $\ell n\tan$ terms is to select one of the two possible values for $W_+ = -\mu \pm h$. There is thus an analogy between $(W_+ + \mu)$ and h in (4.5) and K and μ in (4.4) with $h = 0$. The critical value of h above which W changes is $0.6082\ \bar{\Delta}_0^-$, where $\bar{\Delta}_0^-$ is the effective gap parameter.

In figures 1 and 2 the transition temperature T and K which determines the wavelength of the charge-density wave are plotted as functions of μ for different values of h for an almost half-filled band. For very small $h \leqslant 0.1\,\Delta_0$ not shown in the graphs we find a small change in transition temperature and a negligible change in K. For $h = 0.15$ we see a significant fall in the transition temperature but a small change in K. For larger h, T_c falls even more and, as μ increases, tends towards a common curve independent of h. As μ decreases K approaches zero with a vertical tangent at the Leung point corresponding to that magnetic field (Leung 1975, Doman 1981). For $h \geqslant 0.3\,\Delta_0$, K approaches the straight line $K = \mu - h$. This behaviour together with the limiting behaviour of T_c results from electrons of only one spin direction contributing to the charge-density wave. For $h \leqslant 0.1\,\Delta_0$, the effect of the magnetic field on K is opposite for the two spin directions and this is why K hardly changes with h for small h.

An idea of the size of the magnetic field at which the effects described become significant may be obtained by equating $k_B T_c$ with $\mu_B H$. For TTF-TCNQ taking $T_c = 54K$ this gives a value of H around $70T$. There should, however, be a noticeable reduction in the transition-temperature for fields which are a fraction of this value.

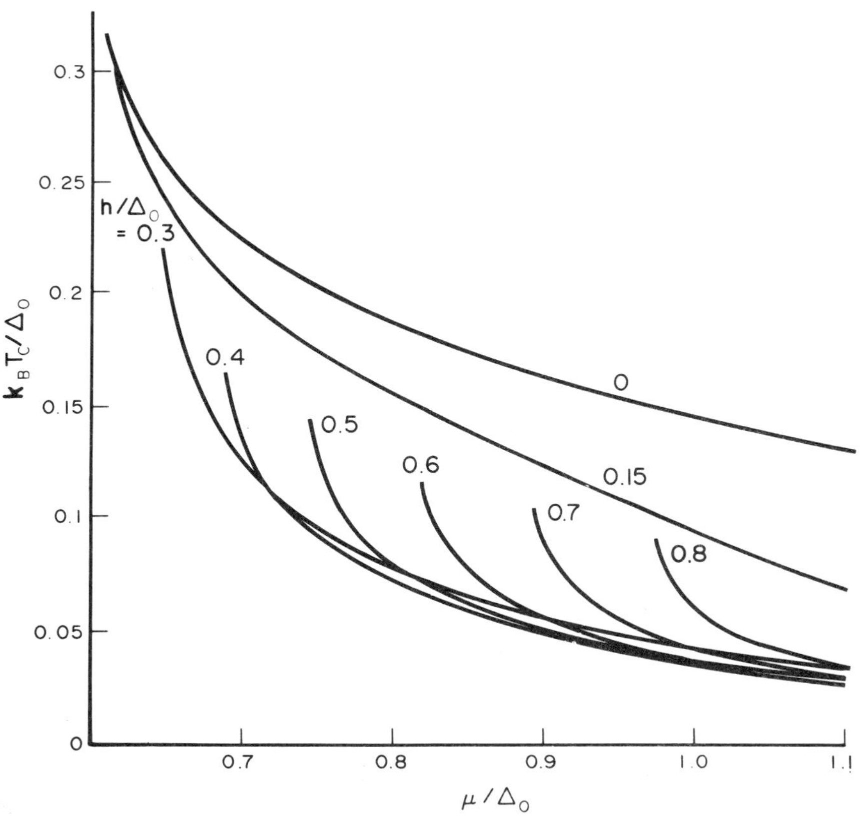

Fig. 1 The transition temperature T_c as a function of μ for $\mu > \mu_L$ and $h/\Delta_0 = 0.15, 0.3, 0.4, 0.5, 0.6, 0.7$ and 0.8

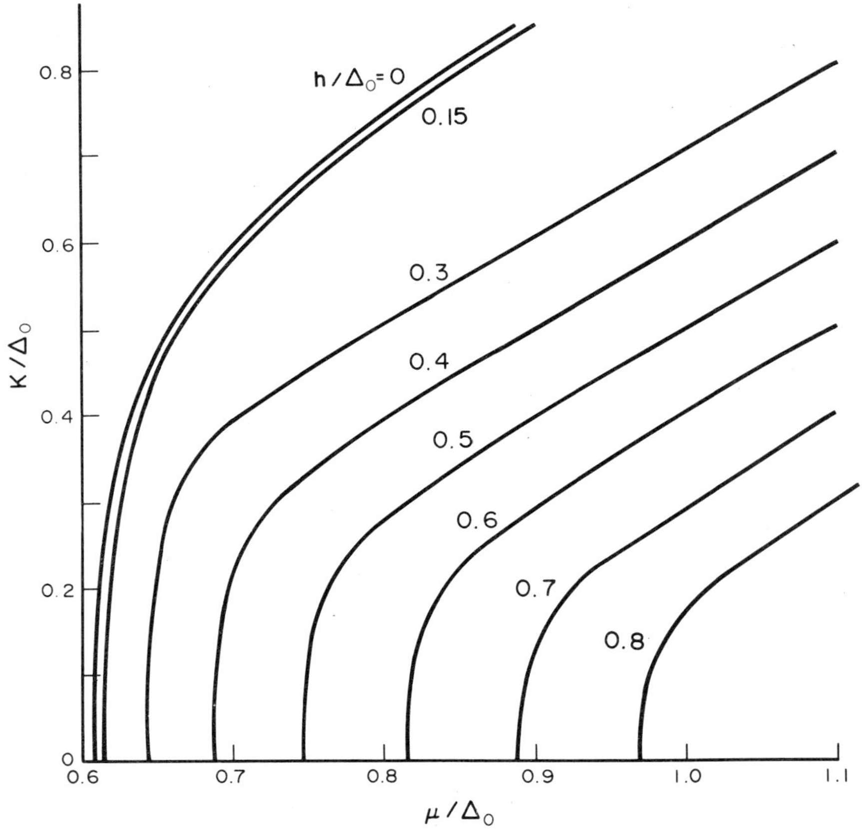

Fig. 2 The wavelength parameter K as a function of μ for $h/\Delta_0 = 0$, 0.15, 0.3, 0.4, 0.5, 0.6, 0.7 and 0.8

5. The Commensurate Incommensurate Transition near the Leung point

In the previous section we examined the properties of the system at a temperature just below the normal — incommensurate CDW transition. We derived the conditions for a gap Δ_m to open up and produce an incommensurate CDW with wavevector $p = 2\pi m/na$ under the assumption that $|\Delta_{mq}| \ll \varepsilon_{Q/2}$. We noted that $\Delta_{(m\pm1)Q}$ would be of order Δ^2_{mQ}, and $\Delta_{(m\pm r)Q}$ of the order Δ^r_m. At slightly lower temperatures, (4.2) will be satisfied for $\Delta_{(m\pm1)Q}$, and if n is large enough for $\varepsilon_{p+Q} - \varepsilon_p < kT$. At such temperatures $\Delta_{(m\pm1)Q}$, $\Delta_{(m\pm2)Q}$ etc. can not be neglected as they could be of the same order of magnitude as Δ_m.

It was first shown by Kotani (1977) using Green's functions that only the gap parameters with wavevector an odd multiple of the incommensurate CDW wavevector are significant, while those with a wavevector $2rp = 2rmQ$ will be of a higher order and therefore negligible. We can adapt Kotani's argument starting from (2.5) and (2.7). Firstly let us now regard Q as the incommensurate CDW wavevector. Equations (2.5) and (2.7) remain true with this change in interpretation. We are interested in $\langle k + mQ|\lambda, k\rangle$ for $k \sim Q$. From (2.5), we see that if $Q \sim \frac{\pi}{a}$ then $\varepsilon_{\mathbf{k+mQ}} \sim E_{\lambda,\mathbf{k}}$ for m odd, but $|\varepsilon_{\mathbf{k+mQ}} - E_{\lambda,\mathbf{k}}| \sim W$, the band width, for m even. Thus for m an odd number $\langle k + mQ|\lambda, k\rangle$ will be of the same order of magnitude, while for even m $\langle k + mQ|\lambda, k\rangle$ will be of order $\Delta_{\mathbf{Q}}$. Turning now to (2.7), the even number $\Delta_{\ell\mathbf{Q}}$ will be given by a product of $f(E_{\lambda,\mathbf{k}} + h\sigma - \mu)$ and a sum of products of two odd number m $\langle k+mQ|\lambda, k\rangle$ and two even number m $\langle k+mQ|\lambda, k\rangle$. The product of the two odd number m $\langle k + mQ|\lambda, k\rangle$ gives a very small contribution on integrating, leaving $\Delta_{2\ell\mathbf{Q}} \sim \Delta_{\mathbf{Q}}^2$ or higher.

We write the free energy F in the form

$$F = -k_{\mathbf{B}}T \sum_{\sigma} \int dE\, \rho(E)\, \ln\left\{1 + \exp\left[-\beta\left(E - h\sigma - \mu\right)\right]\right\} + \frac{1}{\lambda} \sum_{\ell = \mathrm{odd}} |\Delta_{\ell p}|^2\,, \tag{5.1}$$

where we are summing over the gap parameters for the odd harmonics of the incommensurate CDW vector, and ρ the density of states per site and per spin is related to the diagonal elements of the Green's function $G_{\mathbf{kq}}(z) = \langle k|(z - \mathcal{H})^{-1}|q\rangle$ by

$$\rho(E) = -\frac{1}{\pi}\,\mathrm{Im} \sum_{\mathbf{k}} G_{\mathbf{k},\mathbf{k}}(E + i0)\,. \tag{5.2}$$

For small values of the gap parameters $\Delta_{\ell\mathbf{p}}$ we can write $G_{\mathbf{k},\mathbf{k}}$ as a perturbation expansion, see also Kotani (1977)

$$\begin{aligned} G_{\mathbf{k},\mathbf{k}}(z) = g_{\mathbf{k}} + g_{\mathbf{k}}^2 \Bigg[&\sum_{\ell = \mathrm{odd}} \Delta_{\ell\mathbf{p}}\, \Delta_{-\ell\mathbf{p}}\, g_{\mathbf{k}+\ell\mathbf{p}} + \\ &+ \sum_{\ell_1 \ell_2 \ell_3 = \mathrm{odd}} \Delta_{\ell_1\mathbf{p}}\, \Delta_{\ell_2\mathbf{p}}\, \Delta_{\ell\mathbf{p}}\, \Delta_{-(\ell_1+\ell_2+\ell_3)\mathbf{p}} \times \\ &\times g_{\mathbf{k}+\ell_1\mathbf{p}}\, g_{\mathbf{k}+(\ell_1+\ell_2)\mathbf{p}}\, g_{\mathbf{k}+(\ell_1+\ell_2+\ell_3)\mathbf{p}} + \ldots \Bigg]\,. \end{aligned} \tag{5.3}$$

In this expansion we only have an even number of factors of Δ owing to summing over the odd harmonics. The Green's function $g_{\mathbf{k}}$ is the normal state Green's function $g_{\mathbf{k}} = 1/(z - \varepsilon_{\mathbf{k}})$, and if we write $p = \frac{\pi}{a} + \kappa$

$$g_{\mathbf{k}+\ell\mathbf{p}} \approx \frac{1}{z \pm (\varepsilon_{\mathbf{k}} + \ell\hbar v_{\mathbf{k}}\kappa)} = \frac{1}{z \pm \varepsilon_{\mathbf{k}}} \pm \frac{\ell\hbar v_{\mathbf{k}}\kappa}{(z \pm \varepsilon_{\mathbf{k}})^2} + \ldots\,, \quad \begin{matrix} \ell \text{ odd} \\ \ell \text{ even} \end{matrix} \tag{5.4}$$

where $v_{\mathbf{k}}$ are the electron velocities.

Following Mertsching and Fischbeck (1980), we can express the free energy in terms of the polygamma functions. Up to terms of order Δ^6 and κ^4 we find

$$
\begin{aligned}
f - f_{\mathbf{N}} = \pi W(F - F_{\mathbf{N}}) &= c_1 \sum_{\ell} \Delta_{\ell p} \Delta_{-\ell p} \\
&+ c_2 \sum_{\ell_1 \ell_2 \ell_3} \Delta_{\ell_1 p} \Delta_{\ell_2 p} \Delta_{\ell_3 p} \Delta_{-(\ell_1 + \ell_2 + \ell_3)p} \\
&+ c_3 \sum_{\ell_1 \ell_2 \ell_3 \ell_4 \ell_5} \Delta_{\ell_1 p} \Delta_{\ell_2 p} \Delta_{\ell_3 p} \Delta_{\ell_4 p} \Delta_{\ell_5 p} \Delta_{-(\ell_1 + \ell_2 + \ell_3 + \ell_4 + \ell_5)p} \\
&+ (\hbar v_{\mathbf{k}} \kappa)^2 \Big\{ c_2 \sum \ell^2 \Delta_{\ell p} \Delta_{-\ell p} - 5 c_3 \sum_{\ell_1 \ell_2 \ell_3} \ell_1 \ell_2 \Delta_{\ell_1 p} \Delta_{\ell_2 p} \Delta_{\ell_3 p} \Delta_{-(\ell_1 + \ell_2 + \ell_3)p} \Big\} \\
&+ \tfrac{1}{2} (\hbar v_{\mathbf{k}} \kappa)^2 c_4 \sum_{\ell} \ell^4 \Delta_{\ell} \Delta_{-\ell} + \dots , \qquad (5.5)
\end{aligned}
$$

with $v_{\mathbf{k}} = aW/\hbar$, and

$$
\begin{aligned}
c_1 &\approx \ln \frac{4\pi}{\beta \Delta_0} + \mathcal{R}e\, \psi \left(\tfrac{1}{2} + \frac{i \beta \mu}{2\pi} \right) , \\
c_2 &\approx - \tfrac{1}{8} \left(\frac{\beta}{2\pi} \right)^2 \mathcal{R}e\, \psi'' \left(\tfrac{1}{2} + \frac{i \beta \mu}{2\pi} \right) , \qquad (5.6) \\
c_3 &\approx \frac{1}{192} \left(\frac{\beta}{2\pi} \right)^4 \mathcal{R}e\, \psi^{(4)} \left(\tfrac{1}{2} + \frac{i \beta \mu}{2\pi} \right) .
\end{aligned}
$$

In deriving (5.4) we have used the relations

$$
\begin{aligned}
&- \frac{2}{\beta} \int \ln \left\{ 1 + e^{-\beta(E - \mu)} \right\} \left[-\frac{1}{\pi} \operatorname{Im} \frac{1}{(z - \varepsilon)^{\mathrm{m}} (z + \varepsilon)^{\mathrm{n}}} \right]_{z = E + i0} dE \, d\varepsilon \\
&= \frac{1}{(m-1)! \, (n-1)!} \left(-\tfrac{1}{2} \right)^{\mathrm{m+n-3}} \frac{\partial^{\mathrm{m+n-3}}}{\partial \mu^{\mathrm{m+n-3}}} \int_{-\mathrm{W}}^{\mathrm{W}} \frac{f(\varepsilon - \mu)}{\varepsilon} \, d\varepsilon , \qquad (5.7)
\end{aligned}
$$

$$
\int_{-\mathrm{W}}^{\mathrm{W}} \frac{f(\varepsilon - \mu)}{\varepsilon} \, d\varepsilon \approx \ln \frac{2\pi}{\beta W} + \mathcal{R}e\, \psi \left(\tfrac{1}{2} + \frac{i \beta \mu}{2\pi} \right) . \qquad (5.8)
$$

We are now in a position to discuss the stability of the commensurate CDW. The normal — commensurate CDW phase transition curve will be determined by $c_1 = 0$, $c_2 > 0$. As the Fermi energy μ increases, c_2 decreases until we reach the triple point the Leung point (Leung 1975) where $c_2 = 0$. In the incommensurate phase, where $c_2 < 0$, thermodynamic stability is assured by the higher order terms $\sim c_3 > 0$.

6. The Commensurate Incommensurate Transition below the Leung Point

In this and the following sections we shall examine the properties of the incommensurate state close to the commensurate – incommensurate phase boundary. We shall closely follow the work of Mertsching and Fischbeck (1981) which itself is an extension to finite temperatures of the ground state calculations of Brazovski, Gordyunin and Kirova (1980). We shall for the most part take the external magnetic field to be zero. The generalisation to finite magnetic field is for the most part quite straightforwaed (Doman 1983).

For an incommensurate wave with wave-vector $p=\pi/a+\kappa$, any wave-vector in the first Brillouin zone can be written as $k=k_0+np$ modulo $2\pi/a$ with some fixed k_0, thereby determining n. The Hamiltonian (2.1) in the mean field approximation may then be written

$$\mathcal{H}=\sum_{\mathbf{n},\sigma}(\varepsilon_{\mathbf{n},\sigma}-h\sigma)\,c^+_{\mathbf{n}\sigma}c_{\mathbf{n}\sigma}+\sum_{\ell}\sum_{\mathbf{n},\sigma}\Delta_{2\ell+1}\,c^+_{\mathbf{n}+2\ell+1,\sigma}c_{\mathbf{n}\sigma}+\frac{N}{\lambda}\sum_{\ell}\left|\Delta_{2\ell+1}\right|^2 \qquad (6.1)$$

where $c_{\mathbf{n},\sigma}=c_{\mathbf{k}_0+\mathbf{np},\sigma}$, and $\varepsilon_{\mathbf{n}}=\varepsilon_{\mathbf{k}_0+\mathbf{np}}$. We now define the Fourier transformed quantities

$$\phi_\sigma(x)=\phi_\sigma(x+2\pi/\kappa)=\frac{1}{\sqrt{L}}\sum_{\mathbf{n}}c_{\mathbf{n},\sigma}\,e^{\mathbf{i}\mathbf{n}\kappa}\,, \qquad (6.2)$$

$$\Delta(x)=-\Delta(x+\pi/\kappa)=\sum\Delta_{2\ell+1}\,e^{\mathbf{i}(2\ell+1)\kappa} \qquad (6.3)$$

The chain length $L=Na$ is presumed to be an integral multiple of the incommensurate CDW wave length $2\pi/\kappa$. The Hamiltonian becomes

$$\mathcal{H}=\frac{\kappa L}{2\pi}\sum_{\sigma}\int_0^{2\pi/\kappa}dx\,\phi^+_\sigma(x)\,\varepsilon\!\left(k_0-i\frac{d}{dx}\right)\phi_\sigma(x+\pi/\kappa)$$
$$+\frac{\kappa L}{2\pi}\sum_{\sigma}\int_0^{2\pi/\kappa}dx\,\Delta(x)\,\phi^+_\sigma(x)\,\phi_\sigma(x)+\frac{N\kappa}{\pi\lambda}\int_0^{\pi/\kappa}dx\,\Delta^2(x)\,, \qquad (6.4)$$

which can be written in a spinor notation

$$\Phi_\sigma(x)=\frac{1}{\sqrt{2}}\begin{pmatrix}\phi_\sigma(\mathbf{x})\\ \phi_\sigma(\mathbf{x}+\pi/\kappa)\end{pmatrix}$$

as

$$\mathcal{H}=\frac{\kappa L}{\pi}\sum_{\sigma}\int_0^{\pi/\kappa}dx\,\Phi^+_\sigma(x)\left[\varepsilon\!\left(k_0-i\frac{d}{dx}\right)\sigma_1+\Delta(x)\,\sigma_3\right]\Phi_\sigma(x)-\frac{N\kappa}{\pi\lambda}\int_0^{\pi/\kappa}\Delta^2(x)\,dx\,, \qquad (6.5)$$

where σ_1 and σ_3 are Pauli matrices. We now make the continuum approximation by linearising $\varepsilon(k)$ at $k_0=\frac{\pi}{2a}$; i.e. $\varepsilon\!\left(k_0-\frac{id}{dx}\right)\approx i\hbar v\,\frac{d}{dx}$ with $\hbar v=\frac{\partial\varepsilon_{\mathbf{k}}}{\partial k}$ at $k=k_0$. We also rotate $\Phi_\sigma=\mathrm{U}\Psi_\sigma$ with $\mathrm{U}=\frac{1}{\sqrt{2}}\begin{pmatrix}1&i\\i&1\end{pmatrix}$. This gives

$$\mathcal{H} = \sum_{\sigma} \int_0^L dx\, \Psi_\sigma^+(x) \left[-i\hbar v \sigma_1 \frac{d}{dx} - \sigma_2 \Delta(x)\right] \Psi_\sigma(x) + \frac{1}{\lambda a} \int_0^L \Delta^2(x)\, dx \tag{6.6}$$

We now write the free energy (5.4) in terms of the transformed Δ_s. Measuring lengths in units of a and energy, $\hbar v/a$, we find

$$F - F_N = \frac{\kappa}{\pi^2} \int_0^{\pi/\kappa} \left[c_1\Delta^2 + c_2\Delta^4 + c_3\Delta^6 + (c_2 + 5c_3\Delta^2)\Delta'^2 + \frac{c_3}{2}\Delta''^2 + \dots\right] dx \tag{6.7}$$

The Euler-Lagrange equation for the Δ which minimises F is

$$c_3\left(\Delta^{(4)} - 10\Delta\Delta'^2 - 10\Delta^2\Delta'' + 6\Delta^5\right) - 2c_2(\Delta'' - 2\Delta^3) + 2c_1\Delta = 0\,. \tag{6.8}$$

This is satisfied by the solution

$$\Delta'' + 2A\Delta - 2\Delta^3 = 0, \tag{6.9}$$

whose first integral is

$$\Delta'^2 = \Delta^4 - 2A\Delta^2 + a_0 = (\Delta^2 - \Delta_1^2)(\Delta^2 - \Delta_2^2)\,, \tag{6.10}$$

where $\Delta_{1,2}^2 = A \pm \sqrt{A^2 - a_0}$, provided that the two constants $A \geqslant \sqrt{a_0} \geqslant 0$ fulfil the condition

$$c_3(a_0 + 2A^2) + 2c_2A + c_1 = 0\,. \tag{6.11}$$

Integrating (6.10) and inverting the elliptic integral gives

$$\Delta(x) = \Delta_2\, sn(\Delta_1 x, k_1) \tag{6.12}$$

where sn is the Jacobian elliptic function with modulus $k_1 = \Delta_1/\Delta_2$. The integral of (6.7) extends over a half period π/κ of $\Delta(x)$ which is given by

$$\frac{\pi}{\kappa} = \frac{2K(k_1)}{\Delta_1}, \tag{6.13}$$

where $K(k_1)$ is the complete elliptic integral of the first kind.

7. The Electron Spectrum

The Schrödinger equation of the Hamiltonian (6.6) is

$$\sum_{\beta} \left(-i\hbar\sigma_1 \frac{d}{dx} - \sigma_2\Delta(x)\right)_{\alpha,\beta} \psi_\beta(x) = \mathcal{E}\psi_\alpha(x)\,, \tag{7.1}$$

where $\psi_\alpha(x) = \psi(x + \alpha\pi/\kappa)$ and $\alpha = 0, 1$.

Using the periodicity of $\Delta(x)$ in (6.3) and denoting ψ_0 by v and ψ_1 by $u = v(x + \pi/\kappa)$, the equations for u and v decouple to give

$$v'' + (\mathcal{E}^2 - Q(x))v = 0\,, \quad Q(x) = \Delta^2(x) - \Delta'(x)\,, \tag{7.2}$$

$$u'' + (\mathcal{E}^2 - P(x))u = 0\,, \quad P(x) = \Delta^2(x) + \Delta'(x)\,. \tag{7.3}$$

Differentiating $Q(x)$ twice and using (6.9) and (6.10), we find

$$Q'' - 3Q^2 + 2AQ = -a_0 \tag{7.4}$$

whose first integral is

$$Q'^2 = 2(Q^3 - AQ^2 - a_0 Q + Aa_0) \tag{7.5}$$

where the integration constant Aa_0 is determined as the value of Q'^2 at the zeros of Q.

We can reduce (7.5) to Weierstrass's normal form by putting $2t = Q - A/3$ to obtain

$$t'^2 = g(t) \equiv 4(t-e_1)(t-e_2)(t-e_3) \; ; \tag{7.6}$$

$$e_1 = A/3 \;, \quad e_2 = -A/6 + \sqrt{a_0}/2 \;, \quad e_3 = -A/6 - \sqrt{a_0}/2 \;.$$

Integrating gives

$$x = \int_{e_3}^{t} \frac{d\tau}{\sqrt{g(\tau)}} = \left(\int_{\infty}^{t} - \int_{\infty}^{e_3}\right) \frac{d\tau}{\sqrt{g(\tau)}} = \int_{\infty}^{t} \frac{d\tau}{\sqrt{g(\tau)}} - \omega' \tag{7.7}$$

so that

$$Q(x) = 2t + e_1 = e_1 + 2\wp(x+\omega') \tag{7.8}$$

is given by Weirstrass's $\wp$ function, the real and imaginary half periods of which are

$$\omega = \frac{\pi}{\kappa} = \int_{e_3}^{e_2} \frac{dt}{\sqrt{g(t)}} = \frac{K(k)}{\sqrt{e_1 - e_3}} \;, \quad \omega' = \int_{\infty}^{e_3} \frac{d\tau}{\sqrt{g(\tau)}} = \frac{i\,K(k')}{\sqrt{e_1 - e_2}} \tag{7.9}$$

where

$$k^2 = \frac{e_2 - e_3}{e_1 - e_3} \;, \quad k'^2 = 1 - k^2 = \frac{e_1 - e_2}{e_1 - e_3} \tag{7.10}$$

are the squares of the modulus and its complement.

Equation (7.2) now appears as a Lamé equation in the Weierstrassian form

$$v'' + (\mathcal{E}^2 - e_1 - 2\wp(x+\omega'))\,v = 0 \;. \tag{7.11}$$

We have a one dimensional Bloch problem. and its physically interesting solutions are quasi-periodic. Therefore we write

$$v(x) = w(x)\exp\left[ic\int_0^x \Phi(x')\,dx'\right] \; ; \quad w(x+2\omega) = w(x), \quad \Phi(x+2\omega) = \Phi(x) \;. \tag{7.12}$$

This transforms (7.11) into

$$w'' + (\mathcal{E}^2 - e_1 - 2\wp(x+\omega') - c^2\Phi^2)\,w + ic(\Phi' w + 2\Phi w') = 0 \;. \tag{7.13}$$

In eqn (7.10) there are two functions. This means we can specify one of them so as to satisfy $\Phi' w + 2w\Phi' = 0$, i.e. $\Phi = 1/w^2$. Then (7.13) becomes

$$w'' + (\mathcal{E}^2 - e_1 - 2\wp(x+\omega'))\,w - c^2/w^3 = 0 \;. \tag{7.14}$$

It is convenient to change the independent variable to $t = \wp(x+\omega')$ in (7.12). On using (7.6) this gives

$$g(t)\ddot{w} + \tfrac{1}{2}\dot{g}(t)\dot{w} + (\mathcal{E}^2 - e_1 - 2t)w - c^2/w^3 = 0\,, \tag{7.15}$$

where dots denote differentiation with respect to t. Firstly, let us consider $c = 0$. In this case (7.15) is the algebraic form of Lamé's equation. The prefactor of t is usually written $n(n+1)$ and determines the degree n of the Lamé equation. Here, $n = 1$. For integer n, there are $2n+1$ terminating solutions — the Lamé polynomials (Erdelyi *et al.* 1958). It is easily verified that $w_{oi} = (t - e_i)^{\frac{1}{2}}$, $(i = 1, 2, 3)$ are the three terminating solutions for $c = 0$. The corresponding energy eigenvalues are

$$\mathcal{E}_i^2 = e_1 - e_i. \tag{7.16}$$

In the same way, $w = (t-e)^{\frac{1}{2}}$ is a solution of (7.15) if

$$e = e_1 - \mathcal{E}^2\,, \quad c(\mathcal{E}) = \pm\sqrt{-g(e)}/2 = \pm\,\mathcal{E}\sqrt{(\mathcal{E}^2 - \mathcal{E}_2^2)(\mathcal{E}^2 - \mathcal{E}_3^2)}\,. \tag{7.17}$$

It is now clear that $\mathcal{E}_2$ and $\mathcal{E}_3$ are band edges. There is a band gap for $\mathcal{E}_2^2 < \mathcal{E}^2 < \mathcal{E}_3^2$ for which c is imaginary. These are the only gaps in the spectrum.

Substituting into (7.12) we have

$$v(x) = \left(\frac{\wp(x+\omega') - e}{2L(\overline{\wp} - e)}\right)^{\frac{1}{2}} \exp\left(ic\int_0^x \frac{dx'}{\wp(x'+\omega') - e}\right) \tag{7.18}$$

The normalization has been defined so that

$$\int_0^L (|u|^2 + |v|^2)\,dx = 1\,,$$

so that

$$\overline{\wp} = \frac{1}{\omega}\int_0^{\omega} \wp(x+\omega')\,dx = \frac{1}{\omega}\int_{e_3}^{e_2} \frac{t\,dt}{\sqrt{g(t)}} = e_1 - (e_1 - e_3)\frac{E(k)}{K(k)}$$

where $E(k)$ is the complete elliptic integral of the second kind, N.B. $e_3 < \overline{\wp} < e_2$. To determine the wave number of the Bloch function (7.10), we use the relation $v(x+2\omega) = e^{2iq\omega}v(x)$ to obtain (Gröbner and Hofreiter 1958)

$$q(\mathcal{E}) = \frac{c}{2\omega}\int_0^{2\omega}\Phi\,dx = \frac{c}{\omega}\int_0^{\omega}\frac{dx}{\wp(x+\omega') - e} = \frac{c}{\omega}\int_{e_3}^{e_2}\frac{dt}{(t-e)\sqrt{g(t)}}$$

$$= \pm\frac{1}{K(k)}\sqrt{\frac{(e_1-e)(e_2-e)}{e_3-e}}\;\Pi\left(\frac{e_2-e_3}{e_3-e}, k\right), \tag{7.20}$$

where $\Pi(s,k)$ is the complete elliptic integral of the third kind. With $e = e_1 - \mathcal{E}^2$ this is the inverse of the energy band function $\mathcal{E}(q)$. The splitting into two bands becomes explicit on representing the complete elliptic integral of the third kind in terms of complete and incomplete elliptic integrals of the first and second kinds (Abramowitz and Stegun 1964). Thus

$$q(\mathcal{E}) = \begin{cases} \dfrac{\delta}{K(k)}\, S\left(\arc\sin\left(\dfrac{|\mathcal{E}|}{k'\delta}\right),\, k\right) & \text{for} \quad |\mathcal{E}| \leqslant k'\delta \\ \sqrt{\dfrac{(\mathcal{E}^2-\delta^2)(\mathcal{E}^2-k'^2\delta^2)}{\mathcal{E}^2}} + \dfrac{\delta}{K(k)}\, S\left(\arc\sin\left(\dfrac{\delta}{|\mathcal{E}|}\right),\, k\right) & \text{for} \quad |\mathcal{E}| \geqslant \delta . \end{cases} \tag{7.21}$$

Here we have introduced the parameter $\delta = (e_1 - e_3)^{\frac{1}{2}}$ so that

$$e_1 - e_3 = \delta^2 , \quad e_1 - e_2 = k'^2\delta^2 , \quad e_2 - e_3 = k^2\delta^2 \tag{7.22}$$

and the function $S(\phi,k)$ defined by

$$S(\delta,k) = (E(k) - K(k))\, F(\phi,k') + K(k)\, E(\phi,k') \tag{7.23}$$

where $F(\phi,k')$ and $E(\phi,k')$ are the incomplete integrals of the first and second kind. From the Legendre relation (Abramowitz and Stegun 1964) we have $S(\pi/2\ , k) = \pi/2$. Therefore at the band edges $\mathcal{E} = \mathcal{E}_2 = k'\delta$ and $\mathcal{E} = \mathcal{E}_3 = \delta$, we get

$$q(\mathcal{E}_2) = q(\mathcal{E}_3) = \pm\, \frac{\pi\delta}{2K(k)} = \pm\, \frac{\pi}{2\omega} = \pm\, \frac{\kappa}{2} \tag{7.24}$$

as the boundaries of the first Brillouin zone.

We obtain the density of states per unit length and spin from (7.21) and (7.19)

$$\rho(\mathcal{E}) = \frac{2}{2\pi}\left|\frac{dq}{d\mathcal{E}}\right| = \frac{|\mathcal{E}^2 - \wp - e_1|}{\pi\sqrt{(\mathcal{E}^2-\delta^2)(\mathcal{E}^2-k'^2\delta^2)}}\,[\theta(\mathcal{E}^2-\delta^2) + \theta(k'^2\delta^2-\mathcal{E}^2)] \tag{7.25}$$

This expression holds in both bands with θ the step function.

8. The Gap Equation and the Free Energy

The main objective of this section is to derive the conditions which determine the values of the parameters δ and k, (7.20). These values will be determined by minimising the free energy F given by

$$F = -\,\frac{k_{\mathbf{B}}T}{L}\, \ln\, Tr\left(e^{-\beta(\mathcal{H}-\mu N)}\right) \tag{8.1}$$

where μ is the chemical potential and N the particle number operator. We obtain the gap equation by writing

$$\delta F/\delta\Delta(x) = \left\langle \delta\mathcal{H}/\delta\Delta(x)\right\rangle = 0 ,$$

where the angular brackets denote the grand canonical average. Using the Hamiltonian (6.6), we find

$$\begin{aligned} \Delta(x) &= \frac{\lambda}{2} \sum_{\sigma} \left\langle \Psi^+_{\sigma}(\alpha)\, \sigma_2\, \Psi_{\sigma}(x)\right\rangle \\ &= \lambda \sum_{\mathbf{q}} f(\mathcal{E}_{\mathbf{q}} - \mu)\; i(u^*(x)\, v(x) - v^*(x)\, u(x)) \qquad (8.2) \\ &= -\lambda \sum_{\mathbf{q}}{}' \frac{f(\mathcal{E}_{\mathbf{q}} - \mu)}{\mathcal{E}_{\mathbf{q}}} \left(\frac{d}{dx} + 2\,\Delta(x)\right) v^*(x)\, v(x) , \end{aligned}$$

where we have used the fact that the solutions of (7.1) are electron eigenfunctions. Substituting for $v(x)$ from (7.16), and using (7.8) and (7.2), we see that (8.2) is identically satisfied provided that

$$\frac{1}{\lambda} = -\frac{1}{L}\sum_{\mathbf{q}} \frac{\mathcal{E}_{\mathbf{q}}\, f(\mathcal{E}_{\mathbf{q}} - \mu)}{\mathcal{E}_{\mathbf{q}}^2 + \wp - e_1}\,. \tag{8.3}$$

Separating this expression into terms which depend on the temperature and those which don't gives the form

$$\ell n\left(\frac{\delta_0}{k\delta}\right) = \int_{\delta}^{\infty} \Big(f(\mathcal{E}+\mu) + f(\mathcal{E}-\mu)\Big)\frac{\mathcal{E}\,d\mathcal{E}}{\sqrt{(\mathcal{E}^2-\delta^2)(\mathcal{E}^2-k'^2\delta^2)}}$$
$$+ \int_{0}^{k'\delta} \Big(f(\mu-\mathcal{E}) - f(\mu+\mathcal{E})\Big)\frac{\mathcal{E}\,d\mathcal{E}}{\sqrt{(\delta^2-\mathcal{E}^2)(k'^2\delta^2-\mathcal{E}^2)}}\,. \tag{8.4}$$

Having linearised the dependence of the free particle energy on its momentum (6.4), it is now necessary to introduce a momentum cut-off at $q_{\mathbf{s}}$ which was $\pi/2$ in the lattice tight binding model, $q_{\mathbf{s}} \gg \delta$. This results in a lower bound for the energy spectrum $-|\mathcal{E}_{\mathbf{s}}| \leqslant \mathcal{E}_{\mathbf{q}}$. The gap at $T = 0$ is related to $q_{\mathbf{s}}$ by

$$\delta_0 = 2q_{\mathbf{s}}\, e^{-\pi/\lambda}\,. \tag{8.5}$$

Equation (8.4) gives us one of the conditions necessary to determine k and δ. The other condition comes from minimising F, the free energy per unit length given by

$$F = -\frac{2k_{\mathbf{B}}T}{L}\sum_{\mathbf{q}} \ell n\left[1 + e^{-\beta(\mathcal{E}_{\mathbf{q}} - \mu)}\right] + \frac{1}{\lambda L}\int_0^L \Delta^2(x)\,dx\,, \tag{8.6}$$

which is evaluated using the density of states (7.25) and the gap function (6.12). The lattice part, written in terms of δ and k using Landen's transform of $E(k_1)$ and $K(k_1)$ is

$$\frac{1}{\lambda L}\int_0^L \Delta^2(x)\,dx = \frac{\Delta_1^2}{\lambda}\left(1 - \frac{E(k_1)}{K(k_1)}\right) = \frac{\delta^2}{\lambda}\left(1 + k'^2 - 2\,\frac{E(k)}{K(k)}\right). \tag{8.7}$$

The electronic part $F_{\mathbf{e}}$ can be divided into terms which depend on T and those which don't

$$F_{\mathbf{e}} = -2k_{\mathbf{B}}T\left\{\int_{-\mathcal{E}_{\mathbf{s}}}^{k'\delta} \left[-\beta(\mathcal{E}-\mu) + \ell n\Big(1 + e^{\beta(\mathcal{E}-\mu)}\Big)\right]\rho(\mathcal{E})\,d\mathcal{E}\right.$$
$$\left. + \int_{\delta}^{\infty} \ell n\Big(1 + e^{-\beta(\mathcal{E}-\mu)}\Big)\rho(\mathcal{E})\,d\mathcal{E}\right\}$$

$$= -\frac{\mu}{\pi}(2q_{\mathbf{s}}+\kappa) + 2\int_{-\mathcal{E}_{\mathbf{s}}}^{-\delta} \mathcal{E}\rho(\mathcal{E})\,d\mathcal{E}$$

$$-2k_{\mathbf{B}}T\left\{\int_{-\mathbf{k}'\delta}^{\mathbf{k}'\delta} \ln\left(1+e^{\beta(\mathcal{E}-\mu)}\right)\rho(\mathcal{E})\,d\mathcal{E}\right.$$

$$\left.+\int_{\delta}^{\infty} \ln\left[\left(1+e^{-\beta(\mathcal{E}+\mu)}\right)\left(1+e^{-\beta(\mathcal{E}-\mu)}\right)\right]\rho(\mathcal{E})\,d\mathcal{E}\right\}, \tag{8.8}$$

where $-\mathcal{E}_{\mathbf{s}} = -|\mathcal{E}_{q_{\mathbf{s}}}|$ is the cut off energy. The first integral is easily performed

$$2\int_{-\mathcal{E}_{\mathbf{s}}}^{-\delta} \mathcal{E}\rho(\mathcal{E})\,d\mathcal{E} = -\frac{2}{\pi}\left[\delta^2\left(1-\frac{k^2}{2}-\frac{E}{K}\right)\ln\frac{2\mathcal{E}_{\mathbf{s}}}{k\delta} + \frac{\mathcal{E}_{\mathbf{s}}^2}{2} - \frac{1+k'^2}{4}\delta^2\right] \tag{8.9}$$

where by inversion of (7.21)

$$\mathcal{E}_{\mathbf{s}} = q_{\mathbf{s}} + \frac{\delta^2}{q_{\mathbf{s}}}\left(\frac{1+k'^2}{2} - \frac{E}{K}\right) + \ldots \tag{8.10}$$

Here we have omitted the argument k of the elliptic functions E and K, and we shall continue to do this. The correction to $q_{\mathbf{s}}$ in this expression is only important in the $\mathcal{E}_{\mathbf{s}}^2$ term in (8.9). With this substitution, we find for the total free energy at $T=0$

$$F(0,\mu) = -\frac{1}{\pi}\left\{q_{\mathbf{s}}(q_{\mathbf{s}}+2\mu) + \frac{\pi\mu\delta}{K} + \delta^2\left[\left(1+k'^2-2\frac{E}{K}\right)\left(\ln\left(\frac{\delta_0}{k\delta}\right)+1\right) - \frac{1+k'^2}{2}\right]\right\} \tag{8.11}$$

On integrating by parts and using (7.24) we find for the temperature dependent part of F

$$F(T,\mu) - F(0,\mu) = \frac{\delta}{\beta K}\ln\frac{\left(1+e^{-\beta(\delta+\mu)}\right)\left(1+e^{-\beta(\delta-\mu)}\right)}{\left(1+e^{-\beta(\mu-k'\delta)}\right)\left(1+e^{-\beta(\mu+k'\delta)}\right)}$$

$$+\frac{2}{\pi}\int_0^{k'\delta}\left(f(\mu-\mathcal{E}) - f(\mu+\mathcal{E})\right)q(\mathcal{E})\,d\mathcal{E}$$

$$-\frac{2}{\pi}\int_{\delta}^{\infty}\left(f(\mathcal{E}+\mu) + f(\mathcal{E}-\mu)\right)q(\mathcal{E})\,d\mathcal{E} \tag{8.12}$$

where $q(\mathcal{E})$ is given by (7.21).

We can now obtain the second condition by minimising F with respect to variations in k and δ. It is not difficult to show that if (8.4) is satisfied, $\partial F/\partial k = 0$ is equivalent to $\partial F/\partial\delta = 0$. In particular, using (8.4) we find that

$$\frac{\partial F}{\partial\delta} = \frac{1}{K}G(T,\mu,\delta,k'), \tag{8.13}$$

where

$$G(T,\mu,\delta,k') = -\mu + \frac{2}{\pi}\delta E + \frac{1}{\beta}\ln\frac{\left(1+e^{-\beta(\delta+\mu)}\right)\left(1+e^{-\beta(\delta-\mu)}\right)}{\left(1+e^{-\beta(\mu-\mathbf{k}'\delta)}\right)\left(1+e^{-\beta(\mu+\mathbf{k}'\delta)}\right)} +$$

$$+\frac{2}{\pi}\int_0^{\mathbf{k}'\delta}\left(f(\mu-\mathcal{E}) - f(\mu+\mathcal{E})\right)\left[\mathcal{S}\left(\arcsin\left(\frac{\mathcal{E}}{k'\delta}\right), k\right) + \right.$$

$$\left. + \mathcal{E}\delta\,\frac{E-(1+k'^2-\mathcal{E}^2/\delta^2)K}{\sqrt{(\delta^2-\mathcal{E}^2)(k'^2\delta^2-\mathcal{E}^2)}}\right]d\mathcal{E} -$$

$$-\frac{2}{\pi}\int_\delta^\infty\left(f(\mathcal{E}+\mu) + f(\mathcal{E}-\mu)\right)\left[S\left(\arcsin\left(\frac{\delta}{\mathcal{E}}\right), k\right) - \right.$$

$$\left. - \mathcal{E}\delta\,\frac{E-(k'^2\delta^2/\mathcal{E}^2)K}{\sqrt{(\mathcal{E}^2-\delta^2)(\mathcal{E}^2-k'^2\delta^2)}}\right]d\mathcal{E} \qquad (8.14)$$

In eqn. (8.13) we can either have that $1/K = 0$ in which case $k = 1$, $k' = 0$ which leads to the commensurate phase, with (8.4) reducing to the familiar finite temperature gap equation, or of $k \neq 1$, $G = 0$ together with (8.4) gives us the incommensurate phase.

If we now impose a magnetic field on the system, we have effectively two fermi energies, $\mu + h$ for the electrons whose spins are up, and $\mu - h$ for those with spin down. The modification of the equations in this and the two previous sections is quite straightforward (Doman 1983); wherever an expression involving μ appears, this is replaced by the average of this expression with μ replaced by $\mu + h$ and with μ replaced by $\mu - h$. We can see from (8.4) and (8.11) that for small h, the change in k and δ will be very small.

The magnetisation per unit length is given by

$$m = \int_\delta^\infty \frac{f(\mathcal{E}\pm\mu-h) - f(\mathcal{E}\pm\mu+h)}{\sqrt{(\mathcal{E}^2-\delta^2)}\sqrt{(\mathcal{E}^2-k'^2\delta^2)}}\left(\mathcal{E}^2-\delta^2 E/K\right)\,d\mathcal{E}$$

$$+\int_0^{\mathbf{k}'\delta}\frac{f(\mu\pm\mathcal{E}-h) - f(\mu\pm\mathcal{E}+h)}{\sqrt{(\delta^2-\mathcal{E}^2)}\sqrt{(k'^2\delta^2-\mathcal{E}^2)}}\left(\delta^2 E/K - \mathcal{E}^2\right)\,d\mathcal{E}. \qquad (8.15)$$

From this expression we see that if $h \ll \mu - k'\delta$ and $h \ll \delta - \mu$, the magnetisation and the susceptibility will be very small, while at $T = 0$, h must exceed a critical value h_c in order that $m \neq 0$ — a conclusion in agreement with Brazovski *et al.* (1981).

Another interesting result arises from the symmetry between the dependence of k and δ on μ and h. For $\mu = 0$ the conditions determining k and δ have the form (8.4) and (8.11) with μ replaced by h. It follows then that for large enough h, we have a soliton band appearing in incommensurate phase. The physical interpretation of this is that this band will be occupied by spin up electrons only. This is because the effective fermi energies of the spin up and spin down electrons are $+h$ and $-h$ respectively.

At $T=0$ we have from (8.11) with $q_s = \pi/2$ the electron density per unit length

$$n = -\frac{\partial F}{\partial \mu} = 1 + \frac{\delta}{K(k)} = 1 + \frac{1}{\omega} = 1 + \frac{\kappa}{\pi}. \tag{8.16}$$

Thus the excess number of electrons $L(n-1)$ is equal to the number of states per spin in the middle bands and equal to the number of periods of $\Delta(x)$. The electron charge density $|\phi(x)|^2$ is obtained from $v(x)$ and $u(x)$, see eqns. (7.2), (7.3), (6.5), (6.6), (7.18) and (7.8)

$$|\phi_0(x)|^2 = |v(x)|^2 + |u(x)|^2 = \frac{\Delta^2(x) - 3e_1 + 2\mathcal{E}^2}{4L(\bar{\wp} - e_1 + \mathcal{E}^2)}. \tag{8.17}$$

From (6.12) and (6.10) we have

$$\Delta^2(x) < \Delta_2^2 = A - \sqrt{A^2 - a_0} = 3e_1 - 2\sqrt{(e_1-e_2)(e_1-e_3)} = 3e_1 - 2k'\delta^2 \tag{8.18}$$

So that for $\mathcal{E}^2 < k'^2\delta^2$ the numerator (and the denominator) in (8.17) are negative. This means that for these energies the charge density is greatest where $|\Delta|$ is smallest, i.e., at the kinks. For the other bands where $\mathcal{E}^2 > \delta^2$, the charge density is greatest where $|\Delta(x)|$ is largest. For small h the extra electrons if $n > 1$, or the holes for $n < 1$ will be located at the kinks. If $\mu = 0$ and $h \neq 0$, the kink band will only be occupied by spin up electrons. We may regard this as a state with extra spin up electrons and spin down holes localised on the kinks.

9. The Commensurate Incommensurate Phase Boundary

For the incommensurate phase at $T=0$ we have for $h < \mu - k'\delta$ and $h < \delta - \mu$

$$\delta_0 = k\delta \quad \text{and} \quad \mu = \frac{2}{\pi}\delta E(k). \tag{9.1}$$

This tells us that the energy needed to create another kink is $(2/\pi)\,\delta E(\delta_0/\delta)$. This is always less than δ, the energy to put an electron in the upper band since for $0 \leqslant k \leqslant 1$, $\frac{\pi}{2} \geqslant E(k) \geqslant 1$.

From (7.9) we see that as $k \to 1$, $\kappa \to 0$, and (8.4) goes continuously over to the gap equation for the commensurate phase. This means that the commensurate incommensurate phase transition is continuous. This is true also when $h \neq 0$. It follows then that on the phase boundary, from eqns. (8.14), $G(T, \mu, \delta = \delta_c(T,\mu), k' = 0) = 0$, where $\delta_c(T,\mu)$ is the commensurate gap for T and μ as given by (8.4) with $k' = 0$. We have

$$2\,\ell n(\delta_0/\delta_c) = \int_{\delta_c}^{\infty} d\mathcal{E} \sum f(\mathcal{E} \pm \mu \pm h)/\sqrt{\mathcal{E}^2 - \delta_c^2} \tag{9.2}$$

and

$$0 = -\mu + 2\delta_c/\pi + \frac{1}{2\beta}\sum \ell n\left(1 + \exp\left[-\beta(\delta_c \pm \mu \pm h)\right]\right)$$
$$-\frac{1}{\beta}\sum \ell n\left(1 + \exp\left[-\beta(\mu \pm h)\right]\right)$$
$$-\frac{1}{\pi}\sum\int_{\delta_c}^{\infty} d\mathcal{E}\, f(\mathcal{E} \pm \mu \pm h)\left[\text{arc sin}\left(\frac{\delta_c}{\mathcal{E}}\right) - \frac{\delta_c}{\sqrt{\mathcal{E}^2 - \delta^2}}\right] \tag{9.3}$$

where the sums are over all possible combinations of $\pm$ signs.

In the figures we see the complete phase diagram for zero magnetic field and for $h = 0.5\,\delta_0$, $0.625\,\delta_0$ and $h = \delta_0$. These figures show the effect of the magnetic field in depressing the transition temperature. For $h = 0.5\,\delta_0$ we see the commensurate-incommensurate phase boundary move out to higher μ. For $h > 0.6082\,\delta_0$ we find two Leung points for $\mu > 0$. At first we have a narrow band of temperatures in which the system is in the incommensurate phase. As h increases, this band broadens until for $h \gtrsim 0.637\,\delta_0$ the commensurate-incommensurate phase boundary detaches itself from the $\mu = 0$ line. For much larger values of h see Fig. 5, this phase boundary becomes almost vertical like the larger μ one.

At $T = 0$ Brazovski *et al.* (1981) have shown that for $h \neq 0$ there is a gap for each of the two fermi wave-vectors of the up and down spin electrons. This will give four gaps altogether with five energy bands. However, using energy considerations we can see that transition to such a state when $h \leqslant \delta_0$ will be at a much lower temperature. The Peierls transition is driven by the reduction of the energy of the electron gas on the formation of the gap at the expense of an increase in the lattice energy by the distortion. The energy reduction on the formation of the gap will be large if the slope of $\varepsilon_{\mathbf{k}}$ is large. However, if the extra CDW has a wavelength close to an existing CDW, the slope of the $\varepsilon_{\mathbf{k}}$ near to its band edge for the existing CDW will be small. This means the energy gained by the electron gas on forming the gap will be small too.

10. Conclusion

We have examined the static properties of a CDW system in the mean field approximation. It would be very interesting also to examine the magnetic field effects on the transport properties of the solitons. The kinetic properties of the solitons for CDW have already been studied by Artemenkov and Volkov (1980, 1981).

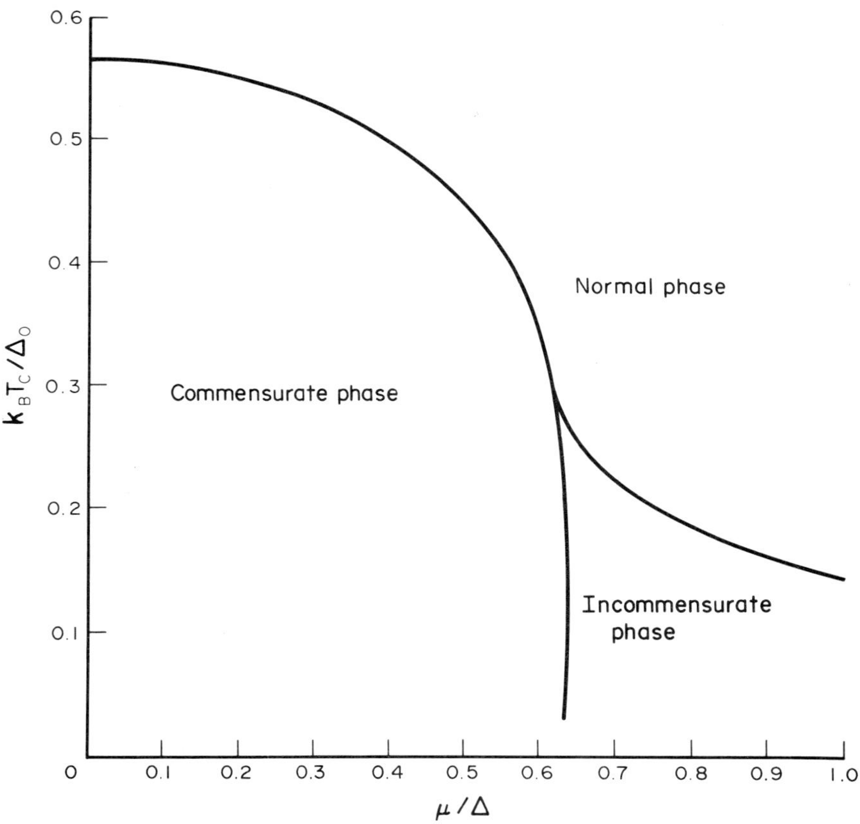

Fig. 3 The phase diagram for $h = 0$

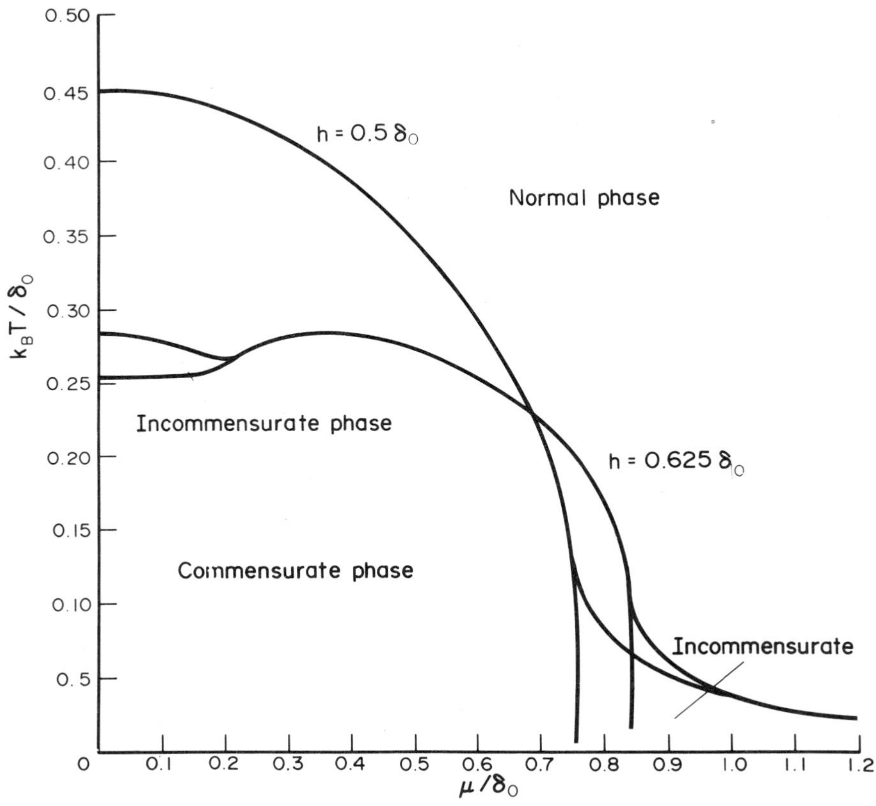

Fig. 4 The phase diagrams for h = 0.5 and h = 0.625

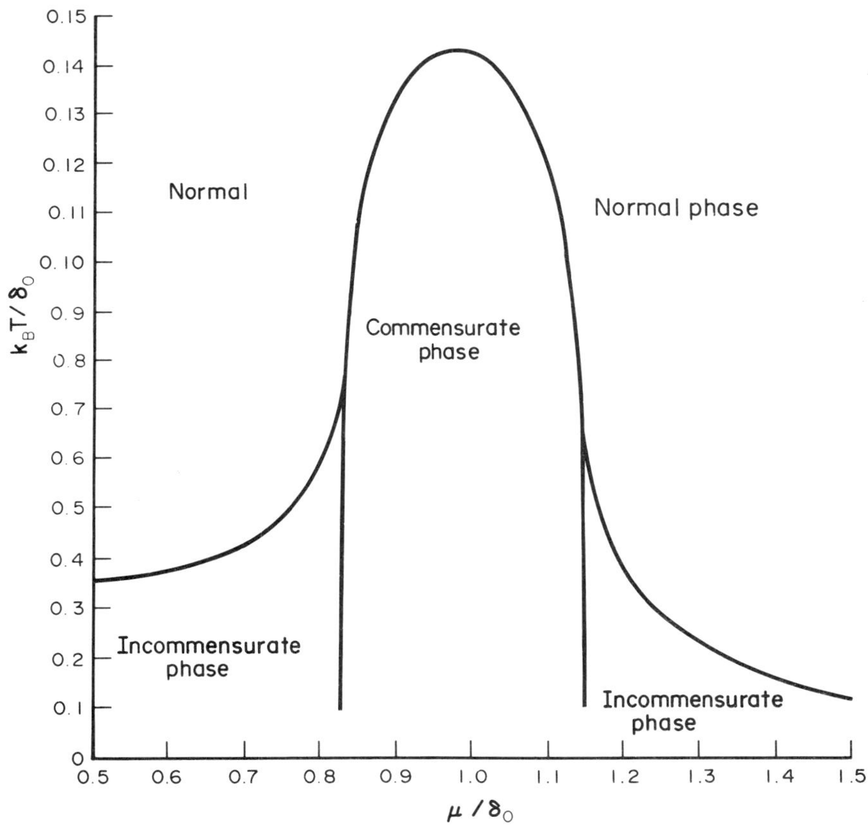

Fig. 5 The phase diagram for $h = 1$

References

Abramovitz, M. and Stegun, I.A., 1970. *Handbook of Mathematical Functions*, Dover Publications Inc. New York.
Artemenkov, S.N. and Volkov, A.F., 1980. Zh. Eksp. Teor. Fiz., **80**, 2018; (Sov. Phys. JETP, **53**, 1050).
Artemenkov, S.N. and Volkov, A.F. 1981. Zh. Eksp. Teor. Fiz, **81**, 1872; (Sov. Phys. JETP, **54**, 992).
Brazovski, S.A., Dzyaloshinski, I.E. and Kirova, N.N., 1981. Zh. Eksp. Teor. Fiz. **81**, 2279; (Sov. Phys. JETP, **54**, 1209).
Brazovski, S.A., Gordyunin, S.A. and Kirova, N.N., 1980, Plasma Zh. Eksp. Teor. Fiz., **31**, 486. (Sov. Phys. JETP Lett., **31**, 456)
Doman, B.G.S., 1981a, J. Phys. C: Solid State Phys., **14**, 1457.
Doman, B.G.S., 1981b, J. Phys. C: Solid State Phys., **14**, 3393.
Doman, B.G.S., 1983, J. Phys. C: Solid State Phys., **16**, 2069.
Erdelyi, A., Magnus, W., Oberhettinger, F. and Tricomi, F., 1958, *Higher Transcendental Functions*, Bateman Manuscript Project, McGraw-Hill Publ. Co., New York; Toronto; London.
Fröhlich, H. 1954, Proc. Roy. Soc., **A233**, 296.
Gröbner, W. and Hofreiter, N., 1958, *Integraltafel*, Springer Verlag Wien, Innsbruck.
Kotani, A., 1977, J. Phys. Soc. Japan, **42**, 408.
Kotani, A. and Harada, I., 1980, J. Phys. Soc. Japan, **49**, 535.
Leung, M.C., 1975, Phys. Rev. **B11**, 4272.
Mertsching, J. and Fischbeck, H.J., 1980, Phys. Stat. Sol. (b), **99**, 51.
Mertsching, J. and Fischbeck H.J., 1981, Phys. Stat. Sol. (b), **103**, 783.
Peierls, R. 1956, *Quantum Theory of Solids*, Oxford University Press.

On the Electrodynamics of Two-dimensional (Surface) Superconductors

V. L. Ginzburg
P. N. Lebedev Physical Institute of the USSR Academy of Sciences, Moscow, USSR

1. Ordering and phase transitions in two-dimensional systems, for example on the surface of three-dimensional specimens, are undoubtedly of interest in a number of key problems in the physics of the condensed state. One such problem, which is the subject of the present paper, is two-dimensional (surface) superconductivity [1]. In ref. 1 we put forward the question of the possibility of the existence of superconductivity for a 'surface metal', i.e. in the case of a partially filled degenerate band of surface electron (Tamm's) levels Such a band may appear, in principle, on the surface of a three-dimensional metal and a three-dimensional dielectric. It is quite obvious how interesting it is to investigate a metallic and, particularly, a superconducting surface of a dielectric substance.

If we approach the problem in the spirit of the ordinary theory of superconductivity, in particular in the BCS approximation, then such a two-dimensional type of superconductivity is quite admissible as noticed in ref. 1. By analogy, a surface magnetic ordering (ferromagnetic or antiferromagnetic) and, say, surface ferromagnetism for a substance paramagnetic in a bulk, is possible in principle [2].

Papers 1, 2 published in 1964 discussed the problem only in the self-consistent field approximation, i.e. fluctuations were disregarded. Soon it was noticed, however, that when one takes into account fluctuations in two-dimensional, to say nothing of one-dimensional systems, in the case of a degenerate order parameter (of the type of macroscopical wave function $\psi = |\psi| e^{i\phi}$ for superfluidity or superconductivity), the ordinary long-range order cannot exist in equilibrium. At the same time, as is well known, superconductivity is observed also in very thin films and wires, as well as in layers of metal in layers compounds (see ref. 4, Ch.6, and the literature cited there). A large number of subsequent papers discussed this problem and it was eventually established that two-dimensional ordering (superconductivity, superfluidity, ferromagnetism, etc.) is possible even when fluctuations are taken into account (see ref.4, Ch. 1, 6, and ref.5, and the literature cited there).

2. The discussion of the properties of two-dimensional (surface) superconductors thus has a real physical meaning. In particular, this concerns the electrodynamics of surface superconductors (in this connection, see, e.g., ref.6). At first glance there is no problem here, and one can expect that a surface superconductor (say, when superconductivity is realized for surface-level electrons) will be, in an electrodynamic respect, quite similar to the ordinary superconducting film, but with a thickness of $2d \sim a \sim 3\cdot 10^{-8} - 10^{-7}$ cm, which corresponds to the surface state localization region in the z-direction perpendicular to the surface. This was just the conclusion of reference 1.

Here appears some doubt however, or, if you like, a paradox. Even in a very thin film of the three-dimensional type† the current density (in particular, the density of a superconducting current) may depend on z. This becomes especially obvious for electrons in a film, which are considered in a one-electron appproximation. Meanwhile for a two-dimensional 'film' the wave function of an electron in the same approximation has the form:

$$\psi = \psi_q(z)\, u(x,y) e^{-i(E/\hbar)t} , \tag{1}$$

where for free electrons

$$u(x,y) = Ae^{i(K_x x + K_y y)} , \quad E = E_q + \frac{\hbar^2 (K_x^2 + K_y^2)}{2m} .$$

Here the coordinates x and y lie in the plane of the film, and the index q characterizes the state of 'transverse' motion; apparently, in the lowest state $q = 0$, the function $\psi_0(z)$ has no nodes and is non-negligible only in the region $z \sim a$. In the state (1) the momentum is directed along $\boldsymbol{K}(K_x, K_y)$ and does not depend on z. The same seems to be true of the current density. On the other hand, the total current in a layer (film)

$$\mathcal{J} = \int j(z)\, dz = \frac{c}{4\pi} (H_{2t} - H_{1t}) ,$$

---

† **This means that quantization of levels in the z-direction normal to the field may be neglected; practically, this occurs at $d \gg a$.**

where $H_{1,2,t}$ are the values of the tangential component of the magnetic field strength respectively on either side of the film. If the total (transport) current $\mathcal{J} = 0$, the magnetic field H_t is continuous, and in the 'film' itself the field does not depend on z. Meanwhile for an ordinary superconducting film and at $\mathcal{J} = 0$, the magnetic field depends on z, being minimal at $z = 0$ (in the centre of the film), due to current in the film, which has a different sign at $z > 0$ and at $z < 0$. So, for a thin superconducting film (plate) of thickness $2d$ in the external field H_0 (parallel to the film) the field and the current density in the film have the form [7]

$$H = H_0 \frac{ch(\psi_0 z/\delta_0)}{ch(\psi_0 d/\delta_0)}, \quad j = -\frac{c}{4\pi}\frac{\psi_0}{\delta_0}\frac{sh(\psi_0 z/\delta_0)}{ch(\psi_0 d/\delta_0)}, \tag{2}$$

where $\psi_0^2 = \psi^2/\psi_\infty^2$, $\psi_\infty^2 = n_s$ is the concentration of 'superconducting electrons' in the field $H_0 \to 0$ and $\delta_0 = \sqrt{\frac{mc^2}{4\pi e^2 n_s}}$ is the field penetration depth at $H_0 \to 0$ when $\psi_0^2 = 1$; the condition that the film should be thin and ψ^2 should be z-independent reduces, practically, to the inequality $(d/\delta_0)^2 \ll 1$ (for more details see refs. 7,8; the ψ-theory of superconductivity[8] is used here).

3. Thus, the impression is produced that a two-dimensional (surface) superconductor differs from a thin superconducting film already qualitatively. This impression deepens in the light of the existing theory of layered superconducting compounds [9,10] (see also ref. 4, Ch.6), in which the function ψ_q is introduced for each layer as the order parameter. Differentiation is carried out with respect to x and y, and the general scheme is similar to ref. 8. In the z-direction (across the layer) the components of the field $\boldsymbol{H}$ are considered to be continuous (in the absence of transport current), and interaction between layers is taken into account by introducing into the free energy ψ_r—dependent term with different values of r.

Such an approach is justified in a certain approximation, and we would like only to emphasize here that a two-dimensional superconductor does not differ qualitatively from a three-dimensional superconducting film[†]. The point is that when speaking above of the surface state (with the wave function (1)). we disregard the diamagnetic current. The density of this current in a state with a zero momentum (i.e. at $\boldsymbol{K} = 0$) is equal to

$$\boldsymbol{j} = -\frac{e^2}{mc}\psi_q^2 \boldsymbol{A}. \tag{3}$$

[†] I take this opportunity to express my gratitude to L.N. Bulayevsky and G.F. Zharkov for discussion of the questions touched upon in the present paper.

All the differences from the theory [8] in which a macroscopic ψ-function is used are reduced in this case to the substitution in (3) of ψ_q for ψ. For a thin film (at $(d/\delta_0)^2 \ll 1$) in a weak field, in (2) one can put $\psi_0 = 1$ and $j_x \approx -(cH_0 z/4\pi\delta_0^2)$ (if the field $\boldsymbol{H}$ is directed along the y-axis, in a first approximation $A_x = H_0 z = -(4\pi\delta_0^2/c)\, j_x$, $H_0 = H_y$). Then the magnetic moment of unit film surface is given by

$$\mu_y = \frac{1}{2c}\int [\boldsymbol{rj}]_y d\boldsymbol{r} = -\frac{1}{12\pi}\left(\frac{d}{\delta_0}\right)^2 dH_0 = -\frac{e^2 n_s d^3 H_0}{3mc^2}\,. \tag{4}$$

If one proceeds from expression (3) disregarding the paramagnetic part of the current, then in a similar approximation

$$\mu_y = -\frac{e^2 H_0}{2mc^2}\int \psi_0^2 z^2\, dz\,,$$

i.e. the role of $n_s d^3$ is played by the quantity $\frac{3}{2}\int \psi_0^2 z^2 dz$, whose meaning becomes particularly clear when the normalization $\int \psi_0^2\, dz = n_s^*$ is used, where n_s^* is the concentration of 'superconducting electrons' in surface levels per unit surface area. Integration over z has been formally carried out above from $z = -\infty$ to $z = +\infty$ but in effect we mean integration over the region $|z| \sim a$, where the function $\psi_0(z)$ is not yet too small; it is also assumed that the ground level $q = 0$ is being considered. To the penetration depth $\delta_0 \sim 3\cdot 10^{-6}$ there corresponds the value

$$n_s = \frac{mc^2}{4\pi e^2 \delta_0^2} \sim 3\cdot 10^{22}\ \mathrm{cm}^{-3}\,.$$

Apparently, the maximum reasonable value $n_s \sim n_s a \sim 10^{15} - 10^{16}\ \mathrm{cm}^{-2}$.

4. To the moment (4) there corresponds the mean magnetic susceptibility per unit volume

$$\chi = \frac{\mu_y N}{H_0} = -\frac{1}{12\pi}\left(\frac{d}{\delta_0}\right)^2 dN\,,$$

where N is the number of layers parallel to the field in a unit volume. At $d \sim 3\cdot 10^{-8}$ cm and $\delta_0 \sim 3\cdot 10^{-6}$ cm, for example, $\chi \leqslant -10^{-6}$ since $N \leqslant \frac{1}{2d} \sim \frac{1}{a}$. Thus, the susceptibility corresponds to a more or less usual diamagnetic susceptibility (spin effects are and will be neglected). This is the more so as the superconducting transition usually does not at all affect the value of the diamagnetic contribution to χ. Indeed, the distance between the levels of 'transverse motion' in a layer

$$(E_1 - E_0) \sim \frac{\hbar^2}{2ma^2} \sim 10^{-13} - 10^{-12}\ \mathrm{erg}\,,$$

whereas the superconducting gap $\Delta \sim K_B T_c \sim 10^{-14}$ erg even at $T_c \sim 100$ K.

Nevertheless, even in a strictly parallel field, a surface superconductor is far from being equivalent to a corresponding layer in a normal state (we do not speak of the superconductivity for transport current, but have in mind the diamagnetic contribution to the magnetic susceptibility). The validity of this conclusion, which is the main subject of the present note, becomes obvious if one considers a surface superconductor of a cylindrical form.

For definiteness we shall consider the cylinder to be circular with a radius R and introduce cylindrical coordinates z, r, ϕ (the coordinate z is directed along the cylinder axis). The Schrödinger equation for the one-electron function $\psi_{\mathbf{qm}}(r,\phi)$ has the form

$$\left\{-\frac{\hbar^2}{2m}\left[\frac{1}{r}\frac{\partial}{\partial r}\left(r\frac{\partial}{\partial r}\right)+\frac{1}{r^2}\frac{\partial^2}{\partial\phi^2}\right]+V(r)+\right.$$

$$\left.+\frac{ie\hbar}{mcr}A_\phi\frac{\partial}{\partial\phi}+\frac{e^2}{2mc^2}A_\phi^2\right\}\psi_{\mathbf{qm}}=E_{\mathbf{qm}}\psi_{\mathbf{qm}}\,, \tag{5}$$

where the magnetic field

$$H_0=H_{\mathbf{z}}=\frac{1}{r}\frac{\partial}{\partial r}(rA_\phi)$$

is considered to be directed along the z-axis, so that one has taken into account only the component A_ϕ of the vector potential $\boldsymbol{A}$. In (5) $V(r)$ is some effective potential, which provides the surface level localization. The ground state $\psi_0(r)\equiv\psi_{00}(r)$, equivalent to the state $\psi_0(r)$ introduced in (1), satisfies the equation

$$\left(-\frac{\hbar^2}{2m}\frac{1}{r}\frac{\partial}{\partial r}\left(r\frac{\partial}{\partial r}\right)+V(r)-E_0\right)\psi_0=0\,.$$

To the states (1) with $K\neq 0$ there correspond the solutions of Eqn (5) that contain the phase factors $e^{\mathbf{i}\mathbf{m}\phi}$. The current density in the state $\psi_{\mathbf{qm}}$ has the component

$$j_\phi=\frac{ie\hbar}{2mr}\left(\psi_{\mathbf{qm}}\frac{\partial\psi^*_{\mathbf{qm}}}{\partial\phi}-\psi^*_{\mathbf{qm}}\frac{\partial\psi_{\mathbf{qm}}}{\partial\phi}\right)-\frac{e^2}{mc}\left|\psi_{\mathbf{qm}}\right|^2A_\phi\,.$$

In the ground state evidently,

$$j_\phi=-\frac{e^2}{mc}\psi_0^2A_\phi\,.$$

Assuming the field screening to be negligibly small, one can put

$$A_\phi=\tfrac{1}{2}H_0r\,.$$

In this case the magnetic moment per unit cylinder length is given by

$$\mu_{\mathbf{z}}=\frac{1}{2c}\int j_\phi\cdot 2\pi r^2\,dr=-\frac{\pi e^2H_0}{2mc^2}\int\psi_0^2(r)\,r^3\,dr=$$

$$=-\frac{\pi e^2R^3H_0}{2mc^2}\int\psi_0^2(r)\,dr=-\frac{\pi e^2n^*R^3H_0}{2mc^2}\,. \tag{6}$$

Here we assume that in the level $q=0$ there exist n^* electrons (we think of a unit surface, i.e. the normalization $\int \psi_0^2(r)\,dr = n^*$ is used, integration being carried out over all the thickness of the surface state, or practically, over the region with a size of about a around the radius R). For bosons such a state is really possible, and a large diamagnetism would take place [11], but electrons cannot, of course, concentrate on the ground level. For higher levels the paramagnetic term is essential in the expression for current. As a result we deal with almost a complete compensation, and if the spin is disregarded, there remains only a weak Landau diamagnetism.

In a superconducting state, however, the situation changes radically, and one can think that in a weak field there exists only a diamagnetic term (6) with a replacement of n^* by n_s^*, which is a surface concentration of 'superconducting electrons' (of course, instead of n_s^* and the already used volume concentration n_s one can introduce corresponding half as much concentrations for Cooper pairs with a charge $2e$ and mass $2m$).

Thus, in a superconducting state the indicated cylinder has an effective susceptibility

$$\chi = \frac{\mu_z}{\pi R^2 H_0} = -\frac{e^2 n_s^* R}{2mc^2} \sim 10^{-13} n_s^* R. \tag{7}$$

Evidently, at $n_s^* \sim 10^{15}$ cm^{-2} the value of $|\chi|$ will be high ($|\chi| > 10^{-2}$), when the above approximation is already invalid, but diamagnetism can undoubtedly be large. True, in the general case one should consider the possibility of the appearance of states with an azimuthal quantum number $m \neq 0$, and, accordingly, the quantization of flux (usually [12,13] the quantum number m introduced above is denoted by the letter n: for a thick cylinder the number m gives the number of trapped flux quanta). It is natural to solve the corresponding problems, at least those near the critical temperature T_c, as in the case of thin cylindrical films [12,13]. A corresponding film of thickness $2d \sim a$ is characterized by the effective penetration depth

$$\delta_{0,\text{eff}} = \sqrt{\frac{mc^2}{4\pi e^2 n_{s,\text{eff}}}}$$

with $n_{s,\text{eff}} \sim (n_s^*/a)$. For such a film the characteristic parameter

$$\mu = \frac{2Rd}{\delta^2_{0,\text{eff}}}$$

(see ref.13) will be large even for very thin cylinders (for example, at $\delta_{0,\text{eff}} = 3\cdot 10^{-6}$ cm and $d = 3\cdot 10^{-8}$ cm the parameter $\mu \sim 10^4 R$ and $\mu \gg 1$ already at $R \sim 10^{-3}$ cm). At $\mu \gg 1$ the cylinder in a metastable state may possess large diamagnetism up to fields H_0, which can be noticeable.

The behaviour of cylindrical and, generally, multiply connected two-dimensional superconductors, as is clear from what has been said above, even in a field strictly parallel to a superconducting surface, differs from the behaviour of such a surface conductor in a normal state. This also applies to some extent to superconducting layered compounds with closed (multiply connected) layers. One should consider also two-dimensional superconductors that can undergo, besides a superconducting, also a magnetic transition (in particular, from a paramagnetic to a ferromagnetic state). On the whole, the corresponding range of questions requires, in our opinion, a more detailed analysis.[†]

In conclusion I would like to express my delight in the possibility of participating in the present volume and thus to congratulate also in this form Dirk ter Haar upon his jubilee and to wish him all the best.

References

1. Ginzburg V.L. and Kirzhnits D.A., ZhETP, **46**, 397 (1964); Sov. Phys. JETP, **19**, 269 (1964); see also, Ginzburg, V.L., Physics Letters, **13**, 101 (1964).
2. Bulaevsky L.N. and Ginzburg V.L., Fiz. Metal. Metallov., **17**, 631 (1964).
3. Hohenberg, P.C., Phys. Rev., **158**, 383 (1967).
4. *High Temperature Superconductivity*. (Ed. by V.L. Ginzburg and D.A. Kirzhnits), Consultants Bureau, (1982).
5. Eggington M.A. and Leggett, A.J. Collective Phenomena, **2**, 81 (1975).
6. Ginzburg V.L., Solid State Commun., **39**, 991 (1981).
7. Ginzburg V.L., Doklady Acad. Nauk USSR, **83**, 385 (1952); see also **118**, 464 (1958).
8. Ginzburg V.L. and Landau L.D., ZhETP, **20**, 1064 (1950); English Transl. see in: D. ter Haar, *Men of Physics*, L.D. Landau, Pergamon Press, p.138 (1965)
9. Lawrence W. and Doniach S., Proc. LT-12, Kyoto, 361 (1970).
10. Bulaevsky, L.N., ZhTEP, **64**, 2241 (1973); **65**, 1278 (1973); Sov. Phys. JETP, **37**, 1133 (1973); **38**, 634 (1973); Uspekhi Fiz. Nauk, **116**, 449 (1975); Sov. Phys. Usp. **18**, 514 (1975).
11. Ginzburg, V.L., Uspekhi Fiz. Nauk, **48**, 26 (1952); Forschritte der Physik, **1**, 101 (1953).
12. Ginzburg, V.L., ZhTEP, **42**, 229 (1962); Sov. Phys. JETP, **15**, 207 (1962).
13. Arutunian R.M. and Zharkov, G.F., ZhETP, **78**, 1530 (1980); Sov. Phys. JETP, **51**, 768 (1980).

† This paper was written and sent to England in the middle of 1982. Afterwards L.N. Bulaevsky, G.F. Zharkov and the present author had clarified some of the corresponding questions, (see ZhEJP, **85**, 1707 (1983)).

The Bethe Ansatz and Exact Solutions of the Kondo and Related Magnetic Impurity Models

A. C. Hewson
Department of Mathematics, London, SW7 2BZ

ABSTRACT

We give an elementary introduction to the Bethe Ansatz and its application to one-dimensional models, discussing in particular its recent application to the diagonalization of the Kondo model. We review the success of this approach in giving exact results for this model and other magnetic impurity systems.

1. Introduction

In the last few years exact solutions of the Kondo problem have been found [1,2] coming from an unexpected source, the application of the ansatz first used by Bethe in 1931 in calculating the ground state energy of the Heisenberg spin chain. These results are of great interest to anyone concerned with magnetic systems. They are also of interest to the many-body theorist in general as there are so few many-body problems that are directly applicable to real physical systems for which exact solutions have been found. The results provide very clear examples of such fundamental theoretical concepts as fermi-liquid behaviour, scaling as a function of temperature and magnetic field, cross-overs from a strong coupling to a perturbation regime, etc. The accessibility of this work, however, has been limited as the Bethe ansatz approach is not well known, being largely the province of specialists in one-dimensional systems, and is seldom mentioned in the standard textbooks on many-body theory. I have thought it worthwhile to provide a brief review of the technique here, aimed at the non-specialist, concentrating particular attention on the application to magnetic impurity problems.

In the next section we give an introduction to the Bethe ansatz approach. As the aim is pedagogical we start with the simplest application rather than follow the historical development, and then discuss those applications which are most relevant to a consideration of magnetic impurity problems. The simplest application will be considered in some detail to lay the framework of ideas for the later discussion of applications to more complex systems. In the second part we consider the application of this technique to the diagonalization of the Kondo or *s-d* model, with a very brief summary of earlier work to set it in context. We briefly describe some of the exact results that have been obtained. We then assess the prospect of getting new, and very important results, for other magnetic impurity models, which have applicability to a wider range of magnetic systems.

2. The Bethe Ansatz

Though Bethe first used the ansatz which is now associated with his name to calculate spin wave excitations of the one-dimensional Heisenberg model [3], the simplest example of its application is to a one-dimensional spin-less Bose gas where the particles interact through a delta function interaction. A solution of this problem was given by Lieb and Linniger in 1964 [4]. The Schrödinger equation for this system is

$$-\sum_{i=1}^{N} \frac{\partial^2\Psi}{\partial x_i^2} + c \sum_{\langle i,j\rangle}^{N} \delta(x_i - x_j)\,\Psi = E\Psi\,, \tag{2.1}$$

for N particles, where c is the strength of the interaction (in units of $2m$, the particle mass) which we take to be positive. The common theme in the Bethe ansatz approach is to look at the form of the solutions of the two- or three-particle problem in the hope of generalizing the results to the N-particle case. The two-particle solution of (2.1) is quite straightforward and we shall consider this in some detail as it will serve to introduce some of the basic ideas.

For two particles, coordinates x_1, x_2, the interaction only comes into play along the line $x_1 = x_2$. The coordinate space therefore naturally divides into the two regions $R_1\,(x_1 < x_2)$ and $R_2\,(x_2 > x_1)$. For each of these regions we are dealing with the equation for free particles which has a plane wave solution $e^{i(k_1x_1 + k_2x_2)}$, corresponding to a total energy $E = k_1^2 + k_2^2$. This, however, is degenerate with the plane wave with the k's interchanged $e^{i(k_2x_1 + k_2x_2)}$ so a solution in the form of a linear combination of these plane waves,

$$\Psi(x_1, x_2) = \theta(x_2 - x_1)\,\Psi_{R_1}(x_1, x_2) + \theta(x_1 - x_2)\,\Psi_{R_2}(x_1, x_2) \tag{2.2}$$

where

$$\Psi_{R_i}(x_1, x_2) = A_{12}(i)e^{i(k_1x_1+k_2x_2)} + A_{21}(i)e^{i(k_2x_1+k_1x_2)} \tag{2.3}$$

can be sought subject to the conditions

(i) Bose symmetry $\Psi(x_1, x_2) = \Psi(x_2, x_1)$,

(ii) matching conditions at the interface $x_1 = x_2$, and

(iii) periodic boundary conditions.

Imposition of (i) gives the coefficients in the region R_2 in terms of those in R_1, $A_{\ell m}(2) = A_{m\ell}(1)$ for $m, \ell = 1, 2$ $(m \neq \ell)$. Condition (ii) requires continuity of the wave-function together with a discontinuity in gradient found by integrating (2.1) over the interface.

$$-\left(\frac{\partial}{\partial x_1} - \frac{\partial}{\partial x_2}\right)(\psi_{R_1} - \psi_{R_2}) = 2c\psi_{R_1} \quad \text{on} \quad x_1 = x_2 . \tag{2.4}$$

The result of applying this condition is

$$\frac{A_{21}[1]}{A_{12}[1]} = -e^{-i\Theta_{21}}, \tag{2.5}$$

where

$$\Theta_{21} = \Theta(k_2 - k_1) = 2\tan^{-1}\left(\frac{k_2 - k_1}{c}\right). \tag{2.6}$$

It corresponds in form to a phase shift as the particles scatter through the interaction.

If the system is taken to be a ring of length L then the boundary conditions (iii) take the form

$$\Psi(x_1 + L, x_2) = \Psi(x_1, x_2) = \Psi(x_1, x_2, +L).$$

Applying this last condition gives for the allowed values of k_1 and k_2,

$$k_s L = 2(n_s + 1)\pi - \theta(k_s - k_t) \tag{2.7}$$

for $s, t = 1, 2$ $(s \neq t)$, where n_s is an integer. For $c > 0$ the solutions of (2.7) are real and generate the complete set of two-particle states of the model. As the set of k's is conserved in the scattering at the interface it is possible to generalize this approach to the solution to the general case of N particles. From a set of k's $\{k_i\}$, $i = 1, 2, \ldots, N$, sets of $N!$ plane waves can be constructed from

$$e^{i\sum_s k_s x_s}$$

by permuting the k's as in (2.3), in each of the $N!$ regions. A total wave-function can be constructed from these in the form (2.2, 2.3) with $N! \times N!$ coefficients to be determined by applying the symmetry, matching and periodic boundary conditions. A conjectured solution of this type is the so-called Bethe ansatz. The calculation follows along exactly the same line as in the two-particle case, the only complication that occurs is that the set of conditions at the interfaces is greater than the number of coefficients to be determined so that a check on consistency must be made. This is best illustrated in the three-particle case. For example, if the coefficients of the plane wave $e^{i(k_r x_1 + k_s x_2 + k_t x_3)}$ in the region $x_1 < x_2 < x_3$ (123, from now on we shall follow the usual convention and denote a region $x_p < x_q < x_r$ simply by pqr) is denoted by A_{rst}, applying the symmetry condition gives the coefficients in the other regions in terms of these. Matching conditions at the interface $x_1 = x_2$ give

$$\frac{A_{213}}{A_{123}} = -e^{i\theta_{21}}, \quad \frac{A_{312}}{A_{132}} = -e^{-i\theta_{31}}, \quad \frac{A_{231}}{A_{321}} = -e^{-i\theta_{23}} . \tag{2.8}$$

These illustrate the general rule that coefficients which differ only in a pair of nearest neighbour indices, e.g., A_{rst} and A_{rts}, are related via $A_{rst} = -e^{-i\theta_{st}} A_{rts}$. Problems of consistency can arise because coefficients such as A_{321} and A_{123} can be related by different nearest neighbour paths, $321 \to 231 \to 213 \to 123$ or $321 \to 312 \to 132 \to 123$. However, it is straightforward to show that these two paths give the same results, and that there is consistency in the general case.

The application of the periodic boundary conditions gives equations of the form $A_{123} = e^{ik_1 L} A_{231}$, or equivalently using relations of the form (2.8)

$$A_{123} = e^{ik_1 L} e^{-i(\theta_{31} + \theta_{21})} A_{123} , \tag{2.9}$$

and hence a set of transcendental equations which determine the k's. In the general case these are

$$k_s L = 2\pi n_s - \sum_{t=1}^{N} \theta(k_s - k_t) \tag{2.10}$$

where n_s are integers or half integers (depending on N) and play the role of quantum numbers for the system. The total energy corresponding to this solution is given by

$$E = \sum_{s=1}^{N} k_s^2 .$$

Ordering the k's such that $k_{s+1} > k_s$, and subtracting the equations for k_{s+1} and k_s gives

$$L(k_{s+1} - k_s) = 2\pi m_s - \sum_t (\theta_{s+1,t} - \theta_{s,t}), \tag{2.11}$$

where $m_s = n_{s+1} - n_s$. As $\theta_{s,t}$ is a monotonically increasing function of s, and the solution vanishes if $k_{s+1} = k_s$, $m_s \geq 1$. The ground state clearly corresponds to $m_s = 1$ for all s. On taking the limit $L \to \infty$ (with $N \to \infty$ for finite density) and introducing a density of states

$$\rho(k) = \text{limit}_{L\to\infty}(1/(k_{s+1}-k_s))$$

for the ground state, equation (2.11) goes over to the integral equation

$$2\pi\rho(k) = 1 + \int_{-D_0}^{D_0} \frac{2c\rho(k')}{[c^2+(k-k')^2]}\,dk' , \tag{2.12}$$

with the ground state energy per particle E_0 given by

$$E_0 = \frac{1}{\rho_0}\int_{-D_0}^{D_0} k^2\rho(k)\,dk . \tag{2.13}$$

The limits D_0 are determined by specifying the density ρ_0 through the relation

$$\rho_0 = \int_{-D_0}^{D_0} \rho(k')\,dk' . \tag{2.14}$$

Hence the problem of calculating the ground state energy reduces to finding the solution of (2.12).

A 'particle-hole' type of excitation of this system [5] can be generated by omitting one of the k's, k_h corresponding to an 'occupied' quantum number n_h in the ground state set, and including k_p corresponding to an 'unoccupied' quantum number n_p outside the ground state set. There is a slight readjustment of all the other k's from their ground state values. Remarkably, however, the total energy of this excitation relative to the ground state energy, E_{ex}, can be expressed in an exact quasi-particle form

$$E_{ex} = \varepsilon(k_p) - \varepsilon(k_h) , \tag{2.15}$$

with $\varepsilon(k)$, the quasi-particle energy, determined by the integral equation

$$\varepsilon(k) = k^2 + \frac{1}{\pi}\int_{-D_0}^{D_0} \frac{c\varepsilon(k')\,dk'}{[c^2+(k-k')^2]} \tag{2.16}$$

and includes the adjustment of all the other particles.

What is more, *all* the excitations are of this 'quasi-particle-quasi-hole' type. This is very significant. As shown by Yang and Yang [6], it enables the thermodynamics of the system to be calculated. Although equation (2.10) gives an algorithm for calculating all the energy states, and hence in principle the free energy, the particle-hole structure of the excitations makes this possible in practice. Yang and Yang expressed the free energy in terms of the 'particle' and 'hole' densities, ρ and ρ_h respectively, given by the integral equations,

$$2\pi\rho\left(1 + e^{\varepsilon(\mathbf{k})/T}\right) = 1 + \int_{-\infty}^{\infty} \frac{2c\rho(k')dk'}{[c^2 + (k-k')^2]} \tag{2.17}$$

and

$$\varepsilon(k) = k^2 - \mu_0 \frac{T_c}{\pi} \int_{-\infty}^{\infty} \frac{\log(1 + e^{-\varepsilon(k')/T})}{[c^2 + (k-k')^2]} dk' , \tag{2.18}$$

where $\rho_h/\rho = e^{\varepsilon(\mathbf{k})/T}$, T is the temperature and μ_0 is the chemical potential. For details of the derivation of (2.17),(2.18) and some of the other results we refer the reader to the original papers[4,5,6].

This example of the one-dimensional Bose gas has served to introduce the basic ideas involved in all applications of the Bethe anzatz. The corresponding problem for an electron gas, where spin degrees of freedom must be included, introduces an application to a more complex system. This application, which we shall consider in some detail here, gives results which are directly applicable to the Kondo problem considered in the next section. The general version of this problem was solved by Yang [7] and Gaudin [8]. The approach described here, however, follows rather more closely that of Belavin [9] in a recent application of the Bethe ansatz to an SU(2) field theory (Thirring) model.

If a solution of (2.1) for two electrons, one of spin up (↑) and one of spin down (↓), is sought in the same form as (2.2),(2.3) then,

$$\Psi = \theta(x_2 - x_1)\left(\Psi^{(12)}_{\uparrow\downarrow}(x_1, x_2) + \Psi^{(12)}_{\downarrow\uparrow}(x_1, x_2)\right)$$
$$+ \theta(x_1 - x_2)\left(\Psi^{(21)}_{\uparrow\downarrow}(x_1, x_2) + \Psi^{(21)}_{\downarrow\uparrow}(x_1, x_2)\right) \tag{2.19}$$

where

$$\Psi^{(\mathbf{Q})}_{\sigma_1\sigma_2}(x_1, x_2) = \chi_1(\sigma_1)\chi_2(\sigma_2)\Big(A_{\sigma_1\sigma_2}(Q, 12)e^{i(\mathbf{k}_1\mathbf{x}_1 + \mathbf{k}_2\mathbf{x}_2)}$$
$$+ A_{\sigma_1\sigma_2}(Q, 21)\, e^{i(\mathbf{k}_2\mathbf{x}_1 + \mathbf{k}_1\mathbf{x}_2)}\Big) \tag{2.20}$$

where $Q = 12\,(x_1 < x_2)$ or $21(x_2 < x_1)$, $\sigma_1, \sigma_2 = \uparrow,\downarrow$, and $\chi(\uparrow) = \alpha$, $\chi(\downarrow) = \beta$ are the spin functions.

A similar form can be written in the case of N particles using the set of plane waves that can be constructed from a set of k's $\{k_i\}$, $i = 1,2,\ldots,N$. The coefficients of these plane waves can be denoted by $A_{\sigma_1\sigma_2\ldots\sigma_N}(Q,P)$, where Q, a permutation of the integers $1,2,\ldots,N$, denotes the region, a similar permutation P denotes the set of subscripts of the k's that are combined with $x_1, x_2, x_3, \ldots, x_N$ to form the particular plane wave and $\sigma_1, \sigma_2, \ldots, \sigma_N$ refer to the spin states of the particles 1 to N respectively. The matching conditions across the interfaces (ii) and periodic boundary conditions (iii) are as in the earlier Bose problem but (i) must be replaced by the antisymmetry condition for fermions. Applying (1), $x_1 \leftrightarrow x_2$ $\sigma_1 \leftrightarrow \sigma_2$, relates terms with different spin labels so, we find from (2.19),

$$\Psi^{(12)}_{\uparrow\downarrow}(x_1, x_2) = -\Psi^{(21)}_{\downarrow\uparrow}(x_2, x_1) \qquad (2.21)$$

or

$$A_{\uparrow\downarrow}(12,12) = -A_{\downarrow\uparrow}(21,21), \quad A_{\uparrow\downarrow}(12,21) = -A_{\downarrow\uparrow}(21,12). \qquad (2.22)$$

In the general case these conditions reduce the set of unknown coefficients by a factor $N!$ The equations relating the coefficients obtained by applying (i) can be automatically satisfied by introducing a new set defined by

$$A_{\sigma_1\sigma_2\ldots\sigma_N}(Q,P) = (-1)^p A_{\sigma_{Q1}\sigma_{Q2}\ldots\sigma_{QN}}(QP), \qquad (2.23)$$

where QP denotes the product of the two permutations Q and P, and p is the parity of the P permutation [9], and $Q1, Q2, \ldots, QN = Q[12\ldots N]$. Using (2.23) we see, for example, $A_{\uparrow\downarrow}(21, 12) = A_{\downarrow\uparrow}(21)$, $A_{\downarrow\uparrow}(12, 21) = -A_{\downarrow\uparrow}(21)$ and hence the second equation in (2.22) is satisfied.

Applying the matching conditions across the interface $x_1 = x_2$ yields the pair of equations

$$\begin{aligned} A_{\uparrow\downarrow}(21) &= b_{12}A_{\uparrow\downarrow}(12) + a_{12}A_{\downarrow\uparrow}(12) \\ A_{\downarrow\uparrow}(21) &= b_{12}A_{\downarrow\uparrow}(12) + a_{12}A_{\uparrow\downarrow}(12) \end{aligned} \qquad (2.24)$$

where

$$a_{ij} + b_{ij} = 1, \quad a_{ij} = a(k_i - k_j) = \frac{k_i - k_j}{k_i - k_j + ic}, \quad i,j = 1,2, \quad i \neq j.$$

These can be written as an operator or matrix equation

$$A_{\sigma_1\sigma_2}(21) = \left(b_{12}I + a_{12}\hat{P}_{\sigma_1\sigma_2}\right) A_{\sigma_1\sigma_2}(12), \qquad (2.25)$$

where $\hat{P}_{\sigma_1\sigma_2}$ interchanges the spin labels $\sigma_1\sigma_2$. This equation corresponds to (2.5). Similar equations arise in the general case relating coefficients which differ only in a pair of nearest neighbour indices. For example, $A_{\sigma_1\sigma_2\sigma_3}(321)$ and $A_{\sigma_1\sigma_2\sigma_3}(231)$ are related via

$$A_{\sigma_1\sigma_2\sigma_3}(321) = (b_{23}I + a_{23}P_{23})\, A_{\sigma_1\sigma_2\sigma_3}(231) \qquad (2.26)$$

where $\hat{P}_{23}$ must be given the more general definition as an operator which permutes the spin labels in the positions *corresponding* to the indices 2 and 3, i.e. it interchanges σ_1 and σ_2 as indicated in (2.26). If ξ_R denotes the column vector $(A_{\sigma_1\sigma_2\sigma_3}(R))$ in 2^3 dimension spin space then

$$\xi_{321} = \hat{K}_{23}\,\xi_{231}$$

where

$$\hat{K}_{ij} = (b_{ij}I + a_{ij}\hat{P}_{ij}), \qquad (2.27)$$

is equivalent to the set of equations (2.26). The operator $\hat{K}_{ij}$ is only defined to operate on a vector ξ_R which has i and j as nearest neighbour indices in R. A similar consistency problem arises as in the Bose case, ξ_{321} and ξ_{123} can be related via alternative routes $321 \to 231 \to 213 \to 123$ or $321 \to 312 \to 132 \to 123$. Consistency is achieved in this case, however, only if

$$\frac{a_{21}}{b_{21}} + \frac{a_{32}}{b_{32}} = \frac{a_{31}}{b_{31}} . \tag{2.28}$$

This is satisfied with a_{ij} and b_{ij} defined as in equation (2.24). This is a critical condition for the success of a solution by the Bethe ansatz.

The application of periodic boundary conditions gives equations such as

$$A_{\sigma_1\sigma_2\sigma_3}(123) = e^{ik_1L} A_{\sigma_2\sigma_3\sigma_1}(231) \tag{2.29}$$

which relate different components of ξ_{123} and ξ_{231}. It is then useful to introduce the X'_{ij} operator of Yang [7], defined by $X'_{ij} = \hat{K}_{ij}\hat{P}_{ij}$, so one can write $A_{\sigma_2\sigma_3\sigma_1}(231) = X'_{13}A_{\sigma_2\sigma_1\sigma_3}(213)$, $A_{\sigma_2\sigma_1\sigma_3}(213) = X'_{12}A_{\sigma_1\sigma_2\sigma_3}(123)$, and hence express (2.29) in the form

$$e^{-ik_1L}\xi_{123} = X'_{13}X'_{12}\xi_{123} . \tag{2.30}$$

In the general case the periodic boundary conditions lead to the set of equations

$$T_j\xi_I = \mu_j\xi_I, \quad T_j = X'_{j,j-1}X'_{j,j-2}\cdots X'_{j,1}X'_{1,N}\cdots X'_{j,j+1} \tag{2.31}$$

where $\mu_j = e^{-ik_jL}$ and I denotes the identity permutation.

The problem has now been reduced to the diagonalization of the operators T_j which act on the spin degrees of freedom only. At this stage the Bose problem gave a scalar equation and was essentially solved. With spin degrees of freedom we still have a difficult problem to solve.

The operators X'_{ij} are easier to deal with than the $\hat{K}_{ij}$ operators because their definition can be extended so that $X'_{ij}\xi_I = (a_{ij} + b_{ij}\hat{P}_{ij})\xi_I$ for *any* indices i,j not just nearest neighbours, and the combined transformation is the product of these operators. The ancillary problem (2.1) can be interpreted in terms of *fixed* spins at position i round a ring of length L. It can be expressed in terms of the usual spin operators using

$$\hat{P}_{ij} = \tfrac{1}{2}(I + 4\,\mathbf{S}_i\cdot\mathbf{S}_j) .$$

The commutivity of the operators T_i and T_j follows from the relation

$$X'_{ij}X'_{i\ell}X'_{j\ell} = X'_{j\ell}X'_{i\ell}X'_{ij} \tag{2.32}$$

which is equivalent to the path consistency condition and is satisfied if (2.28) holds. Hence the set of operators $\{T_i\}$ can be simultaneously diagonalized.

This ancillary problem is solved in Yang's paper [7] by considering spin deviation excitations from the reference state $|0\rangle$ in which all the spins are fully aligned. Quasi-momenta λ_r, $r = 1,2,\ldots,M$, are associated with the excitations corresponding to M reversed spins, and a solution is obtained by yet another application of the Bethe ansatz, in a generalized form. The transcendental equations for the two types of quasi-moment λ_r, k_i are

$$\sum_{i=1}^{N} \theta(2\lambda_r - 2k_i) = 2\pi J_r + \sum_{s=1}^{M} \theta(\lambda_r - \lambda_s) \tag{2.33}$$

and

$$k_j L = i \log \mu_j = 2\pi n_j - \sum_{s=1}^{M} \theta(2k_j - 2\lambda_s) \tag{2.34}$$

where n_j and J_r are integers or $\frac{1}{2}$ integers depending on M and N. The original paper of Yang is very brief without details of the derivation of (2.33),(2.34) but these have been given in a recent paper by Fung[10].

This ancillary problem can be seen in the wider context of soluble lattice models. These include not only spin chains but also two-dimensional ice and ferroelectric models, and the more general six and eight vertex models of Baxter, all of which can be solved by the Bethe ansatz. For discussion of these problems we refer the reader to the reviews of Kasteleyn[11], Gaudin [12] and Lieb and Wu [13] which contain references to the original papers. More recently the Bethe ansatz has been successfully applied to a number of relativistic field theoretic models, the Massive Thirring model [14], the Gross-Neveu model [15] and the SU(2) Thirring model. Side by side with this has been the development of the inverse scattering method for quantum systems. The inverse scattering method has proved to be a general technique for solving certain classical nonlinear equations, giving insight into the connection between the existence of an infinite number of conserved quantities and soliton solutions. It has proved possible to extend the approach to just those quantum mechanical systems that can be solved by the Bethe ansatz[16-19]. The method provides an alternative approach to the solution of these models and a framework that links classical and quantum mechanical systems. For a survey of these developments there is a comprehensive review by Thacker[20] which also includes discussion of some of the earlier applications of the Bethe ansatz to statistical mechanical models. To give a flavour of the elegant algebraic techniques that have been devised in the quantum inverse scattering method, we outline the solution of the ancillary problems (2.31) using this approach. The full details can be found in the paper of Kulish and Reshetkhin[21].

Rather than work directly with the T operators as defined by (2.31), a more generalized operator $\tilde{T}(u)$ is introduced, defined by

$$\tilde{T}_c(u) = X'_{cN}(u-k_N)X'_{cN-1}(u-k_{N-1})\dots X'_{c1}(u-k_1) \qquad (2.35)$$

where u is a variable. The operator is defined within an enlarged spin space $V \otimes C$, V is the original product spin space and C a two-dimensional auxiliary spin space, $X'_{cj}(u) = a(u) + b(u)\hat{P}_{cj}$, where $\hat{P}_{cj}$ exchanges the spin states at j and the auxiliary site c. Using the states of C as a basis, $\tilde{T}(u)$ can be expressed as a 2×2 matrix operator,

$$\tilde{T}(u) = \begin{bmatrix} \hat{A}(u) & \hat{B}(u) \\ \hat{C}(u) & \hat{D}(u) \end{bmatrix} \qquad (2.36)$$

The operators T_j are related to $\tilde{T}(u)$ via $T_j = \mathrm{Trace}\,(\tilde{T}_c(k_j)) = \hat{A}(k_j) + \hat{D}(k_j)$. The matrix (2.36) is the quantum equivalent of the scattering coefficients of the classical inverse scattering problem and is commonly known as the monodromy matrix. It is closely related to the transfer matrices of Baxter[22] whose work has been a major influence on the development of this approach. One of the key relations is the so-called Yang-Baxter equation,

$$X'_{cd}(u-v)\; X'_{cj}(u)\; X'_{dj}(v) = X'_{dj}(v)\,X'_{cj}(u)\,X'_{cd}(u-v)\,, \qquad (2.37)$$

where a second auxiliary space D has been added which can be thought of as an additional site d. This equation is equivalent to (2.32) and ensures the commutivity of the $\tilde{T}(u)$'s.

The philosophy of the quantum inverse scattering approach is to attempt to find creation operators that generate the required eigenstates. That such operators exist has been shown for the Bose gas model (2.1)[16,17]. In its field theoretic form (2.1) corresponds to the Hamiltonian

$$\mathcal{H} = \int dx \left(\frac{\partial \phi^+}{\partial x} \frac{\partial \phi}{\partial x} + c\phi^+ \phi^+ \phi\phi \right) \qquad (2.38)$$

where the field operators ϕ satisfy the commutation relation

$$[\phi(x), \phi^+(x')] = \delta(x - x')\ .$$

This Hamiltonian has a classical counterpart, and the classical equation of motion of ϕ is the nonlinear Schrödinger equation. This *nonlinear* equation can be solved by finding the scattering data for a suitable *linear* problem. From this scattering data and the inverse scattering transform a solution for the time development from some initial state for the nonlinear equation can be generated. The main interest in this context, however, is that the scattering data (operators in the quantum mechanical case) have a simple Poisson bracket (commutator) with the Hamiltonian. These operators generate the required eigenstates of the quantum mechanical problem. To find these one is really concerned with the direct rather than the inverse part of the scattering problem. For the case considered here the operators $B(u_\alpha)$ acting on the reference state $|0\rangle$ create the required eigenstates for a suitable choice of the argument u_α. To give an explicit proof of this certain commutation relations are required. These can be obtained by re-expressing the Yang-Baxter equation (2.37) as an equation relating the operators $\tilde{T}_c(u)$ and $\tilde{T}_d(u)$,

$$X'_{cd}(u-v)\,\tilde{T}_c(u)\,\tilde{T}_d(v) = \tilde{T}_d(v)\,\tilde{T}_c(u)\,X'_{cd}(u-v)\ , \qquad (2.39)$$

which can be written as a matrix equation using the product of the auxiliary spin spaces C and D as a basis.

$$R(u-v)\ \tilde{T}(u) \otimes \tilde{T}(v) = \tilde{T}(v) \otimes \tilde{T}(u)\ R(u-v)\ , \qquad (2.40)$$

where $\otimes$ denotes a direct product of the matrices given in (2.35) and $R(u) = a(u)\,P_{cd} + b(u)\,I$. Equating individual matrix elements in (2.40) gives the communtation relations,

$$\hat{A}(u)\ \hat{B}(u_1) = \frac{1}{a(u_1-u)}\ \hat{B}(u_1)\ \hat{A}(u) - \frac{b(u_1-u)}{a(u_1-u)}\ \hat{B}(u)\ \hat{A}(u_1) \qquad (2.41)$$

$$\hat{D}(u)\ \hat{B}(u_1) = \frac{1}{a(u-u_1)}\ \hat{B}(u_1)\ \hat{D}(u) + \frac{b(u_1-u)}{a(u_1-u)}\ \hat{B}(u)\ \hat{D}(u_1) \qquad (2.42)$$

and

$$\hat{B}(u)\ \hat{B}(v) = \hat{B}(v)\ \hat{B}(u)$$

Operating with $\hat{A}(u)+\hat{D}(u)$ on the state $\hat{B}(u_1)|0\rangle$ and using (2.41), (2.42) one finds,

$$\Big(A(u)+D(u)\Big)\ B(u_1)|0\rangle = B(u_1)\left(\frac{1}{a(u_1-u)}\ A(u) + \frac{1}{a(u-u_1)}\ D(u)\right)|0\rangle$$
$$+ \frac{b(u_1-u)}{a(u_1-u)}\ B(u)\ \Big(D(u_1)-A(u_1)\Big)|0\rangle\ . \qquad (2.43)$$

The second term on the right-hand side of (2.43 can be made to vanish for an appropriate choice of u_1 so $\hat{B}(u)|0\rangle$ is then an eigenstate of $\hat{A}(u)+\hat{D}(u)$. From

$$\hat{A}(u)|0\rangle = |0\rangle \quad \text{and} \quad D(u)|0\rangle = \prod_{i=1}^{N} a(u-k_i)\,|0\rangle\ ,$$

which can be shown directly from (2.35), the second term vanishes if

$$\prod_{i=1}^{N}\left[\frac{u_1-k_i}{u_1-k_i+ic}\right] = 1\ . \qquad (2.44)$$

Using the definition of $a(u)$ which is given after equation (2.24) and setting $u=k_j$ in (2.43) gives the eigenvalue μ_j of $\hat{T}_j$,

$$\mu_j = \left[\frac{u_1-k_j+ic}{u_1-k_j}\right]\ . \qquad (2.45)$$

On taking logarithms, these results correspond to those of Yang (2.33),

(2.34) for the case $M = 1$ with $\lambda_1 = u_1 - ic/2$. The general case can be proved by operating with

$$\hat{A}(u) + \hat{D}(u) \quad \text{on} \quad \prod_{\alpha=1}^{M} B(u_\alpha)|0\rangle ,$$

manipulating the terms using the commutation rules in a similar way and choosing the u_α to make the extra terms, such as those in (2.43), vanish. A very similar approach can be used to generate the eigenstates of the lattice ice and ferroelectric models [19]. This elegant and powerful approach may prove to be more general and more comprehensive than the Bethe ansatz itself, integrating as it does the classical and quantum mechanical systems in a single framework.

We have now covered enough ground to be able to consider the application of the Bethe ansatz to the magnetic impurity models, which we turn to in the next section.

3. Diagonalization of the Kondo Model

Before embarking on the details of the diagonalization of the $s-d$ or Kondo model, it will be useful to give a very brief history of the main work on this model covering the period from the calculations of Kondo in 1964 to the applications of the Bethe ansatz in 1980.

The Kondo or $s-d$ model describes the interaction of a localized d-moment with non-interacting conduction electrons. In its usual second quantized form it is

$$\mathcal{H} = 2\frac{J_c}{N}\sum_{\mathbf{k},\mathbf{k}'}\left\{S^+ c^+_{\mathbf{k}\downarrow} c_{\mathbf{k}'\uparrow} + S^- c^+_{\mathbf{k}\uparrow} c_{\mathbf{k}'\downarrow} + S_z\left(c^+_{\mathbf{k}\uparrow} c_{\mathbf{k}'\uparrow} - c^+_{\mathbf{k}\downarrow} c_{\mathbf{k}'\downarrow}\right)\right\} \tag{3.1}$$

where S is the spin operator of the localized d state. The perturbation calculation to third order in the exchange coupling J_c of Kondo in 1964 [23] gave a $\log T$ contribution to the resistance, which, when combined with the regular impurity scattering term, successfully explained the puzzling resistance minimum that had been observed in simple metals with magnetic impurities. The breakdown of the perturbation approach at low temperatures due to the divergence of the $\log T$ term was also found to occur when the series of leading order divergent terms was summed, then breaking down at a finite temperature $T_\mathbf{k}$, the Kondo temperature, for an antiferromagnetic coupling. The challenge of finding a description of the low temperature regime $T < T_\mathbf{k}$ stimulated an enormous amount of theoretical activity in the late 60's and early 70's. The problem became the testing ground for a wide variety of many-body techniques. The physical picture that emerged from this work was a ground state with the impurity moment magnetically screened by the conduction electrons, a transitional region at temperatures of the order of T_k and an impurity moment Curie law in the high temperature regime $T \gg T_k$. Really definitive results, however, were not established until 1974 when Wilson [24] applied a numerical renormalization group technique to the $S=\frac{1}{2}$ model. The renormalization group provided a more precise realization of the scaling approach applied earlier by Anderson, Yuval and Hamnn [25]. In the scaling argument high energy virtual excitations are removed to map the original problem on to the same, or a more generalized model with renormalized couplings. The lower cut-off for the energy of the excitations that can be removed depends on the temperature. At very low temperatures the effective coupling of the antiferromagnetic model becomes so strong that the impurity forms a singlet with an electron and the impurity moment is magnetically screened. In the numerical approach devised by Wilson the behaviour in the cross-over region from high to low temperatures could be calculated. The important results obtained by Wilson were a universal function $F(T/T_k)$ for the temperature dependence of the impurity susceptibility $\chi(T)$, $T\chi(T)=F(T/T_k)$, and certain universal ratios. The first ratio R essentially relates the impurity susceptibility at $T=0$ to γ, the linear coefficient of the impurity contribution in the specific heat,

$$R = \frac{4\pi^2 k_B^2 \chi(0)}{3g^2 \mu_B^2 \gamma} \tag{3.2}$$

which was estimated to have the value 2. A second ratio W, relating $\chi(0)$ to the Kondo temperature T_k, as defined from the asymptotic form of $\chi(T)$ in the high temperature regime,

$$\chi(0) = \frac{Wg^2\mu^2}{4k_B T_k} \tag{3.3}$$

was estimated to have the value 0.4128 ± 0.002. The calculation of the form of the specific heat with temperature proved to be too demanding on computer resources for direct computation via the numerical renormalization group. It has, however, been calculated via this method more recently by Oliveira and Wilkins [26] by making use of an equivalence between the Kondo model and a certain resonant level model established by Wiegmann and Finkelstein [27]. The transformation to the resonant level model effectively removes the spin degrees of freedom, so the computations for this model are significantly less demanding. The curve has a peak at $T \sim 0.5\,T_k$.

Soon after Wilson's calculations Nozières [28], taking for granted the strong coupling limit $J_c \to \infty$ as the relevant fixed point for the scaling trajectories at $T \to 0$, deduced a phenomenological Fermi-liquid description of the low temperature regime. With some simple assumptions Wilson's result, $R = 2$, was deduced.

The success of the numerical renormalization group approach has led to its extension to the Anderson model by Krishna-Murphy, Wilkins and Wilson[29]. The Anderson model is a more general magnetic impurity model which does not assume a well defined localized d-spin, though this does emerge in a certain parameter range. The important parameters of the model are the position of the impurity d-levels, ε_d, the overlap matrix element V associated with the hybridization of the d and conduction electron states, and U the Coulomb interaction between electrons in the impurity d levels. For integral occupation of the d level, and in the limit $U \to \infty$, the model maps into the Kondo model with $J_c \sim |V|^2/|\varepsilon_d|$, ε_d measured relative to the Fermi level. Yosida and Yamada[30] have carried out extensive perturbation calculations in powers of U, and then deduced results for the Kondo model on taking $U \to \infty$. The ratio R can be deduced as a function of U and ranges from 1 in the non-interacting case $U = 0$ to the value 2 in the limit $U \to \infty$. An interesting feature of these results is a sharp peak at the Fermi level of width $\sim T_k$ that exists in the d electron density of states in the approach to the Kondo limit. This was not revealed in the renormalization group calculations of Wilson for the thermodynamic quantities as it requires the calculation of a dynamic quantity, the one-electron Green's function for the d electrons. For further results, and details of the calculations, we refer the reader to the original papers and return to the application of the Bethe ansatz.

To apply the Bethe ansatz the conventional Hamiltonian (3.1) is cast into the form of an equivalent one-dimensional model. This is possible if the dispersion of the conduction electrons $\varepsilon(\boldsymbol{k})$ is taken to be spherically symmetric and linear in k. Expanding $\varepsilon(\boldsymbol{k})$ about the Fermi-momentum k_F gives a leading term $\varepsilon(\boldsymbol{k}) = v_F k$, where energy and momentum are measured relative to ε_F and k_F, and v_F is the Fermi velocity. This was the form assumed by Wilson with a cut-off $|\boldsymbol{k}| \leqslant D$. This is a reasonable simplification because it is the particle-hole excitations about the Fermi level that are relatively the most important, in fact the numerical renormalization group calculation deals with the conduction states as packets over energy intervals that are narrower and narrower as the Fermi level is approached to reflect this. To obtain the required one-dimensional model this cut-off must be removed and linear dispersion assumed at all energies. An inverse Fourier transform is then taken with respect to a coordinate x. In first quantized form this Hamiltonian can be written as

$$H = \sum_{j=1}^{N} \left(-i \frac{\partial}{\partial x_j} + 4J\delta(x)\mathbf{S}_0 \cdot \mathbf{S}_j\right) \tag{3.4}$$

where the impurity spin is located at the origin $x = 0$, x_j is a conduction electron coordinate and J is a dimensionless coupling constant, $\mathbf{S}_0$ and $\mathbf{S}_j$ are the spin operators for impurity and conduction electron respectively. The model is now in a more promising form for the applicability of the methods of section 2.

Diagonalization of the linear dispersion Kondo model by the Bethe ansatz was demonstrated independently by Andrei [1] and Wiegmann [2]. The formulations, however, are rather different. Andrei modified the model to make it similar to the field theoretic Gross-Neveu model which had previously been solved by application of the Bethe ansatz. This involved introducing a variable number of impurities, whose positions were not fixed though they were taken to have infinite mass. A purity label, $\alpha = 0$ for an impurity, $\alpha = 1$ for an electron, was imposed and an antisymmetry requirement with respect to an exchange of these particles. Wave packets of the solutions of this model are required to describe a fixed impurity. Wiegmann's approach used the model as given in (3.4) with the impurity fixed at the origin. The application of the Bethe ansatz to the model in this form is not completely straightforward because if a solution in terms of plane waves is constructed as in (2.20) the coefficients are not completely determined by the matching, antisymmetry and boundary conditions. A subset of these states was used, and a completeness argument is required to show that this set is sufficient. Here we describe an alternative derivation [31], which avoids the expansion in plane waves, and reduces the problem in a few lines to the ancillary problem of Yang (2.31) considered in the last section. The key to this approach is to recognize that (3.4) has a Schrödinger equation which is a differential equation in the centre-of-mass coordinate

$$z_1 = \frac{1}{N} \sum_{i=1}^{N} x_i$$

only. For example, for two electrons, introducing the coordinates

$$z_1 = (x_1 + x_2)/2 , \quad z_2 = (x_1 - x_2)/2 ,$$

the Schrödinger equation corresponding to (3.4) is

$$\left\{-E - i \frac{\partial}{\partial z_1} + 4J\mathbf{S}_1 \cdot \mathbf{S}_0 \delta(z_1 + z_2) + 4J\mathbf{S}_2 \cdot \mathbf{S}_0 \delta(z_1 - z_2)\right\} \psi_{\sigma_1 \sigma_2 \sigma_0} = 0 \tag{3.5}$$

where σ_1, σ_2, σ_0 denotes the spin states of electrons 1,2 and the impurity respectively. The interaction comes into play only along the lines $x_1 = 0$, $x_2 = 0$. Within the quadrants I $x_1, x_2 < 0$, II $x_1 > 0$, $x_2 < 0$, III $x_1 < 0$, $x_2 > 0$, and IV $x_1 > 0$, $x_2 > 0$,

$$\psi^{R}_{\sigma_1\sigma_2\sigma_0} = e^{iEz_1} f^{Q}_{\sigma_1\sigma_2\sigma_0}(z_2) , \tag{3.6}$$

where $Q =$ I, II, III or IV, and $f^{Q}(z_2)$ is an arbitrary function. The boundary conditions can be obtained by integrating over the interfaces. For example, across $x_1 = 0$, $z_2 > 0$,

$$\begin{aligned} \psi^{II}(-z_2+, z_2) &= \hat{R}_{10}\psi^{I}(-z_2-, z_2) \\ \psi^{IV}(z_2+, z_2) &= \hat{R}_{20}\psi^{II}(z_2-, z_2) , \end{aligned} \tag{3.7}$$

where $\hat{R}_{n0} = e^{iJ}(\cos(2J) - i\sin(2J)\hat{P}_{n0})$ and $\hat{P}_{n0}$ permutes the spin states of impurity and nth electron. The spin indices have been dropped and the wavefunction is taken to be a column vector in spin space. Periodic boundary conditions are taken for convenience. In applying the boundary condition $\psi(x_1 + L, x_2 + L) = \psi(x_1, x_2)$ we can start with a solution of the form

$$e^{iEz_1} f(z_2)$$

in region IV ($z_2 > 0$) and generate the solution at $x_1 + L$, $x_2 + L$ by moving along the line $z_1 \to z_1 + L$, $z_2 =$ constant. This cuts across the two interfaces $x_1 = 0$, $x_2 = 0$, and gives

$$\hat{R}_{20}\hat{R}_{10} f(z_2) = e^{iEL} f(z_2) \tag{3.8}$$

$z_2 > 0$. Hence the diagonalization of $\hat{R}_{20}\hat{R}_{10}$ is sufficient to determine the total energy E. A similar equation can be generated starting from the solution in the region $z_2 < 0$

$$\hat{R}_{10}\hat{R}_{20} f(z_2) = e^{iEL} f(z_2) \tag{3.9}$$

which at first sight appears to be incompatible with (3.9) as $\hat{R}_{10}\hat{R}_{20} \neq \hat{R}_{20}\hat{R}_{10}$. However, the antisymmetry condition relates the functions in the region $z_2 < 0$ and $z_2 > 0$ via $f(z_2) = -\hat{P}_{12} f(-z_2)$, and the two equations are in fact consistent as $\hat{R}_{20}\hat{R}_{10} = \hat{P}_{12}\hat{R}_{10}\hat{R}_{20}\hat{P}_{21}$.

The approach can be generalized to the N particle problem by introducing the coordinates,

$$z_\ell = \frac{1}{N}\sum_{r=1}^{N} x_r \, e^{(2\pi i(r-1)(\ell-1))/N} \tag{3.10}$$

Applying the boundary condition $x_r \to x_r + L$, $r = 1, 2, \ldots, N$, which corresponds to $z_1 \to z_1 + L$, $z_\ell \to z_\ell$, $\ell = 2, 3, \ldots, N$, gives

$$\begin{aligned} \hat{R}_{N,0}\hat{R}_{N-1,0}\hat{R}_{N-2,0}, \ldots, \hat{R}_{2,0}\hat{R}_{1,0} f(z_2, z_3, \ldots, z_N) &= \\ = e^{-iEL} f(z_2, z_3, \ldots, z_N) \end{aligned} \tag{3.11}$$

where $f(z_2, z_3, \ldots, z_N)$ is an arbitrary function associated with the solution in the region $x_N < x_{N-1} \dots x_2 < x_1 < 0$. The operators act on the spin degrees of freedom only and the problem is equivalent to (2.31) if we identify $\hat{X}'(k_0 - k_n) = e^{iJ}\hat{R}_{n0}$ with the choice $k_0 = -1$, $k_n = 0$ for all $n \neq 0$, and $c = \tan 2J$. Hence we can write down the solution to (3.11) immediately. From (2.31) the transcendental equation for the λ's, for M reversed spins, is

$$N\theta(2\lambda_r) + \theta(2\lambda_r - 2) = 2\pi J_r + \sum_{s=1}^{M} \theta(\lambda_r - \lambda_s) \tag{3.12}$$

and the energy E in terms of these is

$$E = \frac{2\pi n}{L} - \frac{N}{L} \sum_{s=1}^{M} [\theta(2\lambda_s - 2) + \pi] \tag{3.13}$$

using (2.33) and (2.34). The lowest energy is achieved by filling the larger λ states, and in the limit $N \to \infty$ a density $\rho(\lambda)$ of these excitations can be defined, and (3.12) becomes an integral equation,

$$\rho(\lambda) = g(\lambda) - \int_{-B}^{\infty} K(\lambda - \lambda')\, \rho(\lambda')\, d\lambda' \tag{3.14}$$

where

$$K(\lambda) = \frac{c}{\pi(c^2 + \lambda^2)}$$

and $g(\lambda)$ is given by

$$g(\lambda) = \frac{2}{\pi} \left\{ \frac{c}{c^2 + 4(\lambda - 1)^2} + \frac{1}{N} \frac{c}{(c^2 + 4\lambda^2)} \right\} \tag{3.15}$$

with the magnetization M_{tot}, and the impurity contribution to the ground state energy ΔE_0 given by

$$M_{tot} = \mu\left(N + 1 - N \int_{-B}^{\infty} \rho(\lambda) d\lambda\right), \quad \Delta E_0 = -\frac{N}{L} \int_{-B}^{\infty} \rho(\lambda)\, [\theta(2\lambda - 2) + \pi]\, d\lambda \tag{3.16}$$

In the absence of a magnetic field $M_{tot} = 0$, $B = +\infty$ and the equation can be solved by Fourier transform. In the presence of a magnetic field $\boldsymbol{H}$ the lower limit of integration, $-B$, can be determined by calculating the induced magnetization from (3.16) and (3.15) for the conduction electron gas alone and equating it to the Pauli magnetization making the physically reasonable approximation $\mu H \ll D$, where D is the cut-off imposed on the charge excitations to avoid filling the sea of negative energy states. This can be done analytically as the equations can be solved by the Wiener-Hopf technique. B is found to depend logarithmically on H. By including the impurity driving term in (3.15), down by a factor $1/N$, the impurity magnetization $M_{imp}(H)$ can be calculated for all magnetic field strengths. The results are shown in Fig. 1 and correspond to the curve labelled $j = \frac{1}{2}$. As the ground state is

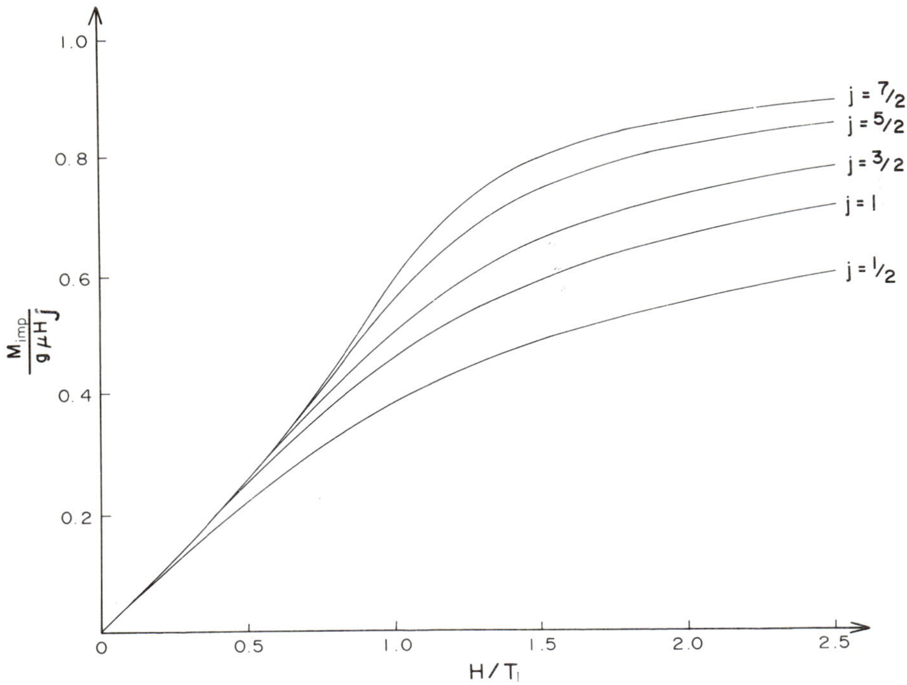

Fig. 1 The impurity magnetic isotherms $M_{\text{imp}}(H)$ scaled by $g\mu j$ for the Coqblin-Schrieffer model in weak and intermediate values of the magnetic field H [43,47]. T_1 is proportional to the characteristic temperature T_0 (essentially T_k).

a singlet the impurity moment is effectively screened by the conduction electrons. In the weak field region the impurity magnetization can be expressed as a power series in H; there is then a cross-over region as the magnetic field breaks up the screening cloud. The full impurity moment is only revealed in very large fields. The approach to saturation is very slow, the deviation from saturation can be expressed as a series in inverse powers of $\log(H/T_1)$. For details we refer the reader to the original papers [2,33].

Calculation of the thermodynamic properties of the model is a more difficult task than for the Bose model considered in section 2, as not all the excitations are of the simple particle-hole type. There are complex solutions of (3.12) for λ of the form

$$\lambda_r^{(n)} = \bar{\lambda}^{(n)} + ic/2(n+1-2r) \tag{3.17}$$

as $N \to \infty$, with $\bar{\lambda}^{(n)}$ real and $r = 1, 2, \ldots, n$, which are known as string excitations. These complex solutions exist for the one-dimensional Heisenberg model and are associated with bound states of the spin waves. Takahashi [34] has set up thermodynamic equations for the Heisenberg model along the lines of Yang and Yang for the Bose problem discussed in section 2. A string excitation density $\rho^{(n)}$ is introduced and a corresponding hole density $\rho_h^{(n)}$. These are determined by an infinite set of coupled integral equations, as there is no limit on the string index number n. The equations are also nonlinear. With minor modifications Takahashi's approach can be used for equation (3.12). The integral equations cannot be solved analytically but certain results can be deduced by expansions in powers of T and H for the low temperature weak field regime. In particular, the Wilson result $R = 2$ for the ratio given in (3.2) can be proved [33,35]. More generally, it can be shown that in the physically relevant region $T \ll D$ the impurity free energy F_{imp} satisfies the scaling law $F_{imp} = -Tf(T/T_0, H/T_0)$, where f is a universal function and $T_0 = De^{-\pi/c}$. The form of T_0 differs from the Kondo temperature T_k deduced either by scaling or high temperature perturbation expressions for the model in its conventional form (3.1), the most important difference being the absence of a term $(J)^{\frac{1}{2}}$ in the prefactor multiplying the exponential in T_0. However, quite different procedures are adopted for introducing the cut-off D in the conventional and Bethe ansatz approaches. The linear dispersion had to be used at *all* energies to convert the model (3.1) into the equivalent one-dimensional form (3.4). Only after solving this equation could the cut-off be imposed in contrast to the usual calculations, such as those by Wilson where the cut-off is applied from the outset. Andrei and Lowenstein [33] have considered this point in some detail. They have calculated the Wilson number W, defined in equation (3.3), to demonstrate that the procedure adopted in the Bethe ansatz approach gives the same universal behaviour in terms of T/T_0 as the conventional procedure gives in terms of T/T_k. This ratio is more significant than the ratio R in characterising the model as it reflects the behaviour in quite different temperature regimes, the low temperature behaviour where the magnetic moment is quenched in the strong coupling regime, and the high temperature perturbational regime where the moment is asymptotically free. Through an ingenious argument Andrei and Lowenstein relate this ratio to the asymptotic forms of the $M_{imp}(H)$ curve in weak and strong fields at $T = 0$ which can be calculated analytically. They give an exact expression for W which when evaluated gives 0.4107, which is in good agreement with Wilson's estimate 0.4128 ± 0.002. Hence they seem to have established beyond reasonable doubt that the difference between T_0 and

T_k in their dependence on the bare coupling and cut-off parameters stems from the different cut-off procedures and is not of any real significance.

To calculate the thermodynamic properties over the whole temperature range, one has to resort to a numerical solution of the coupled integral equations. Several groups [36-38] have carried out such calculations, by truncating the number of strings, ignoring the higher ones, which gives a well controlled approximation and they are in agreement within 1%. The result for the susceptibility curve $F(T/T_k)$ is in excellent agreement with that of Wilson, and there is also reasonably good agreement in the specific heat calculations with those of Oliveira and Wilkins [26] mentioned earlier.

4. Future Prospects

The exact solution of the $S=\frac{1}{2}$ Kondo model using the Bethe ansatz has stimulated similar work for other magnetic impurity models which may be applicable to a wider range of physical systems. One important model that can be solved exactly by this technique [31,39-41] is the Coqblin-Schrieffer model [42]. This model was introduced by Coqblin and Schrieffer in 1969 to describe *Ce* and *Yb* impurities in a simple metallic host. A localized moment associated with the lowest spin-orbit split multiplet of the rare earth ion, with a degeneracy $2j+1$, where j is the total angular momentum quantum number, is coupled via a generalized antiferromagnetic interaction to the conduction electrons which arises from a virtual fluctuation in the number of f-electrons. It has also become known as the SU($2j+1$) Kondo model as it is the natural mathematical generalization of the $s=\frac{1}{2}$ Kondo model which formally corresponds to the case $j=\frac{1}{2}$. The results of this model are of interest not only for rare earth impurity systems but also for certain anomalous rare earth compounds which are paramagnetic down to the lowest temperatures measured and appear to display 'dense' Kondo-like behaviour, and have been termed 'Kondo lattices'. Hewson, Rasul and Newns [43] have calculated the $T=0$ magnetic isotherms for this model, based on equations deduced from its exact diagonalization, for a range of values of j and the results are displayed in Fig. 1. They show an interesting trend in the change of curvature of the low field behaviour with j. There is a positive curvature for $j>1$ in the low field behaviour which is in contrast to the negative curvature of the $s=\frac{1}{2}$ Kondo case ($j=\frac{1}{2}$) and increases with j. This type of behaviour has been observed in the Kondo lattice compound $YbCuAl$ [44], and excellent agreement has been found in fitting the $j=\frac{7}{2}$ curve [Yb^{3+}] to the experimental measurements with a single adjustable parameter [45]. The susceptibility to specific heat coefficient R, defined by equation (3.2) is equal to

$2j/(2j+1)$ for this model. This was first established by the Fermi-liquid arguments of Nozières and Blandin [46] and confirmed by the Bethe ansatz solution[39-41]. This ratio as well as the high temperature susceptibility [47] is also in good agreement with the measurements on *YbCuAl* so that there are quite strong preliminary indications that the impurity model can indeed account for the main features in the behaviour of this compound. This has stimulated work to explain the apparent suppression of the intersite *RKKY* interaction which in normal rare earth systems leads to some form of long range order[48].

The linear dispersion Anderson model has also been solved by the Bethe ansatz (ref.49 and references therein). This model allows charge fluctuations of the impurity as well as spin fluctuations. When the impurity levels are well below (or above) the Fermi level the charge fluctuations are suppressed and, if there is a moment, only the spin fluctuations remain and the behaviour then corresponds to the Kondo model with an antiferromagnetic coupling. When an impurity level lies very close to the Fermi level one enters the regime where charge fluctuations become important. This regime is relevant to intermediate or mixed-valence systems. These are compounds containing rare earth ions which are in a dynamic quantum-mechanical admixture of both $4f^n$ and $4f^{n-1}$ configurations, and which as a consequence have many anomalous properties. The study of these compounds is a very active field at present, both experimentally and theoretically, as more and more such compounds have been found over the last decade. The results for the Anderson model in the charge fluctuation regime should be of great value in helping to understand these systems as it is known that mixed valence impurity and compounds have very similar behaviour. There are only results for the ground state and low temperature properties so far [49] but calculations of the thermodynamic properties over the full temperature range are possible numerically, and are likely to be performed in the near future. Some preliminary results have also been obtained for the degenerate Anderson model [50].

One of the limitations of the Bethe ansatz approach has been the lack of any way of calculating the dynamic correlation functions and Green's functions. Progress has been made in a few special cases [20,51] but no generally applicable technique has been developed. For calculating such interesting physical quantities such as the dynamic susceptibility approximate methods have to be used. However, the approximations involved can be tested by using them to deduce thermodynamic quantities, such as the static susceptibility, and comparing them with the exact results. If the agreement is good, this provides some confidence in the validity of the approximation adopted, which is usually quite uncertain. The same also applies to calculations for more complex systems such as Kondo lattices, or mixed valence compounds. The likelihood that any exact solution can be found for these systems is rather remote, but the exact results for the impurity problem provide a strong special case test for the applicability of any approximate theory. Hence these exact results should be useful for problems beyond their direct sphere of applicability, especially to mixed valence and Kondo lattice systems. Thus a quantitative theory for these systems is a step nearer.

I am happy to acknowledge helpful discussions with Dr J.W. Rasul and Dr D.M. Newns on many of the topics touched on here.

References

1. Andrei, N., Phys. Rev. Letters, **45**, 379 (1980).
2. Wiegmann, P.B., JEPT Lett. **31**, 364 (1980) & J. Phys. C **45**, 1463 (1981).
3. Bethe, H.A., Z. Physik, **71**, 205 (1931).
4. Lieb, E. and Liniger W., Phys. Rev., **130**, 1605 (1963).
5. Lieb, E.H., Phys. Rev. **130**, 1616, (1963).
6. Yang, C.N. and Yang, C.P., J. Math. Phys. **10**, 1115 (1969).
7. Yang, C.N., Phys. Rev. Lett., **19**, 1312 (1967).
8. Gaudin, M., Phys. Letters, **21A**, 55 (1967).
9. Belavin, A.A., Phys. Letters, **87B**, 117 (1979).
10. Fung, M.K., J. Math. Phys., **22**, 2017 (1981).
11. Kasteleyn, P.W., In *Fundamental Problems in Statistical Mechanics III*, (Ed. E.D.G. Cohen, North-Holland, Amsterdam 1975) p.103.
12. Gaudin, M., Saclay Reports CEA-N-1559 (1) and (2), 1973.
13. Lieb, E.H. and Wu, F.Y., In *Phase Transitions and Critical Phenomena*, (Ed. C. Domb and M.S. Green, Academic Press, London 1972) p.331.
14. Bergnoff, H. and Thacker, H.B., Phys. Rev. D., **19**, 3666 (1979).
15. Andrei N. and Lowenstein, J.H., Phys. Rev. Lett., **43**, 1698 (1979).
16. Sklyanin E.K. and Faddeev, L.D., Sov. Phys. Dokl., **23**, 902 (1978); (Dokl. Akad. Nauk. SSSR **243**, 1430).
17. Thacker, H.B. and Wilkinson, D., Phys. Rev., **D19**, 3660 (1979).
18. Sklyanin, E.K., Takhtajan, L.A. and Faddeev, L.D., Theor. Math. Phys., **40**, 688 (1980); (Terr. Mat. Fiz., **40**, 194 (1979).
19. Thacker, H.B., J. Math.Phys., **21**(5) 1115 (1980).
20. Thacker, H.B., Rev. Mod. Phys. **53**, 253 (1981).
21. Kulish, P.P. and Reshetkhin, N.Yu., Sov. Phys-JEPT, **53**, 108 (1981). (Zh. Eksp. Teor. Fiz., **80**, 214).
22. Baxter, R.J., Ann. Phys., **70**, 193 & 323 (1972).
23. Kondo, J., Prog. Theor. Phys., **32**, 37 (1964).
24. Wilson, K.G., Rev. Mod. Phys., **47**, 773 (1975).
25. Anderson, P.W., Yuval, G. and Hamman, D.R., Phys. Rev. **B1**, 4464 (1970).
26. Oliveira, L.N. and Wilkins, J.W., Phys. Rev. Lett., **47**, 1553 (1981).
27. Wiegmann P.B. and Finkelstein, A.M., Sov. Phys. JEPT, **48**(1), 102 (1978). (Zh. Eksp. Teor. Fiz. **75**, 204)
28. Nozieres, P., J. Low Temp. Phys., **17**, 51 (1974).
29. Krishna-Murthy, H.R., Wilkins, J.W. and Wilson, K.E., Phys. Rev., **B21**, 1003 (1980).
30. Yosida, K. and Yamada, K., Prog. Theor. Phys., **53**, 1286 (1975).
31. Hewson, A.C., J. Phys., **C15**, L611 (1982).
32. Hewson, A.C., T. Phys., **C10**, 4973 (1977).

33. Andrei, N. and Lowenstein, J.H., Phys. Rev. Lett., **47**, 356 (1981).
34. Takahashi, M., Prog. Theor. Phys. **46**, 401 (1971); **52**, 108 (1974).
35. Filyov, V.M., Tsvelick, A.M. and Wiegmann, P.B., Phys. Lett. **81A**, 175 (1981).
36. Rajan, V.T., Lowenstein, J.H. and Andrei, N., Phys. Rev. Letters, **49**, 497 (1982).
37. Desgranges, H.U. and Schotte, K.D., Physics Letters, **91A**, 240 (1982).
38. Melnikov, V.I., JEPT Lett., **35**, 10 (1982).
39. Tsvelick, A.M. and Wiegmann, P.B., J. Phys. C., **15**, 1707 (1982).
40. Rasul, J.W., *Proc. Int. Conf. on Valence Instabilities*, Zurich 1982; (Eds. P. Wachter and H. Boppart, North-Holland 1982). p.49.
41. Andrei, N., Furuya, K., and Lowenstein, J.H., Rev. Mod. Phys., **55**, 331 (1983).
42. Coqblin, B. and Schrieffer, J.R., Phys. Rev., **B5**, 4541 (1969).
43. Hewson, A.C., Rasul, J.W. and Newns, D.M., Physics Letters, **93A**, 311 (1983).
44. Mattens, W.C.M., Ph.D. Thesis, University of Amsterdam (1980).
45. Hewson, A.C., Rasul, J.W. and Newns, D.M., Solid State Comm., **47**, 59 (1983).
46. Nozieres, P. and Blandin, A., J. Physique, **41**, 193 (1980).
47. Hewson, A.C. and Rasul, J.W., (to appear in J. Phys. C. (1983)).
48. Newns, D.M., Doniach, S. and Read, N., Preprint (1983).
49. Wiegmann, P.B. and Tsvelick, A.M., J. Phys. C., **16**, 2281 & 2321 (1983).
50. Schlottman, P., Phys. Rev. Lett., **50**, 1697 (1983).
51. Kulish, P.P. and Smirnov, F.A., Phys. Letters, **90A**, 74 (1982).

Probing the Interiors of Neutron Stars

F. K. Lamb
Departments of Physics and Astronomy, University of Illinois at Urbana-Champaign

ABSTRACT

This paper reviews the information on neutron stars that has been obtained from observations of pulsars and pulsing X-ray stars, with emphasis on results from timing studies. Included are summaries of the most recent theoretical work on neutron star structure and the current evidence concerning neutron star masses and radii. Also included are discussions of large frequency jumps and small-scale timing irregularities, the description of the latter phenomenon in terms of random noise processes, and the use of both as probes of the internal structure and dynamical properties of neutron stars. Finally, some of the most promising directions for future work are briefly noted.

1. INTRODUCTION

Most of our knowledge of neutron stars has come from studies of pulsars and pulsing X-ray sources, and especially from pulse timing observations. It was, after all, the regularity of pulsars that first attracted attention (Hewish *et al.* 1968). Soon after their discovery, timing measurements by Richards and Comella (1969) revealed that the Crab pulsar is slowing down, a result that was crucial in establishing that pulsars are rotating neutron stars (Gold, 1969). The detection of a large jump in the frequency of the Vela pulsar (Radhakrishnan and Manchester, 1969; Reichley and Downs, 1969) provided the first evidence of complex internal structure, while the pioneering work of Boynton *et al.* (1972) showed that the small timing irregularities ("restless behaviour") observed in the Crab pulsar could be described as noise.

The discovery of pulsing X-ray sources (Schreier *et al.*, 1972) and the convincing interpretation of these objects as a second class of rotating neutron stars (Pringle and Rees, 1972; Davidson and Ostriker, 1973; Lamb *et al.*, 1973) opened the way for the study of neutron stars using X-ray techniques.

The present paper discusses our observational knowledge of neutron stars, with emphasis on results from timing studies. After summarizing the present theoretical picture of neutron stars and several new developments in § 2, I briefly review the current evidence on stellar masses and radii in § 3. The evidence provided by short-term pulse frequency variations is described in § 4, which includes a discussion of large frequency jumps, the description of small-scale timing irregularities in terms of random noise processes, and the use of intrinsic timing noise as a probe of neutron star interiors. Reviews of early developments and of topics not discussed here may be found in the articles by Ruderman (1972), Pines, Shaham, and Ruderman (1974), Lamb (1975, 1977, 1979a, 1981a, 1981b), and Pines (1980), and in the monograph by Manchester and Taylor (1977).

2. RECENT THEORETICAL DEVELOPMENTS

Theoretical work on neutron stars has been reviewed recently by Baym and Pethick (1975, 1979). In the period since these reviews, there has been substantial progress in including the three-nucleon interaction in the equation of state of neutron star matter. These efforts have been motivated by several facts. First, the three-nucleon interaction is expected on theoretical grounds. Second, comparison of exact solutions for the triton with experiment shows that the three-nucleon interaction is significant. Third, there is general agreement that without inclusion of the three-nucleon interaction, the calculated equilibrium binding energy and density of nuclear matter are too large.

The latter point is illustrated in Fig. 1, which compares the empirical binding energy of nuclear matter as a function of density with various theoretical curves based on modern variational calculations (Carlson *et al.* 1982). The curve marked v_{14} is typical of results obtained with realistic two-nucleon interaction models. It gives a binding energy that is too small for $k_F < 1.4\ \mathrm{fm}^{-1}$, missing the empirical curve by ~ 2 MeV at $1.13\ \mathrm{fm}^{-1}$, and saturates at too high a density. The curve marked $v_{14}+T$ shows the results of a variational calculation using the Tucson model of the three-nucleon interaction (Coon *et al.* 1979). It misses the empirical curve by only ~ 1 MeV at $1.13\ \mathrm{fm}^{-1}$ but saturates at a still higher density. The curve marked $v_{14}+V$ is the result of a variational calculation with an improved three-nucleon

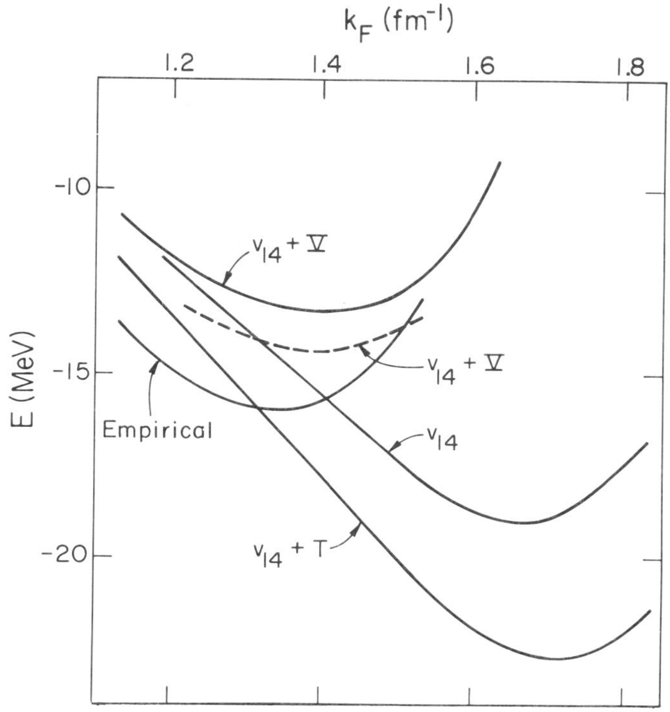

Fig. 1. Comparison of the empirical binding energy of nuclear matter as a function of density with a variety of theoretical curves based on modern variational calculations with realistic two- and three-nucleon interactions. (From Carlson *et al.* 1982.)

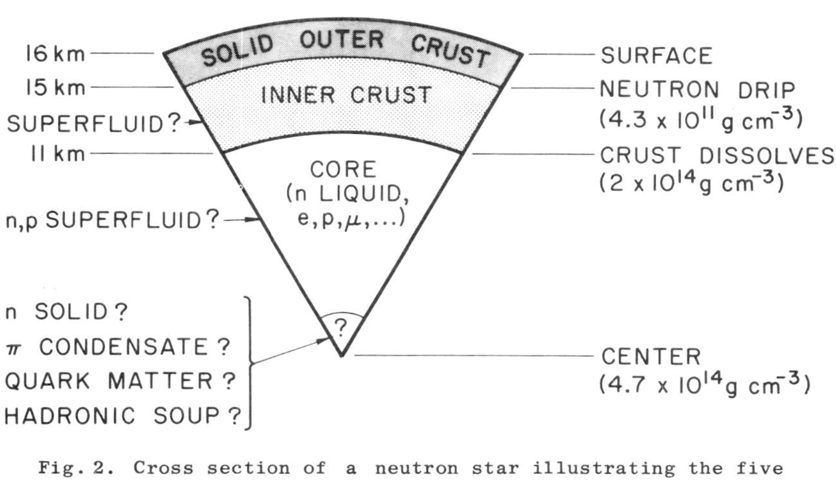

Fig. 2. Cross section of a neutron star illustrating the five regions discussed in the text.

interaction model that accurately describes the properties of the triton as well as the nuclei of ^{3}He and ^{4}He. The dashed curve shows how the result is modified when second-order correlated basis perturbation corrections (Fantoni *et al.* 1982) are included. It gives an equilibrium density in reasonable agreement with the empirical value and a binding energy that is only a little too small.

Phenomenological calculations of the equation of state of neutron star matter incorporating the three-nucleon interaction explicitly (Friedman and Pandharipande, 1981) give a 1.3 $M_\odot$ stellar model with a radius of 11 km and central density of 10^{15} g cm^{-3}. This may be compared with the 16 km radius and 4.7×10^{14} g cm^{-3} central density of a star of the same mass calculated using the earlier tensor-interaction equation of state of Pandharipande and Smith (1975a, 1975b).

In the discussion that follows it will be helpful to have in mind the expected internal structure and dynamical properties of a neutron star. The expected structure is shown in Fig. 2 (the quoted dimensions are those of a 1.3 $M_\odot$ model based on the tensor-interaction equation of state). A neutron star may be conveniently divided into five regions: (1) the *surface*, where the properties of the matter can be strongly affected by the temperature and the magnetic field, (2) the *outer crust*, consisting of a solid lattice of nuclei embedded in a sea of degenerate, relativistic electrons, (3) the *inner crust*, consisting of that part interior to the neutron drip point, (in this region neutrons exist outside the nuclei, so that one has a solid lattice immersed in a sea of electrons and neutrons), (4) the *neutron liquid*, consisting largely of neutrons but with some electrons and protons, and a few muons, (5) possibly a distinct *core*. Such a core could consist of condensed pions, or a solid neutron lattice. Alternatively, the composition of the interior may simply change gradually to a soup of quarks or hyperons as the density increases.

The response of such a star to changes in the rotation rate of the crust is expected to be very complex (see Lamb 1977). From the dynamical point of view it has at least five distinct components: the crust, the neutron fluid, the proton fluid, the electron fluid, and, in heavier stars, possibly a solid core. These components possess a variety of normal modes that can be excited by rotational disturbances, including inertial and hydromagnetic-inertial waves in the electron and proton fluids, Tkachenko oscillations of the neutrons (if they are superfluid), wobble of the crust, and wobble of the solid core (if present). There are also a variety of different relaxation times, including the time required for relative motions of the various components to damp, and the time scale for changes in the crust due to cracking, crumbling, creep, and plastic flow. One of the key challenges facing observers is to devise methods of probing the dynamical behaviour of neutron stars.

3. MASSES AND RADII

The determination of neutron star masses has been recently reviewed by Rappaport and Joss (1981), and I will therefore only summarize the results here. The mass of a neutron star can be measured directly only if it is in a binary system. Thus, all but one of the current estimates come from the study of binary X-ray sources. The seven systems that have provided mass estimates are listed in Table I, which also gives the period, P_{orb}, and the projected semi-major axis, $a_n \sin i$, of the neutron star orbit, and the allowed range of the neutron star mass, M_n.

Table 1

Measured neutron star masses

System		$P_{orb}(d)$	$A_n \sin i$ (lt-s)	M_n ($M_\odot$)
4U0115 − 73	(SMC X-1)[a]	3.892	53.46 ±0.03	0.7 - 1.5
4U0900 − 40	(Vela X-1)[a]	8.965	113.0 ± 0.8	1.6 - 2.5
4U1118 − 60	(Cen X-3)[a]	2.087	39.792±0.005	0.4 - 1.7
4U1538 − 52[a]		3.730	55.2 ± 3.7	0.8 - 3.4
4U1627 − 673[b]		2.88×10^{-2}	<0.04	0.5 - 4.7
4U1653 + 35	(Her X-1)[c]	1.700	13.206±0.006	0.6 - 2.0
PSR 1913 + 16[d]		3.23×10^{-1}	2.3424±0.0007	1.24-1.54

[a] Rappaport and Joss (1981). [b] Middleditch *et al.* (1981).
[c] Middleditch and Nelson (1976), Bahcall and Chester (1977).
[d] Taylor *et al.* (1979).

In four systems (SMC X-1, Vela X-1, Cen X-3, and 1538-52) the mass determination is based on the orbital period and semimajor axis derived from pulse timing, the radial velocity of the companion star derived from measurements of the Doppler shift of its spectral lines, and the eclipse duration, together with some assumptions about the fraction of the critical lobe filled by the companion star and its rate of rotation. So far it has not been possible to determine the radial velocities of the companion stars in Her X-1 and 1627-673 spectroscopically, but orbital solutions have nevertheless been obtained by careful study of the optical pulses from these two systems. The very close binary system PSR 1913 + 16 is a unique case. Assuming that general relativity is the correct theory of gravitation and that both stars behave like point masses, the pulsar mass can be determined from pulse timing alone, with the result given.

The masses listed in Table 1 are consistent with the possibility that all neutron stars have masses in the range 1.2 - 1.6 $M_\odot$, as suggested by current theories of stellar collapse. The maximum stable neutron star mass depends on the equation of state of matter at high densities and the behaviour of gravitational forces. Assuming that general relativity is correct and that the errors quoted in Table 1 reflect the true uncertainties in the masses, the 1.6 $M_\odot$ lower limit on the mass of Vela X-1 is inconsistent with several of the softer equations of state that have been proposed, including those of Leung and Wang (1971), which give maximum masses $\leqslant 0.5\ M_\odot$, Hagedorn's asymptotic equation of state and that of an ideal neutron gas, both of which give a maximum mass $\sim 0.7\ M_\odot$ (see Ruffini, 1974), and the equations of state of Pandharipande (for hyperons), Arponen, and Canuto and Chitre, which give maximum masses in the range 1.36 - 1.46 $M_\odot$ (see Arnett and Bowers, 1977, and references therein). The mass estimates in Table 1 are consistent with the somewhat stiffer equations of state of Bethe and Johnson, Moszkowski, Walecka, Pandharipande and Smith, and Bowers, Gleeson, and Pedigo (again see Arnett and Bowers, 1977), and Friedman and Pandharipande (1981).

New evidence on the masses and radii of neutron stars and the equation of state at high densities is being provided by recent γ- and X-ray observations. The report of apparent emission features in the energy range 300-650 keV in the spectra of several γ-ray bursts may make possible estimates of the surface gravitational redshift and hence M/R, if these can be shown to be due to redshifted 511 keV e^+e^- annihilation radiation from the stellar surface (Mazets *et al.* 1981a, 1981b; but see Fenimore *et al.* 1982, and Laros *et al.* 1982). Comparison of the observed surface luminosities of cooling neutron stars with detailed models is providing new information about the equation of state at high densities (see Glen and Sutherland, 1980; Van Riper and Lamb, 1981; Helfand, 1981; Nomoto and Tsuruta, 1981; Richardson *et al.*, 1982; and Gudmundsson *et al.*, 1982). Measurements of the spectra of X-ray bursts may furnish estimates of neutron star radii if they are produced by thermonuclear flashes in neutron stars (van Paradijs, 1978; Swank, Eardley, and Serlemitsos, 1979), but such estimates may be seriously confused if only part of the stellar surface is involved, if the X-ray photosphere is not at the stellar surface, or if the X-rays interact appreciably with matter above the surface.

4. PULSE FREQUENCY VARIATIONS

Short-term variations as well as long-term changes have been detected in the intrinsic pulse frequencies of many pulsars and pulsing X-ray stars.

The nature of these variations provides clues about the physical processes that are responsible. To the extent that they reflect variations in the angular velocity of the neutron star crust rather than variations in the pulse emission process, they provide information about the internal structure and dynamical properties of the neutron star.

In principle, short-term frequency variations can be continuous, complex and non-repeating. A sequence of such variations would be difficult to analyze in any unified way, In practice, the short-term frequency variations observed in many pulsars and pulsing X-ray stars are consistent with a description in terms of a succession of similar, discrete events ("glitches"). In this case, a unified description may be possible.

The appropriate approach to use depends on whether the individual events have been resolved. If so, detailed modelling can be attempted; if not, a useful description of the statistical properties of the events may still be possible.

4.1 Resolved Events

Sudden, large frequency jumps ("macroglitches") have been detected and at least partially resolved in four pulsars: one each in PSR 1325-43 (Manchester, 1981) and PSR 1641-45 (Manchester *et al.*, 1978), two in the Crab pulsar (see Boynton *et al.*, 1972; Lohsen, 1975), and four in the Vela pulsar (see Downs, 1981). Recently, Downs (1982) has reported two frequency jumps in PSR 0525 + 21. Although other resolved events have been reported, Groth (1975) and Cordes and Helfand (1980) have argued convincingly that these are more likely just local accumulations of unresolved events.

All of these events had a sudden onset (which was not resolved) followed by a slower decay, and are adequately described by the so-called glitch function (Baym *et al.*, 1969)

$$\Omega(t) = \Omega_0(t) + (\Delta\Omega)_0 \left[Qe^{-t/\tau} + (1-Q)\right] ,$$

where $\Omega_0(t)$ is the frequency behaviour extrapolated from the period before the glitch. The parameters in this description are the relative size $(\Delta\Omega/\Omega)_0$ of the initial frequency jump, the fraction Q of the initial jump that eventually decays away, and the time scale τ of the decay. For the Crab and Vela glitches, these parameters have been determined with some precision. Successive glitches in the same star appear similar, but not identical. In the Crab pulsar, $(\Delta\Omega/\Omega)_0 \sim 10^{-8}$, $Q \sim 0.95$, and $\tau \sim 4-15$ d, while in the Vela pulsar, $(\Delta\Omega/\Omega)_0 \sim 10^{-6}$, $Q \sim 0.2$, and $\tau \sim 450$ d.

Large, relatively rapid frequency variations have also been detected in some pulsing X-ray stars, but there is no evidence that these have an onset as sudden as the pulsar macroglitches nor are they described adequately by the glitch function. They may be due to relatively rapid but continuous changes in these systems (see, for example Elsner *et al.*, 1980), or to a large number of smaller, unresolved events (see below).

Pulsar macroglitches have been attributed to a wide variety of causes. These interpretations as well as theoretical studies of the subsequent response of the neutron star have been extensively reviewed elsewhere (see Ruderman, 1972; Pines *et al.*, 1974; Groth, 1975; Lamb, 1977, 1979a; and Pines, 1980) and only a brief summary of the current status of this problem will be given here. Among the explanations considered most promising at present are crustquakes (Ruderman, 1969; Baym *et al.*, 1969; Baym and Pines, 1971), corequakes (Pines *et al.*, 1972; but see Baym *et al.*, 1976), and crust-breaking by vortex pinning (Ruderman, 1976). None of these models is entirely satisfactory. The Vela glitches are too large to be accounted for by crustquakes, while the Crab glitches have too large a Q if the mass of the star is, as expected, $\gtrsim 1\ M_{\odot}$. Corequakes can in principle cause glitches as large as those seen in the Vela pulsar, but may produce too much heat, while the change in the crust moment of inertia in a crust-breaking event, and hence the size of the resulting glitch, is uncertain.

Recently, Greenstein (1979) has pointed out that sudden heating of the neutron superfluid could cause angular momentum to be transferred from the faster core to the slower crust, creating a glitch. In a series of papers, Alpar *et al.* (1981; see also Anderson *et al.*, 1982, and references therein) have developed a model in which $\sim 10\%$ of the pinned vortices in the inner crust break free once every few years. They argue that such events can produce glitches as large as those seen in the Vela pulsar and PSR 1641-45 while releasing only $\sim 10^{39}$ ergs in heat.

Regardless of the cause of a macroglitch, the behaviour of the pulse frequency following the initial jump furnishes information about the internal structure and dynamical properties of the star. In the starquake models, the post-glitch relaxation time τ is the coupling time between the crust and neutron superfluid, while in the thermal-pulse and vortex-unpinning models it reflects cooling of the interior and spin-down of the crust by the external torque, and hence only provides a bound on the crust-superfluid coupling time. In the crust-breaking model of Ruderman, τ is the time scale for plastic flow of the crust, and is unrelated to the crust-superfluid coupling time.

4.2 Unresolved Events

If the individual events cannot be resolved, one may attempt a statistical description of the effect of a large number of events. A natural approach is to try to model the frequency variations by a random noise process. If successful, such modelling allows the observations to be characterized by a small number of parameters, which are the only observables if the process is not resolved.

The pioneering work of Boynton *et al.* (1972; see also Groth, 1975 and Cordes, 1980) showed that the small-scale pulse frequency variations in the Crab pulsar can be adequately described in this way. Soon afterwards Lamb *et al.* (1974, 1976, 1978a, 1978b) argued on theoretical grounds that the frequency variations seen in many pulsing X-ray stars are due to excitation of the neutron star crust by torque noise, and showed that the data then available were consistent with this conjecture. Motivated by this work, Boynton and Deeter (1979) and Boynton (1981) showed that the frequency variations in the pulsing X-ray star Her X-1 are consistent with this hypothesis. Work is in progress to extend this type of analysis to Vela X-1 (see Lamb, 1979b). More recently, Cordes and Helfand (1980) have used the data of Helfand *et al.* (1980) to show that noise models adequately describe the small-scale frequency variations observed in ten pulsars other than Crab.

As examples of the application of noise models, consider the two types of events shown in Fig. 3. A frequency jump caused by a spike in the torque acting on the crust or a sudden change in the crustal moment of inertia could continue indefinitely, giving rise to an event like that shown in Fig. 3a. Alternatively, such a jump might die away partly or completely, due to a subsequent torque on the crust or a change in the moment of inertia of opposite sign to the original, giving rise to an event of the type shown in Fig. 3b. The temporal structure of actual events of either type could be quite complicated.

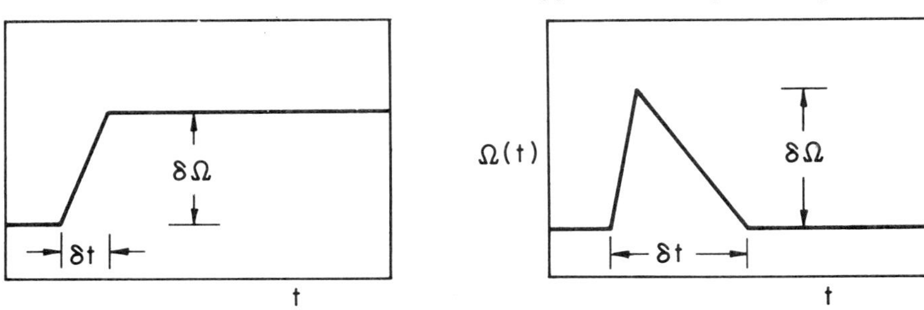

Fig. 3. Two examples of possible small-scale timing irregularities. Variations (a) and (b) can be modelled by white noise in $\dot{\Omega}$ and Ω, respectively (see Lamb, Pines and Shaham, 1978a).

Regardless of their temporal structure, the effect of many such events can, under rather general conditions, be calculated by replacing the actual time history of the event by a simple mathematical expression. Thus, for example, events of the types shown in Fig. 3 can be represented by the two distributions

$$\delta\Omega(t) = \sum_i \delta\Omega_i\, \theta(t - t_i)$$

and

$$\delta\Omega(t) = \sum_i \delta\phi_i \, \delta(t-t_i) \, ,$$

provided that there are a large number of events in the observing interval and none of the events is resolved. Here $\theta(t)$ and $\delta(t)$ are the unit step and delta functions, and $\delta\phi_i = \delta\Omega_i \, \delta t_i$ in terms of the size $\delta\Omega_i$ and duration δt_i of the ith event.

Random noise processes like these can be classified according to their power density spectra or the temporal behaviour of the resulting phase residuals. In practice, determination of the power density spectrum is preferred, for a variety of reasons (see for example Blackman and Tukey, 1958). First, power density spectra are simpler to interpret. Second, they are easier to compare with theory (since one usually goes to the frequency domain in order to compute the variance produced by a given theoretical model, one may as well stay there to make the comparison with observation). Third, fluctuations in the estimated power spectrum caused by sampling effects are generally better behaved than are those in the variance. Finally, in order to make quantitative statements about the observed noise, it is necessary to use statistics whose properties are well-understood. Much of the required work has been done for power density estimators (see Deeter and Boynton, 1982) but is not available for the variance.

I shall call a random noise process that produces a power-law power density spectrum "simple", and one that produces a more complicated spectrum "complex". Unresolved events produce simple noise, which is characterized by just two observables: its spectral index and strength. Table 2 lists a variety of physical mechanisms that can produce simple timing noise in pulsars and pulsing X-ray sources, the type of noise produced, the resulting power density P_ϕ of pulse phase fluctuations as a function of angular frequency ω, the spectrum designation, and the expected behaviour of the rms phase residual $\Delta\phi$ as a function of the length of the observing interval T. Each noise process is identified by specifying the variable in which the process is white noise. It is assumed that individual events are unresolved.

Observed power density spectra are complex, even if the spectrum intrinsic to the source is a power law. This is illustrated in Fig. 4a, which shows schematically the effect on a power-law noise signal (denoted N) of (1) removing polynomial behaviour of degree n from the measured pulse phase (which removes power at $\omega \leqslant \omega_n$), and (2) measurement errors (which produce the white noise in pulse phase denoted ME). Figure 4b shows the same spectrum after prewhitening. The precision with which the spectrum of the noise signal can be determined is governed by the precision ΔP of the local power density estimates and the frequency ω_* above which the power density produced by measurement error exceeds that of the noise signal. In dealing with the extremely steep spectra characteristic of timing noise, great care must be taken to insure that the local power density estimates are not contaminated by leakage of power from distant parts of the spectrum. Further details are given by Lamb (1979a), Boynton and Deeter (1979), Boynton (1981), and Deeter and Boynton (1982).

Table 2

Physical Mechanisms that can Produce Simple Timing Noise

Mechanism	Type of Noise	$P_\phi(\omega)$	Spectrum	$\Delta\phi(T)$
Measurement errors, pulse shape changes	White Noise in ϕ	$\propto \omega^0$	White	$\propto T^0$
Some types of temperature fluctuations	White Noise in Ω	$\propto \omega^{-2}$	First-order Red	$\propto T^{-1/2}$
Crustquakes, other types of temperature fluctuations, vortex unpinning, torque fluctuations	White Noise in $\dot{\Omega}$	$\propto \omega^{-4}$	Second-order Red	$\propto T^{3/2}$
Torque steps	White Noise in $\ddot{\Omega}$	$\propto \omega^{-6}$	Third-order Red	$\propto T^{5/2}$

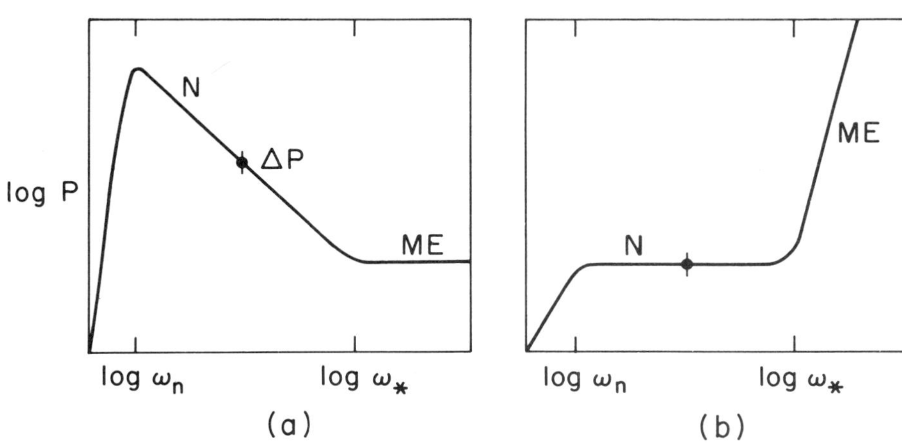

Fig. 4. Schematic illustration of observed power density spectrum (a) and prewhitened spectrum (b).

The spectral index and strength of the observed timing noise signal provides information on the physical processes that are responsible. Thus, Lamb *et al.* (1978a) have argued that the strengths of the timing noise observed in Cen X-3 and Her X-1 are consistent with estimates of the possible strength of accretion torque fluctuations or fluctuations in the torque on the crust caused by vortex unpinning. They have also discussed the noise spectra to be expected from these processes.

Information on the causes of timing noise and on variations in stellar properties can also be obtained by comparing the noise strengths and rms phase residuals of a collection of neutron stars. Table 3 (from Lamb, 1981a) shows how the timing noise produced by various physical processes would scale from star to star if all had similar properties. Lamb (1979a) has argued that the timing noise in pulsing X-ray sources is due to white torque noise which scales as shown. More recently, Cordes and Greenstein (1981) have used the results of Cordes and Helfand (1980) on the spectrum and strength of the timing noise in a collection of eleven pulsars to discuss critically a variety of physical models of the origins of pulsar timing noise.

TABLE 3

Timing Noise Scaling Laws

Physical Process	Noise Strength	$\Delta\phi$
White torque noise	$\propto \dot{\Omega}^2$	$\propto \dot{\Omega} T^{3/2}$
Red torque noise	$\propto \dot{\Omega}^2$	$\propto \dot{\Omega} T^{5/2}$
Vortex unpinning noise [a]	$\propto \Omega\dot{\Omega}$	$\propto \Omega^{1/2} \dot{\Omega}^{1/2} T^{3/2}$
Small crustquakes	$\propto \Omega^3\dot{\Omega}$	$\propto \Omega^{3/2} \dot{\Omega}^{1/2} T^{3/2}$

[a] According to the model proposed by Lamb *et al.* (1978a).

Regardless of its cause, excitation of the stellar crust by simple noise can be used systematically to probe the interior of the star. This possibility was first pointed out by Lamb (1977) and has been developed in detail by Lamb *et al.* (1978a, 1978b). The reason is that the noise "signal" applied to the crust is, in effect, "filtered" by the crust-superfluid system to produce the "output" represented by pulse phase fluctuations. This is expressed mathematically by the relation $P_{obs}(\omega) = F(\omega) P_{ex}(\omega)$ between the power density of the exciting noise process P_{ex} and the observed power density P_{obs}. Here F is the power transfer function of the neutron star, which depends on its internal dynamical properties. The form of the power transfer function is immediately apparent if one works with the variable in which the exciting noise is white, since in this case $P_{\mathbf{ex}}(\omega)$ = const. and hence $P_{\mathbf{obs}}(\omega) \propto F(\omega)$.

The theoretical power transfer function for a two-component neutron star with an internal resonant mode is (Lamb *et al.* 1978a)

$$F(\omega) = \frac{(\omega^2\tau^2 - \beta\omega_0^2\tau^2)^2 + \beta^2\omega^2\tau^2}{(\omega^2\tau^2 - \omega_0^2\tau^2)^2 + \omega^2\tau^2} ,$$

where ω_0 is the frequency of the mode, τ is the crust-superfluid coupling time, and β is the ratio of the inertial moment of the crust to that of the whole star. Three different types of behaviour are illustrated in Fig. 5. The dashed line shows the transfer function for a completely rigid star, whereas the light solid line shows the transfer function for a two-component star with $\beta = 0.36$. In the latter star, the power transfer at $\omega \ll \tau^{-1}$ is smaller than at $\omega \gg \tau^{-1}$ because rotational perturbations at the lower frequency must drive the whole star, whereas those at the higher frequency drive only the crust. The quantity β is a sensitive function of the star's mass and the equation of state. The heavy solid curve shows the power transfer function for a star with an internal resonant mode at the frequency $\omega_0 = 10\ \tau^{-1}$.

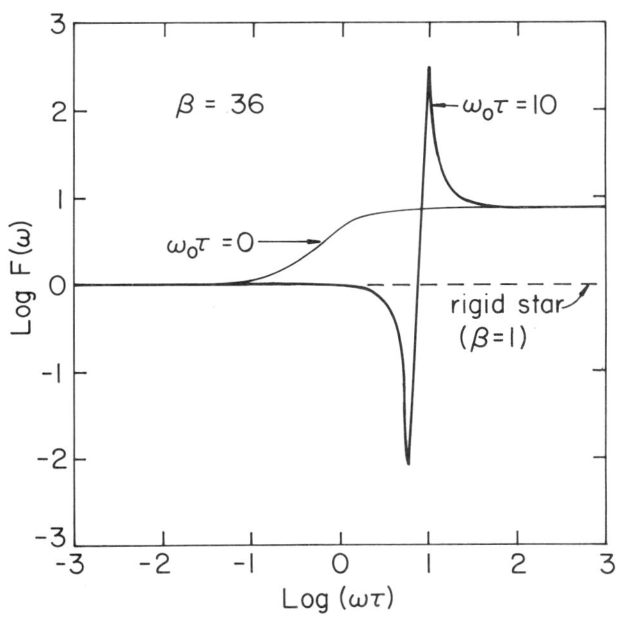

Fig. 5. Theoretical neutron star power transfer functions for a rigid star (dashed line) and a two-component star without (light solid curve) and with (heavy solid curve) an internal resonant mode.

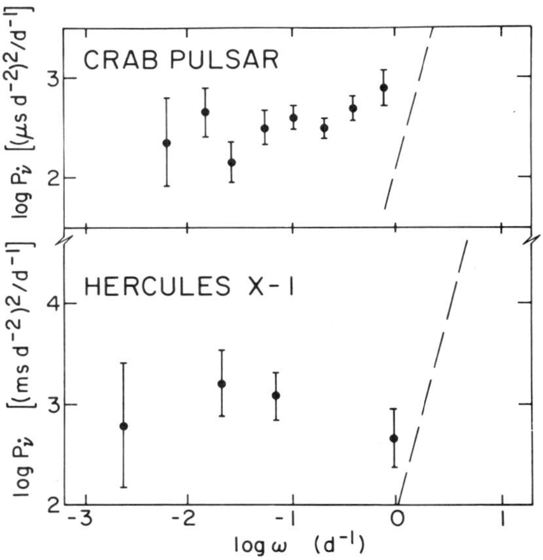

Fig. 6. Prewhitened power density spectra of the pulse timing residuals observed in the Crab pulsar and Her X-1. (From Boynton, 1981.)

Boynton and Deeter (1979; see also Boynton, 1981) have determined the power density spectrum of pulse phase fluctuations in the Crab pulsar, using the Princeton optical timing data, and in Her X-1, using the Uhuru X-ray data. The results are shown in Fig. 6, where the power density in the variable $\dot{\nu} = \dot{\Omega}/2\pi$ is plotted against angular frequency ω. In each case the data points to the left of the dashed line show the timing noise intrinsic to the star, while the dashed line indicates the calculated spectrum of the white phase noise introduced by the measurement errors. The latter has been made 'blue' by the process used to whiten the intrinsic noise spectrum. Both intrinsic noise spectra are relatively flat over more than two decades in frequency, indicating that a description in terms of a simple process (in this case, white noise in $\dot{\Omega}$) is appropriate, and that the neutron star response is approximately that of a rigid body (F = const.). Comparison with the power transfer function for a two-component star excludes τ's in the range 4—14 d, for the Crab pulsar, and 1—9 d, for Her X-1, with 99% confidence, assuming $Q \cong 1-\beta$ is as large as 0.7 (see Fig. 7).

In the case of the Crab pulsar, these results conflict with the parameters derived from analysis of the two macroglitches (contrary to the assertion of Lohsen 1981), implying that the internal dynamics of this neutron

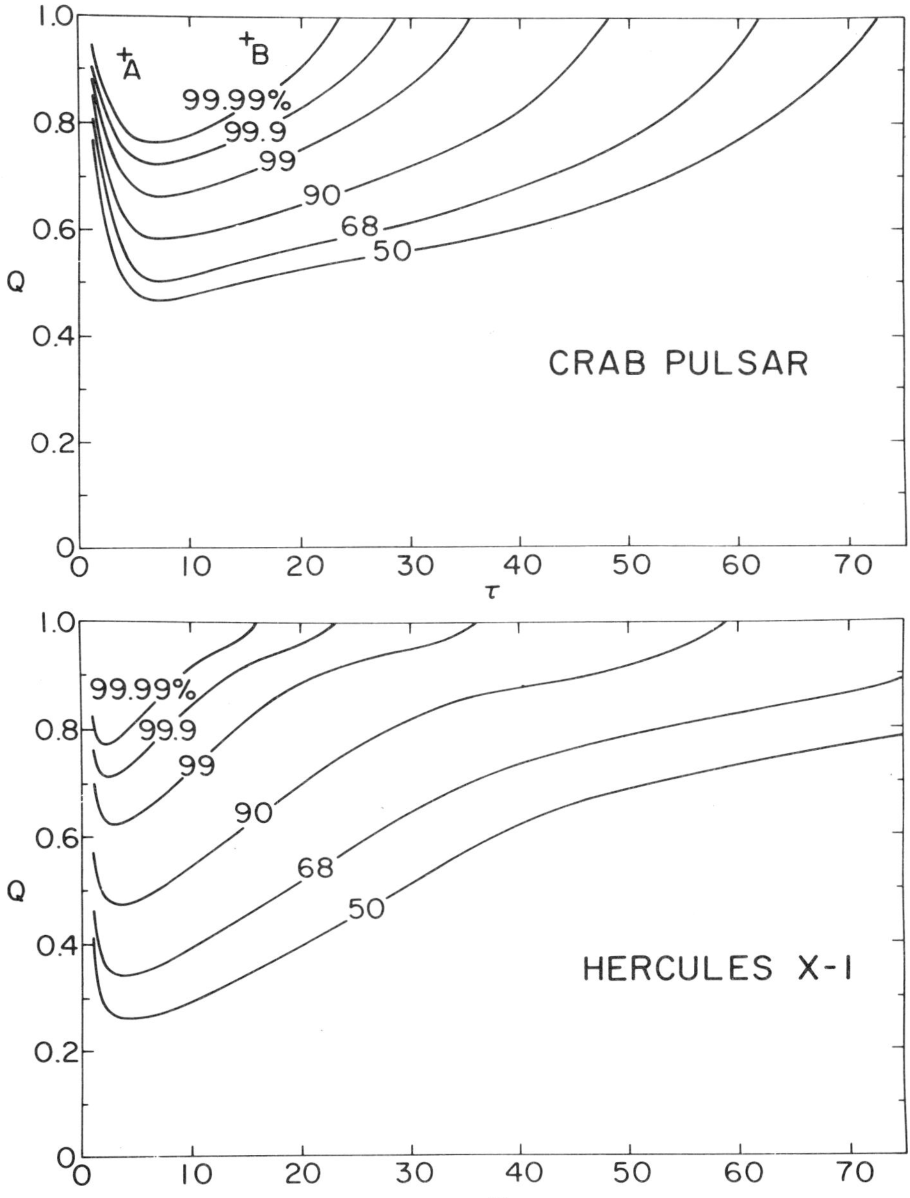

Fig. 7. Chi-square surfaces resulting from a comparison of the theoretical transfer function shown in Fig. 5 with the observed power density spectra of the Crab pulsar and Her X-1 shown in Fig. 6. Regions above the curves can be excluded with the confidence indicated. The two points labelled A and B on the Crab plot show the values of Q and τ derived from the two macroglitches discussed in the text. These values can be excluded with 99.99% confidence. (From Boynton, 1981.)

star are more complicated than can be described by the two-component model. This represents a major advance in the continuing effort to confront theories of neutron star dynamics with observation, and supports theoretical efforts to develop more realistic models. One possibility is that the response of the star depends on the size and rate of events, which are very different for macroglitches and unresolved events. A macroglitch could, for example, heat a portion of the neutron superfluid, thereby reducing τ from a 'normal' value $\gg 40$ d to 4-15 d. Alternatively, macroglitches may release a second component which is not released by the much smaller unresolved events. This would be the case if neutron superfluid vortices in the inner crust remain pinned except when unpinned by a macroglitch. If in addition the inertial moment of the unpinned superfluid is at least comparable to that of the crust component, the behaviour could be consistent with that observed.

6. Concluding Remarks

We have seen that timing observations of pulsars and pulsing X-ray stars have furnished significant information on the properties of neutron stars, including estimates of the masses of seven probable neutron stars. Measurements of macroglitches in four pulsars constrain models of their internal structure, while smaller scale timing irregularities have been detected in more than a dozen pulsars and at least four pulsing X-ray stars. Studies of the latter have provided important information on the causes of timing noise. Regardless of the cause, such irregularities can be used to probe the internal structure of neutron stars, as has been done in the Crab pulsar and Her X-1.

The prospects for future progress are bright. New mass estimates may come from the two most recently discovered binary pulsars (Taylor, 1981) and from additional X-ray stars. X-ray stars are the most promising candidates for studies of neutron star inertial moments and magnetic fields, and for studies of internal dynamical time scales. Such studies are already in progress using data from the HEAO-1 satellite and fresh opportunities will be provided by the launch of the X-ray Timing Explorer in the late-1980's. Pulsars are the most promising candidates for comparisons of macroglitches with smaller scale timing irregularities. Tests of noise strengths, spectra, and scaling laws between sources will provide new constraints on the causes of timing noise and the variation of neutron star properties from source to source.

It is a pleasure to acknowledge useful discussions with Paul Boynton, John Deeter, and Greg Zylstra during the preparation of this review. This research was supported in part by NSF grant PHY 80-25605 and NASA grant NSG 7653.

References

Alpar, A., Anderson, P.W., Pines, D., and Shaham, J., Astrophys. J., **249**, L29 (1981).
Anderson, P.W., Alpar, M.A., Pines, D., and Shaham, J., Phil. Mag., **45**, 227 (1982).
Arnett, W.D., and Bowers, R.L., Astrophys. J. Suppl. **33**, 415 (1977).
Bahcall, J.N., and Chester, T.J., Astrophys. J., **215**, L21 (1977).
Baym, G., Lamb, D.Q., and Lamb, F.K., Astrophys. J., **208**, 829 (1976).
Baym, G., and Pethick, C.J., Ann. Rev. Nucl. Sci., **25**, 27 (1975).
Baym, G., and Pethick, C.J., Ann. Rev. Astron. Astrophys., **17**, 415 (1979).
Baym, G., Pethick, C.J., Pines, D., and Ruderman, M., Nature, **224**, 872 (1969).
Baym, G., and Pines, D., Ann. Phys. N.Y., **66**, 816 (1971).
Blackman, R.B., and Tukey, J.W. *The Measurement of Power Spectra*, Dover, N.Y. (1958).
Boynton, P.E., In *Pulsars*, Proc. IAU Symp. No.95 (eds. W. Sieber and R. Wielebinski), Reidel, Dordrecht, p.279 (1981).
Boynton, P., and Deeter, J., In *Compact Galactic X-ray Sources* (eds. F.K. Lamb and D. Pines), UIUC Phys. Dept., Urbana, Illinois), p.168 (1979).
Boynton, P.E., Groth, E.J., Hutchinson, D.P., Nanos, G.P., Partridge, R.B., and Wilkinson, D.T., Astrophys. J., **175**, 217 (1972).
Carlson, J., Pandharipande, V.R., and Wiringa, R.B., Preprint (1982).
Coon, S.A., Scadron, S.D., McNamee, P.C., Barrett, B.R., Blatt, D.W.E., and McKellar, B.H.J., Nuclear Physics, **A317**, 242 (1979).
Cordes, J.M., Astrophys. J., **237**, 216 (1980).
Cordes, J.M., and Greenstein, G., Astrophys. J., **245**, 1060 (1981).
Cordes, J.M., and Helfand, D.J., Astrophys. J., **289**, 640 (1980).
Davidson, K., and Ostriker, J.P., Astrophys. J., **179**, 585 (1973).
Deeter, J., and Boynton, P., Preprint (1982).
Downs, G.S., Astrophys. J., **249**, 687 (1981).
Downs, G.S., Astrophys. J., **257**, L67 (1982).
Elsner, R.F., Ghosh, P., and Lamb, F.K., Astrophys. J., **241**, L155 (1980).
Fantoni, S., Friman, B., and Pandharipande, V.R., Preprint (1982).
Fenimore, E.E., Laros, J.G., Klebesadel, R.W., Stockdale, R.E., and Kane, S.R. In *Gamma-Ray Transients and Related Astrophysical Phenomena* (eds. R. Lingenfelter, H. Hudson, and D. Worrall), AIP Conf. Proc. No. 77, New York, p.201 (1982).
Friedman, B., and Pandharipande, V.R., Nuclear Phys., **A361**, 502 (1981).
Glen, G., and Sutherland, P., Astrophys. J., **239**, 671 (1981).
Gold, T., Nature, **221**, 25 (1969).
Greenstein, G., Astrophys. J., **231**, 880 (1979).
Groth, E.J., Astrophys. J. Suppl., No. 293 (1975).
Gudmundsson, E.H., Pethick, C.J., and Epstein, R.I., Astrophys. J., **272**, 268 (1982).
Helfand, D.J., In *Pulsars*, Proc. IAU Symp. No. 95 (eds. W. Sieber and R. Wielebinski) Reidel, Dordrecht, p. 343 (1981).
Helfand, D.J., Taylor, J.H., Backus, P.R., and Cordes, J.M., Astrophys. J., **237**, 206 (1980).
Hewish, A., Bell, S.J., Pilkington, J.D.H., Scott, P.F., and Collins, R.A. Nature, **217**, 709 (1968).
Lamb, F.K., In Proc. 7th Texas Symp. Relativistic Astrophys., Ann. N.Y. Acad. Sci., **262**, 331 (1975).
Lamb, F.K., In Proc. 8th Texas Symp. Relativistic Astrophys., Ann. N. Y. Acad. Sci., **302**, 482 (1977).

Lamb, F.K., In *Compact Galactic X-ray Sources* (eds. F.K. Lamb and D. Pines), UIUC Phys. Dept., Urbana, Illinois, p.143 (1979a).
Lamb, F.K., In Proc. HEAO Sci. Symp. (eds. C. Dailey and W. Johnson), NASA, CP 2113, p.223 (1979b).
Lamb, F.K., In *Pulsars*, Proc. IAU Symp. No. 95 (eds. W. Sieber and R. Wielebinski), Reidel, Dordrecht, p.303 (1981a).
Lamb, F.K., In *X-Ray Astronomy in the 1980's* (ed. S. Holt), NASA Technical Memorandum 83848, p.77 (1981b).
Lamb, F.K., Pethick, C.J., and Pines, D., Astrophys. J., **184**, 271 (1973).
Lamb, F.K., Pines, D., and Shaham, J., Talk presented at the 7th Texas Symp. on Relativistic Astrophys. (unpublished) (1974).
Lamb, F.K., Pines, D., and Shaham, J., In *X-Ray Binaries*, Proc. Goddard Conference (eds. E. Boldt and Y. Kondo), NASA, SP-389, p.141 (1976).
Lamb, F.K., Pines, D., and Shaham, J., Astrophys. J., **224**, 969 (1978a).
Lamb, F.K., Pines, D., and Shaham, J., Astrophys. J., **225**, 582 (1978b).
Laros, J.G., Evans, W.D., Fenimore, E.E., and Klebesadel, R.W., Los Alamos National Laboratory Preprint LA-UR-82-936 (1982).
Leung, Y.C., and Wang, C.G., Astrophys. J., **170**, 499 (1971).
Lohsen, E.H.G., Nature, **258**, 689 (1975).
Lohsen, E.H.G., Astron. Astrophys, Suppl., **44**, 1 (1981).
Manchester, R.N., In *Pulsars*, Proc. IAU Symp. No. 95 (ed. W. Sieber and R. Wielebinski), Reidel, Dordrecht, p.267 (1981).
Manchester, R.N., Newton, L.M., Goss, W.M., and Hamilton, P.A., Monthly Notices Roy. Astron. Soc., **184**, 35P (1978).
Manchester, R.N., and Taylor, J.H., *Pulsars*, W.H. Freeman and Co., San Francisco (1977).
Mazets, E.P., Golenetskii, S.V., Aptekar', R.L., Gur'yan, Yu.A., and Il'inskii, V.N., Nature, **290**, 378 (1981a).
Mazets, E.P., Golenetskii, S.V., Il'inskii, V.N., Panov, V.N., Aptekar', R.L., Gur'yan, Yu.A., Proskura, M.P., Sokolov, I.A., Sokolova, Z.Ya., Kharitonova, T.V., Dyatchkov, A.V., and Khavenson, N.G., Astrophys. Spa. Sci., **80**, 3 (1981b).
Middleditch, J., Mason, K.O., Nelson, J., and White, N.E., Astrophys. J. **244**, 1001 (1981).
Middleditch, J., and Nelson, J., Astrophys. J., **208**, 567 (1976).
Nomoto, K., and Tsuruta, S., Astrophys. J., **250**, L19 (1981).
Pandharipande, V.R., and Smith, R.A., Nuclear Phys., **A237**, 507 (1975a).
Pandharipande, V.R., and Smith, R.A., Phys. Letters, **59B**, 15 (1975b).
Pines, D., J. Physique, **C2**, 111 (1980).
Pines, D., Shaham, J., and Ruderman, M.A., Nature, **237**, 83 (1972).
Pines, D., Shaham, J., and Ruderman, M.A., In *Physics of Dense Matter*, IAU Symp. No. 53 (ed. C.J. Hansen), Reidel, Dordrecht, p.189 (1974).
Pringle, J.E., and Rees, M.J., Astron. Astrophys., **21**, 1 (1972).
Radhakrishnan, V., and Manchester, R.N., Nature, **222**, 228 (1969).
Rappaport, S., and Joss, P.C., In *X-Ray Astronomy with the Einstein Satellite* (ed. R. Giacconi), Reidel, Dordrecht, p.123 (1981).
Reichley, P.E., and Downs, G.S., Nature, **222**, 229 (1969).
Richards, D.W., and Comella, J.M., Nature, **222**, 551 (1969).
Richardson, M.B., Van Horn, H.M., Ratcliff, K.F., and Malone, R.C., Astrophys. J., **255**, 624 (1982).
Ruderman, M.A., Nature, **223**, 597 (1969).
Ruderman, M.A., Ann. Rev. Astron. Astrophys., **10**, 427 (1972).
Ruderman, M.A., Astrophys. J. **203**, 213 (1976).
Ruffini, R., In *Astrophysics and Gravitation*, Proc. 16th Solvay Conference, L'Université de Bruxelles, p.349 (1974).
Schreier, E., Levinson, R., Gursky, H., Kellogg, E., Tananbaum, H., and Giacconi, R., Astrophys. J., **172**, L79 (1972).
Swank, J.H., Eardley, D., and Serlemitsos, P.J., Preprint (1979).

Taylor, J.H., In *Pulsars*, Proc. IAU Symp. No. 95 (eds. W. Sieber and R. Wielebinski), Reidel, Dordrecht, p.361 (1981).
Taylor, J.H., Fowler, L.A. and McCulloch, P.M., Nature, **277**, 437 (1979).
Van Paradijs, J., Nature, **274**, 650 (1978).
Van Riper, K.R. and Lamb, D.Q., Astrophys. J., **244**, L13 (1980).

Macroscopic Quantum Tunnelling and all that

A. J. Leggett†
Laboratory of Atomic and Solid State Physics, Cornell University, Ithaca, New York 14853

ABSTRACT

This paper reviews recent experimental and theoretical work on the question of whether the motion of a macroscopic system can be shown to exhibit quantum-mechanical effects, with particular reference to the possible observation of such effects in Josephson systems. It concentrates primarily on the question of quantum-mechanical tunnelling of a macroscopic variable, paying special attention to recent theoretical work on the effects of the dissipation which must characterize the motion of such a variable.

1. Introduction

It is a pleasure to dedicate this paper to Dirk ter Haar, to whose guidance I owe my first steps in condensed-state theory, on the occasion of his sixty-fifth birthday, and to wish him many more happy and productive years of activity in physics.

The subject of this paper is one which has seen considerable advances in the last three or four years on both the experimental and theoretical fronts. Here I shall try to give a picture of our current understanding of it, concentrating mainly on the theoretical side. For reasons of space it is impossible to attempt here an exhaustive survey of the recent literature, and the papers mentioned below represent of course a purely personal selection. I will normally quote results only, referring to the original literature for the details

---

† On leave of absence from University of Sussex, UK and University of Illinois

of the arguments and calculations.†

First, why is the subject interesting ? To me, at least, because it affords an all-too-rare example of a case in which condensed-state physics can be used for a genuine test of the foundations of quantum mechanics.‡ Let's recall that the fundamental "quantum measurement paradox" [3] consists essentially in the observation that in appropriate circumstances extrapolation of the quantum-mechanical description of nature up to the macroscopic level leads to the inescapable conclusion that the state description of a *macroscopic* object (in Schrödinger's [4] original example, a cat) must be undefined or indeterminate until that object is observed. Now the odd thing about this problem is that while it has kept generations of philosophers of science awake at night, most practising physicists dismiss it as of purely 'philosophical' (i.e., no) interest. The reason for this attitude is probably the general belief that the linear superpositions of macroscopically different states which occur in the formulation of the paradox are 'harmless', in the sense that no experimental observation could ever distinguish this description from a classical probabilistic mixture. (I.e., although (very crudely speaking) under appropriate circumstances the technically correct description of Schrödinger's cat is that it is a linear superposition of states of being dead and being alive, the results of all conceivable experiments are allegedly equally well accounted for by an alternative, purely classical description in which the cat 'really is' either dead or alive, but we don't know which until we look.) Since physicists, as opposed to philosophers, generally prefer to discuss hypotheses which can be subjected to an experimental test, the above observation is guaranteed to terminate most physics-department cocktail-party discussions of the question.

But is the observation actually correct ? That is, is it in principle impossible to observe the effects of superposition of quantum-mechanical states corresponding to macroscopically different properties? To some extent this is a matter of definition: just as there are many people who, when faced with incontrovertible evidence that machines can be programmed to perform some tasks normally associated with human intelligence, automatically react by redrawing the boundary of 'intelligent' behaviour so as to exclude these tasks, so the reaction of many physicists to the suggestion that it

---

† Some parts of this paper are in effect an expurgated version of a lengthy paper [1] by A.O. Caldeira and the author which is in press at the time of writing.

‡ For a considerably more extended discussion of the ensuing points, see reference 2.

might be possible to observe the effects of quantum-mechanical interference at the macroscopic level is simply to refuse, as a matter of definition, to apply the term 'macroscopic' to the suggested example: for them, the boundary of the macroscopic worked is *defined* to lie at or beyond the point at which such interference effects vanish. One may conjecture that Nils Bohr, were he alive today, would have reacted in this way: cf. the remarks in ref.5. For people who take this standpoint the questions discussed in this paper will be of purely technical interest, with no relevance to the foundations of physics. The present author takes the opposite point of view, and hopes that at least some of his readers will do likewise.

2. Where to Look?

The most convenient way to discuss the question posed in the last paragraph is to introduce the idea of a macroscopic variable, that is, a variable such that 'appreciably' different values of it correspond to states which are macroscopically distinguishable, and to ask what are the necessary conditions for a quantum-mechanical description of the dynamics of such a variable to lead to results appreciably different from a classical description.† Let us consider for definiteness a mechanical system of macroscopic size such as a ball-bearing; the centre-of-mass coordinate X is then clearly 'macroscopic' in the above sense. Then if the system moves in a potential field $V(X)$ with a characteristic magnitude V_0 and characteristic length scale l, it is easy to convince oneself that to see quantum-mechanical effects one must evade the correspondence limit, that is, attain the order-of-magnitude condition $V_0 \lesssim \hbar\omega_0$, where ω_0 is a typical frequency of classical motion in the potential. Since we have $\omega_0^2 \sim (m^{-1}\,\partial^2 V/\partial X^2) \sim mV_0/l^2$, this condition is equivalent to $V_0 \lesssim \hbar^2/ml^2$. Moreover, if the quantum effects are not to be blurred out by thermal smearing, we also need to satisfy the condition‡ $\hbar\omega_0 \gtrsim k_B T$ (cf. below, section 3); with current technology T cannot be much less than, say, a millidegree. Thus the requirements are very roughly

$$V_0 \lesssim \hbar^2/ml^2\ , \qquad (k_B T)^2/V_0 \lesssim \hbar^2/ml^2\ . \tag{1.1}$$

It is clear that for a mechanical system of the dimensions we would normally call macroscopic, these conditions are quite unattainable.

---

† For further elaboration of the ensuing argument, see ref. 2, section 4 and ref. 1, section 1.

‡ This second condition is somewhat less ineluctable than the first; however, for most practical experiments of the type considered to date it seems necessary.

However, it is possible to attain the condition (1.1) in some systems where the macroscopic variable is not a geometrical coordinate. A promising example [6] for our purposes is a reasonably thick superconducting ring interrupted by a Josephson junction. Such a ring is used as the active element in an r.f. SQUID (superconducting quantum interference device),[7] and we shall refer to it for brevity simply as a SQUID (ring). Here the macroscopic variable is the magnetic flux ϕ trapped in the ring, the 'length' (i.e., flux) scale of the potential energy is the so-called flux quantum

$$\phi_0 \equiv h/2e \cong 2 \times 10^{-15}\ \mathrm{Wb},$$

and its magnitude is of the order of $I_c\phi_0$, where I_c is the critical current of the Josephson junction. Finally, the analogue of the mass in the mechanical case is the capacitance shunting the junction.[6,8] The conditions to see characteristically quantum-mechanical effects are then, very roughly, by analogy with (1.1)

$$CI_c\phi_0^3 \leqslant \hbar^2\,, \qquad k_BT \leqslant \hbar(I_c/C\phi_0)^{\frac{1}{2}}\,. \tag{1.2}$$

Since it is possible to make junctions with capacitances of the order of 10^{-13} F (see also below) and critical currents in the nanoamp to microamp range, these conditions can be attained without excessive difficulty.†

To what extent would a demonstration that the trapped flux in such a SQUID ring is behaving in an essentially quantum-mechanical way constitute direct evidence for the validity of the concept of superposition of macroscopically distinguishable states (rather than merely evidence against classical mechanics at this level)? This is actually quite a subtle question: see ref. 2. As we shall see, the phenomena of interest involve states in which the trapped flux differs by some fraction of the flux quantum ϕ_0. Two questions then arise: (1) are such states correctly described as 'macroscopically different'? (2) to what extent does the fact that the system behaves according to the characteristic predictions of quantum mechanics directly establish the existence of superpositions of such states? As noted above, (1) is to some extent a question of definition: whether one regards a flux difference of the order of ϕ_0 (which, remember, is itself proportional to the quantum constant h) as 'macroscopic' is perhaps a matter of taste. However, from the point of view of the quantum theory of measurement the significant point is that the states involved unarguably differ in the behaviour of a *macroscopic number of particles* (essentially, all or most of the 10^{23}-odd

---

† **Conditions (1.2) are actually somewhat too stringent, since by operating near a point of classical instability one can decrease both the energy and the 'length' scales appreciably.**

electrons in the ring — though see below); for the significance of this feature, see ref. 2. With regard to the second question, observation of the phenomenon of macroscopic quantum coherence (see below) would undoubtedly be the direct evidence required; however, the evidence provided by most of the other quantum-mechanical phenomena to be discussed is more circumstantial. As so often in tests of the foundations of a well-established physical theory, the basic problem is that to 'verify' the standard theory (in this case, quantum mechanics as extrapolated to the macroscopic level) can really only mean to refute a class of alternative theories, and in the case of most of the phenomena to be discussed (in particular, of macroscopic quantum tunnelling) little work has been done to develop alternative theories to the point where meaningful tests become possible. The situation in this regard is closely parallel to the one which was obtained with respect to the EPR paradox [3,9] before the work of Bell [10]: so long as the predictions of quantum mechanics for a two-photon decay process could not be shown to be incompatible with those of a class of intuitively reasonable alternative theories (in this case, local hidden-variable theories or more generally objective local[11] theories) it was difficult to say precisely what (new) was established by experiments which verified the former. In our problem, while it is intuitively clear that the observation of macroscopic quantum coherence would tend to exclude, at least, a class of theories in which the wave function reduces itself at a certain level of complexity or 'disconnectivity' (see ref. 2), no rigorous analogue of Bell's theorem has been proved even for this case, let alone for the other phenomena of interest. Thus, in a sense the most clear-cut, and certainly the most exciting outcome of the experiments would be a *failure* to observe the predicted quantum-mechanical phenomena.

Assuming that we have built ourselves a SQUID ring to the required specifications, what kind of quantum-mechanical behaviour should we look for? To help us answer this question, I show in Fig. 1a the conventional potential energy of the ring as a function of the trapped flux ϕ. In the case plotted the external flux ϕ_x is taken equal to zero. Once we are clearly outside the correspondence limit, almost any kind of experiment, including those on motion in the classically accessible regions of the potential, should show *some* quantum effects; however, in general these will be merely quantitative modifications of the classical behaviour. What we need is some phenomenon which owes its very existence to quantum-mechanical effects. There is, actually, a whole group of such phenomena, all of them resting on the classically unknown possibility of penetration through a potential barrier of height greater than the system's energy. In Fig. 2 I show some of those

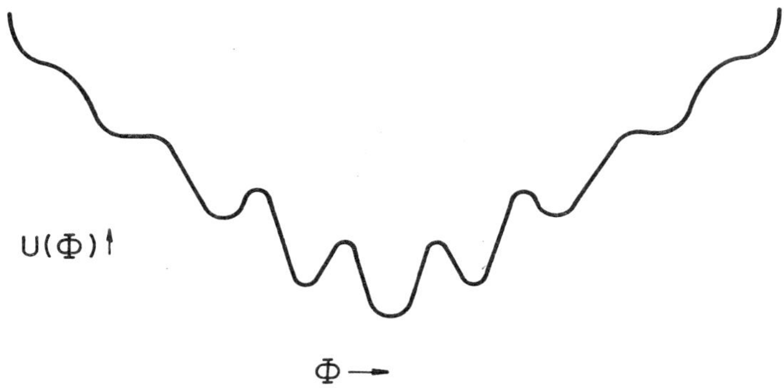

(a)

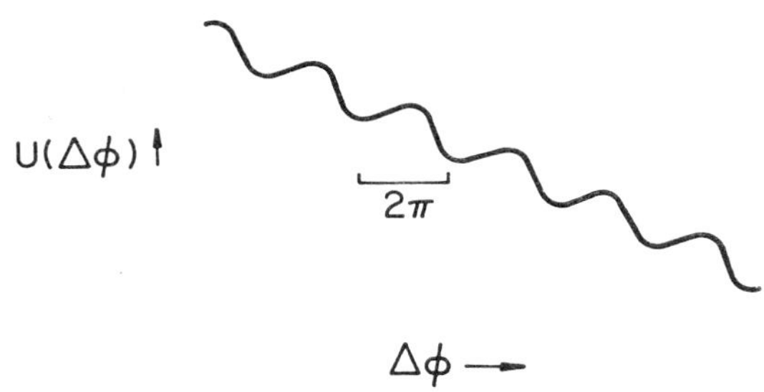

(b)

Fig. 1 (a) The potential energy $U(\phi)$ of a SQUID ring as a function of the trapped flux ϕ. The curve is drawn for external flux ϕ_x equal to zero. For a more general plot see ref. 7.
(b) The (pseudo) potential energy $U(\Delta\phi)$ of a current-biassed junction as a function of phase difference $\Delta\phi$ (the 'washboard potential')

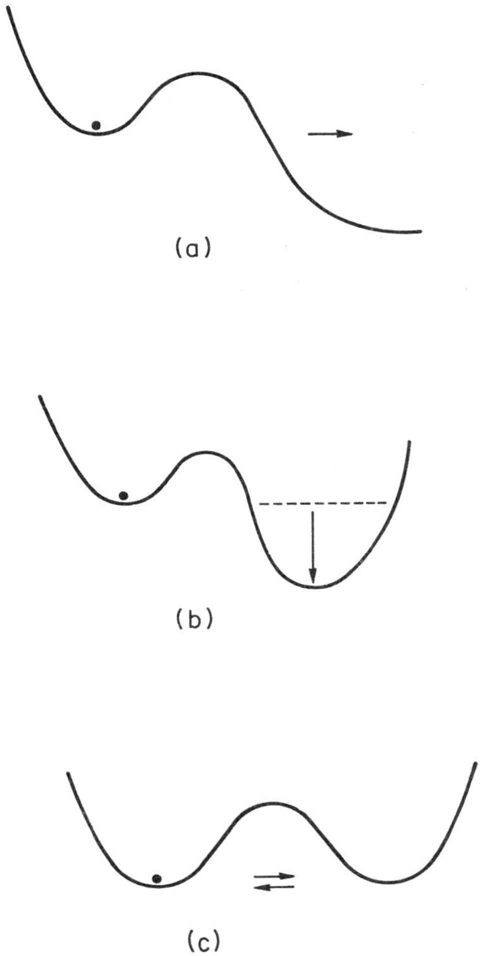

Fig. 2 Various types of phenomena associated with barrier penetration
(a) Quantum tunnelling (decay)
(b) Quantum hopping
(c) Quantum coherence

phenomena. The first, which I call simply quantum *tunnelling* (or quantum decay) is the analogue of the emission of an α-particle from a heavy nucleus: in this case, once the particle has escaped from the metastable potential well, it effectively disappears to infinity and we need keep no further track of it. The second phenomenon, quantum *hopping*, is the analogue of the quantum diffusion of a defect in a solid in the presence (e.g.) of an electric field: in this case the particle does not escape to infinity, but its final state has an energy lower than its initial one, so that no return jump is possible in the absence of excitation by the experimenter or the environment. In this case the system usually emits a real photon (or phonon) in the process of hopping; the inverse process, photo-assisted tunnelling (upwards jumping) is also of interest, cf. below. The third phenomenon, quantum *coherence*, is the analogue of the inversion resonance of the ammonia molecule: the system moves backwards and forwards between two degenerate potential wells separated by a classically impassable barrier. In this case it is essential to keep track of the relative phase of the parts of the wave function which are on different sides of the barrier, whereas for the case of quantum tunnelling this does not play a role (cf. ref. 2). (For the intermediate case, quantum hopping, the situation depends on the details of the parameters: for most cases of practical interest the relative phase is unimportant.) All of these phenomena are in principle expected to occur in a SQUID, given suitable choices of the system parameters.†

Although a SQUID ring is probably the most suitable system for our purposes, some of the phenomena mentioned above can be expected to occur also in other macroscopic systems. In particular, there is one closely related system which has been sued for several recent experiments, namely a simple Josephson junction carrying a fixed externally imposed current.‡ (We shall refer to this system as a current-biassed junction or CBJ.) In this case the relevant macroscopic variable is the phase difference $\Delta\phi$ of the Cooper-pair condensate wave function on opposite sides of the junction. We show the relevant potential energy (the so-called 'washboard potential') in Fig. 1b. Since $\Delta\phi$ is defined only mod 2π, the various potential minima, which corres-

† Strictly speaking, quantum decay (tunnelling) cannot occur in a SQUID since the flux can never escape 'to infinity'. However, in many real-life experimental situations it escapes sufficiently that we have in effect a decay situation. See ref. 1, section 6.

‡ The SQUID ring is obtained from this system by joining up the current leads to form a loop and allowing the system to control its own current dynamics. For the formal relationship between the two systems see ref. 1, section 1.

pond to a phase difference of $2n\pi$, are not physically distinct, and therefore the concepts of quantum hopping and coherence have no direct meaning (though cf. below). However, since in a typical tunnelling situation (Fig. 2a) the width of the barrier is much smaller than 2π, the concept of quantum tunnelling is still well-defined and, if it occurs, leads to the onset of a finite-voltage state of the junction. These points may perhaps be clarified by considering the mechanical analogue of a CBJ, namely a pendulum driven by a constant mechanical torque somewhat less than that needed classically to set it into rotation. (The onset of rotation is the analogue of the onset of finite voltage.) (Cf. also ref. 12.)

One comment needs to be made, in the context of both SQUID and CBJ experiments, concerning the capacitance of the Josephson junction, which in both cases is the analogue[†] of the mass in a mechanical problem. It is clear that, other things being equal, the smaller the capacitance the easier it is to evade the correspondence limit. Now, the geometrical capacitance of Josephson junctions of either the oxide-layer or point-contact type as currently manufactured is rarely less than ~ 0.1 pF, and indeed is often considerably larger than this. However, in the literature[13,14] one sometimes meets estimates of point-contact junction capacitances as small as 10^{-15} F. These estimates are based on a comparison of the experimental current-voltage characteristics of the junctions in question (and in particular the hysteresis properties) with the standard classical analysis,[15] and are justified by the hypothesis that the capacitance C in question refers to a very small region, perhaps with a dimension of the order of a few hundred Å, near the junction itself and acts in series with a substantial inductance L' (perhaps the kinetic inductance of the narrow metallic filaments which the electrons have to pass through to escape from the region in question): see Fig. 3. The whole lumped element is presumably then shunted by the much larger geometrical capacitance C_G, and in the case of a SQUID is in series with the bulk ring inductance L; see ref. 16. The point to bear in mind in the present context is that if this picture is indeed correct, then it is quite possible (and indeed probable) that any quantum tunnelling behaviour observed takes place only in the small region very close to the junction (and to that extent is less unarguably 'macroscopic', since only the electrons in the immediate vicinity are involved); the motion of the genuinely macroscopic variable could then be purely classical.[17] A more general point, in the case of a SQUID, is that if one wants to see quantum phenomena which involve all the

† **Strictly speaking, in the CBJ case it is the quantity $C(\phi_0/2\pi)^2$ which is the analogue of the mass.**

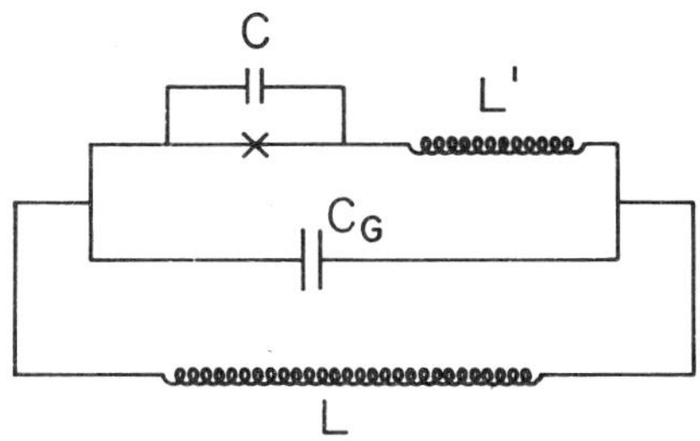

Fig. 3 Effective circuit diagram for a Josephson junction with very small intrinsic capacitance. The junction is, as conventional, represented by a cross.

electrons in the ring, or a substantial fraction of them, then it is no use using capacitances so small that the typical frequencies of relevance in the problem in question are larger than c/l, where l is a typical ring dimension and c is the velocity of light: it is clear that phenomena involving frequencies much larger than this will automatically be localized in a small region close to the junction itself. Moreover, for such phenomena the standard RSJ SQUID dynamics (cf. below) obviously has to be modified substantially.

Other systems besides SQUID's and CBJ's may also be suitable for the observation of macroscopic quantum tunnelling. One system where a detailed theory of such tunnelling has been explicitly invoked [18] to explain the experimentally observed behaviour is the charge density waves believed to exist in some quasi-one- and two-dimensional solids such as $NbSe_3$ and TaS_3 (For a review of these systems see ref. 19.) The tunnelling theory of Bardeen[18,20] has been explicitly formulated by analogy to the theory of Zener tunnelling in solids, but to make contact with the other examples discussed in this paper it may be helpful to try to re-phrase the problem in the following somewhat non-rigorous way†: a charge density wave of wave vector $2k_F$ is described by a phase ϕ which is defined modulo 2π, or equivalently by a 'coordinate' $q = (2k_F)^{-1}\phi$ which is defined modulo πk_F^{-1}. The kinetic energy of the moving wave is

$$\tfrac{1}{2} N_a (m^* + M_F)\, v_0^2 = \tfrac{1}{8} N_a (m^* + M_F)\, k_F^{-2} \dot{\phi}^2 ,$$

† It should be emphasized that the discussion of the CDW problem in this paper represents my own speculative attempt to make contact between recent work in this area and the other topics discussed here; it is not implicit that anyone working in the area necessarily agrees with it. For an alternative approach to the CDW problem which does not involve tunnelling, see ref. 21.

and the periodic potential energy is of the form $N_a U(\phi)$; here m^* and M_F are respectively the electron band mass and Fröhlich mass (see ref. 18), N_a is the number of electrons in a phase-coherent domain, and $U(\phi)$ is a pinning potential per electron which may be very small indeed. In the following we take for definiteness $U(\phi)$ to have a pure cosine form $U = -U_0 \cos\phi$, but this is in no way essential. We see that the small oscillations of the CDW phase correspond to a small-oscillation frequency ω_p which is equal to

$$(4U_0 k_F^2/M^*)^{\frac{1}{2}} \qquad (M^* \equiv m^* + M_F) \; ;$$

hence the spacing of the lowest quantized levels is $\hbar\omega_p$. If now we switch on an external electric field E, then the effect is to add to the potential a linear term $- eEN_a(2k_F)^{-1}\phi$. Thus we get a 'washboard potential' of the type shown in Fig. 1b, and indeed the problem is formally identical to that of the current-biassed junction with the replacements

$$\Delta\phi \to \phi , \quad C(\phi_0/2\pi)^2 \to \tfrac{1}{4} N_a M^* k_F^{-2} ,$$
$$(I_c\phi_0/2\pi) \to N_a U_0 , \quad I_{ext}(\phi_0/2\pi) \to N_a e(2k_F)^{-1} E .$$

However, under practical conditions there is an important difference between the two problems. Namely, while in the CBJ case the dimensionless quantity $K \equiv (32\, CI_c\phi_0^3/\pi^3\hbar^2)^{\frac{1}{2}}$ which determines† the tunnelling probability ($\propto e^{-K}$) between neighbouring wells at zero external current bias is usually fairly large, so that in practice tunnelling is observed only near the point of classical instability ($1 - I_{ext}/I_c \ll 1$), in the CDW case the analogous parameter, namely

$$K' \equiv (64 N_a^2 M^* U_0/\hbar^2 k_F^2)^{\frac{1}{2}} ,$$

might not be very large‡ compared to 1. Thus in the latter case tunnelling could take place even at very small bias (small E); however, such tunnelling can have no physical meaning unless the system ends up in an *excited* state in the 'new' well (which is, of course, physically identical to the old one). Thus, with the 'washboard' of Fig. 1b slightly tilted, the system must actually tunnel through (at least) ν wells, where $\nu = \hbar\omega_p/N_a \pi k_F^{-1} eE$. (One may perhaps think of the Zener analogy as reflecting this feature by putting as it were a fictitious system identical to the real system in each of the (physically identical) wells.)

---

† **Or would determine it, if the wells were physically distinct.**

‡ **The plausibility of this statement of course depends crucially on the details of the microscopic model of the pinning potential $U(\phi)$: note in this context that $U(\phi)$ as defined above may itself be proportional to a negative power of N_a.**

There exist various other systems which have been regarded as possible candidates for the observation of macroscopic quantum tunnelling and/or coherence effects, both in particle physics ('vacuum instability' and 'instantons') and in condensed-state systems: for a partial list see ref. 1, section 1. Apart, possibly, from the ring laser studied in ref. 22, most of them suffer from the disadvantage that either the *classical* dynamics is poorly understood, the expected tunnelling lifetimes are astronomically long, or the effects of such tunnelling show up only in a very indirect way (as in many of the particle-physics examples). For reasons of space I therefore do not discuss them here.

3. Naive Expectations

Consider, say, a SQUID in which the flux ϕ is initially trapped in one of the metastable minima of Fig. 1c. We ask: What is the probability per unit time that the flux 'escapes' from the well into a lower well? If we consider only classical thermal processes. then the probability $P_{\mathbf{c}\ell}$ is given by the well-known Kramers formula[23].

$$P_{\mathbf{c}\ell} = \text{const.}\ \omega_0 \exp - V_0/k_{\mathbf{B}}T\ , \tag{3.1}$$

where ω_0 is the small-oscillation frequency around the metastable minimum and V_0 is the height of the barrier (see Fig. 1a), and the constant is of order 1 for moderate damping. We can, however, also consider the probability of quantum tunnelling through the barrier. At zero temperature the relevant rate $P_{\mathbf{QM}}$ can be expressed in the form (cf. ref. 24)

$$P_{\mathbf{QM}} = \text{const.}\ \omega_0 \left(\frac{V_0}{\hbar\omega_0}\right)^{\frac{1}{2}} \exp - (\alpha V_0/\hbar\omega_0) \tag{3.2}$$

where the constant in front of the prefactor is again of order 1, and where α is a constant which, in the limit as we approach the classical instability (so that the potential in the tunnelling region is well approximated by a quadratic-plus-cubic form) is given by[24] $\alpha = 36/5$. In deriving Eq.(3.2) we assumed that the WKB limit is applicable, i.e., that $\alpha V_0 \gg \hbar\omega_0$. For finite temperature the expression for the escape rate is quite complicated,[25,26] but is not quantitatively very different from the simple sum of (3.1) and (3.2); thus, the logarithm of the lifetime of the metastable state, if plotted against inverse temperature, should be flat at low temperatures and begin to rise linearly at or near the point $k_{\mathbf{B}}T = \hbar\omega_0/\alpha$: see Fig. 4.† Note that formulae (3.1) and (3.2) confirm, for the special case of quantum tunnelling, our previous conditions to see specifically quantum-mechanical behaviour. (If V_0 is *much* larger than $\hbar\omega_0$, tunnelling will of course still occur in

† **See footnote on next page.**

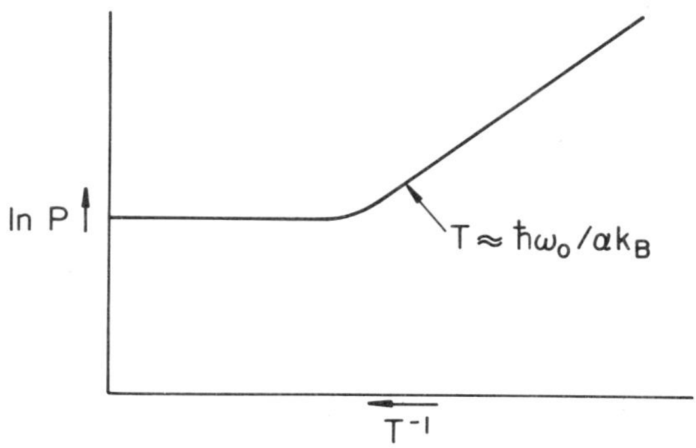

Fig. 4 The logarithm of the decay probability (in arbitrary units) versus inverse temperature for a tunnelling system without dissipation. Note temperature increases to the right.

principle but with unobservably small probability.) Obviously, exactly parallel calculations can be carried out for a CBJ, with results which are formerly identical to the above: the only difference is that in the general case the formulae for ω_0 and V_0 in terms of the system parameters look slightly different. In the case of most practical interest, the limit as the classical instability is approached, the results for the two systems are in exact correspondence.[17]

Using the formulae (3.1) and (3.2) it is straightforward to compute the predicted behaviour in a realistic experiment, for example one in which the external flux (current) is ramped as a function of time and the width of the distribution of values at which the system 'escapes' is plotted.[17] (In the SQUID case this calculation is relevant to the question of the ultimate sensitivity of the SQUID when used as a magnetometer (but see ref. 28).) The results show clearly that specifically quantum-mechanical behaviour should be observable with experimentally realistic values of the system parameters.

It is also straightforward to compute the tunnelling rate in the charge-density-wave system (see end of section 2). Since the system has to tunnel through ν barriers, and the probability of tunnelling through each is $e^{-K'}$,

---

† The result of Affleck[27] is somewhat misleading if applied to the real-life situation of tunnelling in a SQUID (or CBJ), where the parameter $V_0/\hbar\omega_0$ is never more than about 4. The point is that the crossover region between the temperature T_0 above which escape is entirely dominated by above-barrier thermal processes (as described by formula (3.1)) and the lower temperature T_1 below which only tunnelling out of the groundstate is important is appreciable only in the limit $V_0/\hbar\omega_0 \to \infty$: for $V_0/\hbar\omega_0 \leqslant 4$ it is never greater than $\sim 10\%$ of T_0.

where ν and K' are given in section 2, we find the probability of a tunnelling event is

$$P = \text{const. exp} -\left\{(64 N_a^2 M^* U_0/\hbar^2)^{\frac{1}{2}} (\hbar\omega_p k_F/N_a eE)\right\} . \tag{3.3}$$

The factors of N_a cancel, and using the relations $U_0 = M^*\omega^2/4k_F^2$ (see above) and the definitions[18] $e^* \equiv em^*/M^*$, $\varepsilon_g \equiv \hbar\omega_p$, we see that Eq. (3.3) is nearly identical to Bardeen's result

$$P = \text{const. exp} - (\pi\varepsilon_g^2/4\hbar e^* v_F) \tag{3.4}$$

the difference being that the exponent is (3.3) is $(16/\pi^2)$ times that in (3.4). This factor probably reflects subtle differences in the approximations implicitly made in applying the WKB method in the two approaches. (Note that in the above method the numerical factor in (3.3) is anyway a function of the exact shape of the potential $U(\phi)$.)

In the case of a SQUID (but not of a CBJ or CDW: see above) it is also possible to consider the feasibility of observing quantum coherence. To do this, we need to use a 'weakly hysteric' SQUID, that is, one such that the parameter $\beta_L \equiv 2\pi L I_c/\phi_0$ which determines the presence or absence of metastable minima (see ref. 7), is only just greater than 1, and to bias the external flux to half a flux quantum. The the potential for trapped flux ϕ near zero has the form shown in Fig. 5, and in analogy to the case of the ammonia molecule we expect the ground state to split into an even (ψ_+) and an odd (ψ_-) combination of the (approximately harmonic-oscillator) groundstates in the isolated wells, with a splitting $\hbar\Gamma$ between them. These

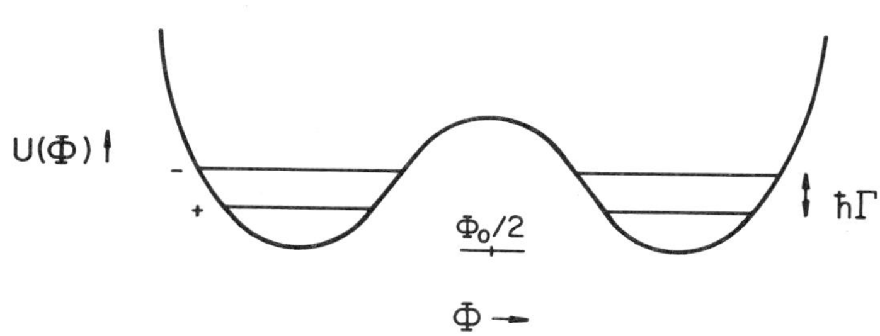

Fig. 5 **Quantum coherence in a SQUID ring. The external flux is biassed to exactly half a flux quantum.**

states are clearly, by any normal definition, linear superpositions of states with macroscopically different properties (cf. above), so unambiguous observation of this effect would adequately answer in the affirmative the question posed in the introduction. In principle one might hope to observe the existence of these superpositions either indirectly,† by resonance absorption of radiation of frequency Γ, or directly, by starting the system in one well and detecting the oscillations in the probability $P(t)$ for finding it there a time t later. In analogy with the NH_3 molecule we would expect

$$P(t) = \cos^2(\Gamma t/2) \tag{3.3}$$

where $\hbar\Gamma$ is, as above, the splitting of the even and odd states. It is crucial to note [2] that in this second type of experiment it is essential <u>not</u> to observe the system continuously: were one to do so, then by the usual arguments of the quantum theory of measurement one would completely destroy all the sought-for interference effects. One would therefore have to observe the system initially to be in a definite well, then switch off the measuring apparatus, wait a time t, and then switch it on again and plot the statistics of $P(t)$, the probability of finding the system in the original well. This would be a difficult experiment but not obviously in principle impossible. Recent calculations by Chakravarty [29] have shown that although under normal conditions the oscillation frequency Γ is very small, there exist windows in the parameter space where it can be made as large as 10^{10} s^{-1} or possibly even larger (cf. ref. 2). However, these estimates are very sensitive to the value of capacitance chosen, and as noted above, values smaller than 0.1 pF should be treated with some caution. Since, in addition, switching one's measuring apparatus on and off in a time of the order of 10^{-10} s may present problems, it may be better for present purposes to err on the side of caution and assume that realistic experiments are likely to involve frequencies of the order of say 10 MHz at most.

Finally, in the SQUID (but not the CBJ) it is possible also to consider 'quantum hopping' and its inverse, photo-assisted tunnelling.[30] The conditions to observe these phenomena are similar to those for observing simple tunnelling, and at first sight it is not clear that they will yield any additional information: see however section 5.

---

† This suggestion actually poses considerable difficulties, since one is dealing with a single system (rather than $\sim 10^{23}$ NH_3 molecules). Naively one would think that the maximum power absorption would be attained for near-critical damping of the resonance oscillation and would then be of order $\hbar\Gamma^2$, which for the maximum tolerable value of $\Gamma(10^{10}\,s^{-1}$: cf. above) is not necessarily undetectably small. However, the considerations introduced in section 5 complicate the discussion considerably.

4. The Effects of Dissipation: Qualitative

All the above discussion is flawed by a serious criticism: we have completely ignored the fact that the macroscopic systems we are talking about are inherently dissipative. Indeed, it is of the very nature of a macroscopic degree of freedom that if we start it off with an energy much greater than its equilibrium thermal (or zero-point) energy it will sooner or later dissipate this into its environment through the familiar mechanisms of friction, ohmic resistance, radiation, etc. In practice we very often have a good phenomenological description of this process in terms of experimentally measured resistances, friction coefficients, etc. without necessarily having a reliable microscopic model from which to calculate them. In particular, the behaviour of many Josephson junctions is believed to be well described by the so-called 'resistively shunted junction' (RSJ) model, which assumes that there is an effective ohmic resistance R in parallel with the junction, so that the circuit diagram looks like Fig. 6.

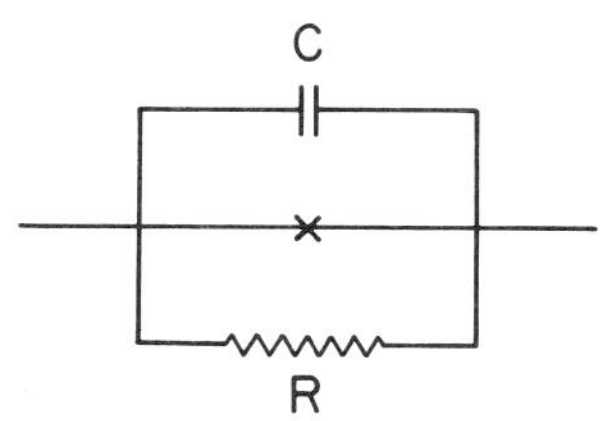

Fig. 6 Equivalent circuit for the RSJ model of a Josephson junction

The classical equation of motion of the phase difference $\Delta\phi$ is then given by

$$\left(\frac{\Phi_0}{2\pi}\right)\cdot\left(C\Delta\ddot{\phi} + \frac{\Delta\dot{\phi}}{R}\right) + I_c \sin\Delta\phi = I_{ext}(t) \tag{4.1}$$

The corresponding equation for the flux Φ threading a SQUID ring is

$$C\ddot{\Phi} + \frac{\dot{\Phi}}{R} + \frac{\Phi}{L} + I_c \sin 2\pi\Phi/\phi_0 = \Phi_{ext}(t)/L\,. \tag{4.2}$$

Moreover, an equation identical in form to (4.1) is usually believed to describe the classical motion of the phase of charge density waves.[19] Note that in all those cases the energy $\dot{W}$ dissipated per unit time is simply proportional to the squared rate of change of the 'coordinate' variable † with a constant coefficient ($\dot{W} = R^{-1}(\phi_0/2\pi)^2(\Delta\dot{\phi})^2$, $R^{-1}\dot{\phi}^2$, etc.) Of course, this is not the only possible type of dissipation. For example, in an ideal tunnel oxide junction the dissipation is due entirely to the normal quasiparticles

---

† **We identify the 'coordinate' as the variable which has a metastable minimum.**

present (at finite temperature) and includes the so-called 'quasiparticle-pair interference term' which is proportional to $\cos\Delta\phi$ (cf. ref. 31); moreover, the coefficient has a substantial frequency-dependence. Nevertheless, this model shares with the RSJ model the property that the dissipation is still proportional to the squared velocity of the coordinate variable, which in general we label $q(t)$. All models sharing this property satisfy an equation of the general form

$$M\ddot{q} + \eta\dot{q} + \frac{\partial V}{\partial q} = F_{\mathbf{ext}}(t) \tag{4.3}$$

where η may in general depend on both q itself and on frequency, but is always positive (and $F_{\mathbf{ext}}$ is an external force which acts only on the system). I shall refer to cases where the dissipative equation of motion is of this type as cases of 'normal' dissipation: almost all cases of current physical interest in the context of macroscopic quantum tunnelling are of this type. A more exotic situation, which I shall refer to as 'anomalous' dissipation, arises when the rate of dissipation $\dot{W}$ is proportional to the squared rate of change of the momentum p conjugate to q.[†] In this case the equation of motion of the coordinate variable takes the form

$$M\ddot{q} + \tau\,\frac{d}{dt}\left(\frac{\partial V}{\partial q}\right) + \frac{\partial V}{\partial q} = F'_{\mathbf{ext}}(t) \tag{4.4}$$

where τ is a (possibly q- and ω-dependent) positive coefficient; however, $F'_{\mathbf{ext}}$ is not a 'true' force (in the sense that it actually couples to the dissipative mechanism as well as to the system). We note that for free motion ($F'_{\mathbf{ext}}(t) = 0$) Eq.(4.4) can be formally cast in the form (4.3), but (in a tunnelling situation) only at the expense of making the coefficient η negative over part of the tunnelling region. An example of such a case, which I believe however is unlikely to be of practical relevance in the context of macroscopic quantum tunnelling or coherence, is given in ref. 32: it consists in effect of a parallel LCR circuit where the capacitance (rather than the inductance, as in the SQUID case) has strongly nonlinear properties. In this case it is the charge Q on the capacitor plates which is the tunnelling variable, while the dissipation is proportional to the squared rate of change of the flux Φ. A similar (but probably equally impractical) example would be a *series* LCR circuit with a nonlinear (ferromagnetic) inductance; here the roles of flux and charge are reversed. One can think of plenty of other such model cases, but it is not clear that any of them are relevant to the real-life macroscopic tunnelling situation.

---

† **Note that in the presence of dissipation we cannot necessarily use the relation $\dot{q} = p/M$, (cf. ref. 1, appendix C), so there is no ambiguity in the classification.**

A case which may conceivably be of somewhat more practical interest in this context is what I shall call the 'mixed' case: this typically occurs when the relaxation processes between the system and its environment are subject to conservation of the momentum p. A formal example of such a case is the motion of the spin-orbit rotation angle Θ in superfluid ^{3}He-B. The classical equation of free motion of the variable Θ is of the general form[33]

$$\dddot{\Theta} + \ddot{\Theta} + \chi_p^{-1}(\partial^2 V/\partial\Theta^2)\,\dot{\Theta} + \chi^{-1}\,\tau^{-1}\,\partial V/\partial\theta = 0\,, \tag{4.5}$$

where τ is a microscopic relaxation time for spin-conserving quasiparticle collisions and χ and χ_p are respectively the total (hydrodynamic) and 'pair' (collisionless) spin susceptibilities (see ref. 33).† If the characteristic frequency ω of the motion is such that $\omega\tau \gg 1$, the motion is nearly collisionless (the 'system' (Cooper pairs) oscillates nearly independently of the 'environment' (normal quasiparticles)) and the relevant 'mass' is the pair susceptibility χ_p; while if $\omega\tau \ll 1$ the pairs and quasiparticles are always nearly in equilibrium with one another and the relevant 'mass' is the hydrodynamic susceptibility χ, which is larger than χ_b. It is easily verified that the dissipation looks 'normal' when viewed from the collisionless end but 'anomalous' when viewed from the hydrodynamic end; this points up the importance, when discussing the effects of dissipation, in always specifying carefully with which nondissipative system the dissipative system is being compared (cf. ref. 34). Although this particular system is not a candidate for the observation of macroscopic quantum effects (since, inter alia, the potential $V(\Theta)$ has no metastable minimum), the problem of the ^{3}He—A → ^{3}He—B transition[35] has many features in common with it and may possibly be of some practical interest. (Cf. ref. 36.)

In the light of the above remarks, it is quite essential, before one can define the experimental conditions needed to look for quantum phenomena at the macroscopic level, to answer the question: *What is the effect of dissipation on quantum tunnelling* (or hopping or coherence)? Much of the progress made in the last three years has come in the understanding of this question, and I shall spend most of the rest of this paper reviewing this.

It is not difficult to attain a qualitative understanding of the likely effects of dissipation on tunnelling (decay). As long as it remains in the metastable well, the system feels an approximately harmonic potential. So

† **Readers unfamiliar with this system may find it easier to consider a mechanical analogue, namely an annular cylinder filled with liquid and suspended on a highly nonlinear torsion fibre so that its angular position has a metastable minimum. Here the moment of inertia of the cylinder alone corresponds to χ_p and that of the cylinder plus liquid to χ; liquid-wall collisions conserve the angular momentum.**

let us consider the effect of dissipation on the groundstate of a simple harmonic oscillator,† for example a simple parallel LCR circuit. In this case the dissipation is equal to $\dot{\phi}^2/R$. Provided that we can be sure that the resistive mechanism is linear in the flux coordinate (which in this case is guaranteed by the microscopic form of the electromagnetic interaction (see ref. 1, appendix A), we can use the quantum-classical correspondence for oscillators and the fluctuation-dissipation theorem to obtain the exact density matrix of the oscillator groundstate (see e.g. ref. 1, appendix B). In this 'damped' groundstate the flux uncertainty $\langle\phi^2\rangle_0$ is *less* than in the corresponding undamped case (i.e., with the resistor removed) while the uncertainty in the charge on the capacitor plates, $\langle Q^2\rangle_0$, is *greater*.‡ For a series LCR circuit, on the other hand, where the dissipation is $\dot{Q}^2R$, the dissipation decreases $\langle Q^2\rangle_0$ relative to the case $R=0$) but increases $\langle\phi^2\rangle_0$. Thus in each case dissipation proportional to the velocity of a given variable decreases the uncertainty in that variable but increases that in the conjugate one (in such a way, of course, that the uncertainty principle is always satisfied). Now, since in the tunnelling problem the tunnelling rate is proportional to the probability of finding the system far outside the classical region, which in turn (for the oscillator) is inversely related to the mean-square uncertainty of the coordinate, we would immediately expect that in the 'normal' case (dissipation proportional to the squared velocity of the tunnelling variable) dissipation would decrease the tunnelling rate. Conversely, we would expect that in the 'anomalous' case the effect would be to increase the rate;$^{\phi}$ and moreover that in the 'mixed' case the actual tunnelling rate will be bracketed between the rates calculated in the two dissipation-free limits, namely the collisionless and hydrodynamic limits.

5. The Effects of Dissipation: Quantitative

To confirm these results and make them quantitative one needs to do two things. First, one must write down a microscopic description (in practice, in the form of a Lagrangian or Hamiltonian) of the interaction between the system and its environment which gives rise to the relevant dissipation.

---

† For a recent review of theoretical work on the quantum damped harmonic oscillator, see ref. 37.

‡ For a model with strictly frequency-independent resistance $\langle Q^2\rangle_0$ diverges. In practice it is proportional to $ln\,\omega_c$, where ω_c is a cutoff frequency of the order of the inverse Drude relaxation time.

ϕ This is proved explicitly in ref. 32 for the particular example studied, but without any consideration of the magnitude of the effect.

Secondly, one must devise a practical scheme for using this description to calculate the quantity of experimental interest. The two stages are effectively independent: stage 1 depends on the nature of the dissipation but is independent of whether one is interested in tunnelling, hopping or coherence, while stage 2 depends strongly on the latter but is more or less independent (at least within the scheme which I shall describe) of the nature of the dissipation.

In some cases stage 1 can be carried out immediately on the basis of an accepted microscopic model. For example, in the case of an oxide-layer tunnel junction the description conventionally used is the standard tunnelling Hamiltonian, which may be shown to produce both the Josephson supercurrent and the dissipative normal current† (see, e.g., ref. 31). However, as noted above, in many cases of practical interest one has a good phenomenological description of the dissipation in terms of experimentally measured friction coefficients, resistances, etc. without necessarily having a detailed microscopic basis for it; in particular this is the case for the so-called RSJ model (see above) which is widely used to describe the behaviour of point-contact and some other Josephson junctions. In such cases stage 1 is by no means trivial. A.O. Caldeira and the author[24,38] considered the case of normal dissipation with constant dissipation coefficient (appropriate to the RSJ model and many other situations), that is, the general case where the classical equation of motion of the tunnelling variable q is of the form (4.3) with η a constant (i.e., independent of amplitude and frequency). Restricting themselves to the zero-temperature case, they showed that a suitable Lagrangian to describe the dissipative system-environment interaction is (cf. ref. 39)

$$L = \tfrac{1}{2}M\dot{q}^2 - V(q) + \tfrac{1}{2}\sum_\alpha m_\alpha(\dot{x}_\alpha^2 - \omega_\alpha^2 x_\alpha^2) + q\sum_\alpha C_\alpha x_\alpha - q^2\sum_\alpha C_\alpha^2/2m_\alpha\omega_\alpha^2 \,. \qquad (5.1)$$

where the spectrum and coupling constants C_α of the 'environmental oscillators' described by the coordinate x_α satisfy the constraint (for ω small compared with some high cutoff ω_c)

$$J(\omega) \equiv \frac{\pi}{2}\sum_\alpha (C_\alpha^2/m_\alpha\omega_\alpha)\,\delta(\omega-\omega_\alpha) = \eta\omega \,. \qquad (5.2)$$

The last term in Eq. (5.1) ('counterterm') is included so that the $V(q)$ which occurs in (5.1) is the same as that occurring in the phenomenological equation

---

† **We note however that there is a difficulty concerning the sign of the so-called 'quasiparticle-pair interference term' which has still apparently not received a totally convincing solution (ref. 31, section 26).**

of motion (4.3); this allows the dissipative and nondissipative systems to be compared at the same value of effective potential. In cases such as the RSJ model the counterterm arises automatically from a correct electrodynamic treatment. These points have received (more than) adequate discussion elsewhere (ref. 34 and ref. 1, appendix A) and I shall not discuss them further here.

While it is easy to show that the Lagrangian (5.1) (with the constraint (5.2)) generates the required classical equation of motion (4.3), it is much less obvious that it is unique in doing so. This question is exhaustively discussed in ref. 1, appendix C, with the following conclusion: Barring, possibly, a class of counterexamples which was not rigorously excluded but appears very pathological, Eq. (4.3) does indeed uniquely constrain the Lagrangian to have the form (5.1) provided either (a) we know on *a priori* grounds that the system-environment interaction is linear in q (as we do, for example, in the RSJ case), or (b) we can exclude on experimental grounds the existence of velocity-dependent reactive terms in the equation of motion (4) which need not necessarily show up at low velocity. If neither of these conditions is fulfilled, it is impossible to exclude a generalization of Eq. (5.1) in which the term $q\sum_\alpha C_\alpha x_\alpha$ is replaced by $\sum_\alpha f_\alpha(q)\, x_\alpha$ (with a similar replacement in the counterterm) and Eq. (5.2) is generalized to read

$$\frac{\pi}{2}\sum_\alpha \left[(\partial f_\alpha(q)/\partial q)^2/m_\alpha \omega_\alpha\right]\delta(\omega-\omega_\alpha) = \eta\omega \; . \qquad (5.3)$$

The interesting point is that while this generalized ('nonfactorizable') coupling produces exactly the same phenomenological classical equation of motion at low velocity as (5.1), it can produce quantitatively different behaviour at stage II of the calculation: in particular, the tunnelling ratios so generated are always higher [1] than those resulting from the use of (5.1) (but still lower than the undamped rates). This may have important implications for the question of the validity of the 'quantum Langevin' approach: see below.

The Langrangian (5.1) can be simply generalized to the case of amplitude- or frequency-dependent dissipation, provided it is still of the 'normal' type: see ref. 1, appendix C. What is perhaps more surprising is that it can also be adapted to the case of anomalous dissipation (Eq. (4.4)). It can be shown† that the equation of motion (4.4) is correctly given by a Lagrangian of the form (5.1) provided that we set $M = 0$ in the first term and take

$$J(\omega) = \frac{M\omega^3\tau}{1+\omega^2\tau^2} \qquad (5.4)$$

† For a discussion of quantum tunnelling in the presence of an arbitrary linear dissipation mechanism, which includes the 'normal' and 'anomalous' cores as special cases, see ref. 40.

Although this form of Lagrangian has the appearance of artificiality, it is enormously convenient when we come to stage II of the calculation. With the exception of the work of Ambegaokar *et al.*[41] on the ideal oxide tunnel junction (which starts at a somewhat more basic level with the tunnelling Hamiltonian and obtains the potential $V(q)$ as part of the output), most recent quantitative work in this area has started from a Lagrangian of the general form (5.1) or the equivalent Hamiltonian. In particular, such a form has been derived and used by Zwerger[42] for the coupling of the supercurrent in a Josephson junction to the radiation field: in this case the q in the coupling term is replaced by $\sin\Delta\phi$, and the quantity $J(\omega)$ is equal to $(\cos\Delta\phi)^{-2}\omega Im\varepsilon(\omega)$ for $\omega>0$, where $\varepsilon(\omega)$ is the $\Delta\phi$-dependent dielectric function of the junction (see ref. 42, Eq. (2.1)).

Before proceeding to stage II of the calculation, it is worthwhile to pause and remark that the description (5.1) has, of course, a long history of use in various areas of physics, including some microscopic tunnelling problems. One such problem which is formally closely related to the problems considered here is that of the quantum hopping or coherence behaviour of a defect in a solid where it is coupled to the lattice phonons. The most sophisticated discussion of this problem is contained in two recent papers by Sethna[43,44] which use techniques closely related to some of those described below. The three most important differences between this problem and ours are that (1) in the defect case the pure tunnelling (decay) problem does not usually arise, even in the approximate sense in which it applies, say, in a SQUID ring (cf. above);† (2) the effect we have elsewhere[1] called 'frequency renormalization' is in the defect case a real physical effect of some importance, while for our problem it is often nonexistent and, even when it does exist in principle, usually physically unobservable (see ref. 1, section 2, and ref. 42); and (3) most importantly, the quantity $J(\omega)$ is in the defect case proportional to ω^3 (because of the three-dimensional phonon density of states) whereas in the most important case of interest for us, linear frequency-independent dissipation, it is proportional to ω. This difference completely alters many aspects of the problem: it means that the defect case is dominated by the high-frequency end of the environment (phonon) spectrum and a description in terms of effective mass corrections is the natural one,[43,44] whereas in our problem it is the low-frequency modes which play the dominant role. This difference is particularly important in the discussion of quantum

† **Thus, what Sethna calculates is an *elastic* tunnelling rate, and his formulae differ from those of ref. 1 by the presence of a Debye-Waller factor.**

coherence: see below.

Once given the Lagrangian (5.1), or the associated Hamiltonian, there are a number of formal calculational techniques by which we can handle the problems of interest to us. One obvious possibility, which is quite helpful to one's qualitative physical intuition but not very convenient for quantitative calculations, is to use the WKB method in the many-dimensional space described by (5.1) (cf. refs. 43, 44). A method which is perhaps less intuitive but gives more clear-cut results, at least for the tunnelling problem, is to use the path-integral formulation of Feynman[45] and explicitly integrate out the environment coordinates (the x_α's) so as to produce an effective action for the system coordinate q alone (cf. ref. 39). There are a number of related ways of doing this, which differ only in the precise quantity they choose to calculate and in whether they work in real or 'imaginary' time or both. Probably the simplest† is the technique used in refs. 38 and 1, which works with 'imaginary' time from the start and calculates the reduced one-particle density matrix $\rho(q,q':\beta)$ of the system variable q at temperature β^{-1}. In the absence of dissipation this quantity would be given by the expression

$$\rho(q,q':\beta) = \int_{q(0)}^{q'(\beta)} Dq(\tau) \exp - S[q(\tau)]/\hbar, \tag{5.5}$$

where the functional $S[q(\tau)]$ has the familiar 'Euclidean' form appropriate to classical motion in an inverted potential

$$S[q(\tau)] = \int_0^\beta [\tfrac{1}{2}M\dot{q}^2 + V(q)]\,d\tau . \tag{5.6}$$

The effect of integrating out the environmental coordinates is remarkably simple: Eq. (5.5) remains valid, but the action functional S acquires an extra term $\Delta S[q(\tau)]$ given by‡

$$\Delta S[q(\tau)] = \tfrac{1}{2}\int_0^\beta d\tau \int_0^\beta d\tau'\, \alpha(\tau-\tau')\,[q(\tau)-q(\tau')]^2 \tag{5.7}$$

where the function $\alpha(\tau)$ is given by the expression

---

† At least in the sense that one can work with real quantities right up to the last moment and hence need not concern oneself with the subtleties involved in analytic continuation.

‡ Expression (16) is strictly valid only if the trajectory $q(\tau)$ is periodically continued for τ outside the range $0 \leqslant \tau < \beta$. See ref. 1, section 4.

$$\alpha(\tau) = \frac{1}{2\pi} \int_0^\infty J(\tau)\, e^{-\omega|\tau|}\, d\omega \ . \tag{5.8}$$

For the case of normal frequency-independent linear dissipation ($J(\omega) = \eta\omega$) we evidently have simply,

$$\alpha(\tau) = \eta/2\pi\tau^2 \ . \tag{5.9}$$

Eqs. (5.7)—(5.8) are however much more general and can be applied even in the 'anomalous' and 'mixed' cases provided that the function $J(\omega)$ is suitable chosen(cf. Eq. (5.4)).

It is also possible to carry out the elimination of the environment coordinates for a general (nonfactorizable) coupling, but the resulting correction ΔS to the effective action will then in general have a more complicated form † which is not bilinear in $q(\tau)$. At this point is should be remarked that Ambegaokar *et al.*[41], starting from the tunnelling Hamiltonian and bypassing the representation (5.1), were able to derive a ΔS which in the limit of the normal phase could be evaluated as

$$\Delta S[\Delta\phi(\tau)] = 8 \int_0^\beta d\tau \int_0^\beta d\tau'\, \tilde{\alpha}(\tau-\tau')\, \sin^2\left\{\tfrac{1}{4}[\Delta\phi(\tau) - \Delta\phi(\tau')]\right\} \tag{5.10}$$

where $\Delta\phi$ is the phase difference across the junction and the function we have called $\tilde{\alpha}(\tau-\tau')$($\hbar/4$ times Ambegaokar *et al.*'s α) is given for $|\tau| \gg \hbar/\mu$ by the expression $\hbar^2/8\pi e^2 R_N$. (R_N is the junction resistance in the normal state and μ the Fermi energy.) To compare this formula with expression (5.7) we take, in the latter, $q(\tau) = \Delta\phi(\tau)$, $\eta = (\Phi_0/2\pi)^2 R_N^{-1}$, and assume frequency-independent dissipation ($J(\omega) = \eta\omega$), so that (5.9) holds. We then find $\tilde{\alpha}(\tau-\tau') \equiv \alpha(\tau-\tau')$, so that in the limit of small excursions in $\Delta\phi$ expressions (5.7) and (5.10) agree exactly. It would be interesting to compare more generally the results which one would get for a tunnel junction using a representation of the (generalized) type (5.1) with those of ref. 41, but at the time of writing this has not been done.

Armed with formulae (5.5)—(5.8) one can calculate the tunnelling rate out of the metastable minimum of $V(q)$ and make at least a start on a discussion of the hopping and coherence problems. The former problem is tackled in ref. 38, using the 'bounce' technique of Langer,[46] Stone,[47] and Callan and Coleman,[48] and a more extended discussion is given in ref. 1, sections 4 and 5. In the latter reference the problem is analyzed in some detail for normal linear dissipation in a cubic potential ($V(q) = \frac{1}{2}M\omega_0^2 q^2 - bq^3$), (which is the

† See ref. 1, Eqs. (4.35)—(4.36).

relevant form for most cases of current experimental interest, cf. above). The general conclusion is that the tunnelling rate has, as in the undamped case, the form

$$P_{QM} = A(\alpha)\ \exp - B(\alpha)/\hbar\ , \tag{5.11}$$

where $\alpha \equiv \eta/2M\omega_0$ is the degree of damping ($= Q^{-1}$) (no connection with $\alpha(\tau-\tau')$ above!) and $A(\alpha)$, $B(\alpha)$ are functions which can be calculated. In particular the correction $\Delta B(\alpha)$ to the undamped WKB exponent B_0 has the general form

$$\Delta B(\alpha) = A_0\, \eta(\Delta q)^2\ , \tag{5.12}$$

where $\Delta q \equiv M\omega_0^2/2b$ is the distance the undamped system would have to travel under the barrier to escape, and the quantity A_0 is only weakly dependent on α, being given by $(12/\pi^3)\ \zeta(3) \cong 0.47$ in the weak-damping limit $\alpha \ll 1$ and by $2\pi/9 \cong 0.70$ in the strong-damping limit $\alpha \gg 1$. The prefactor $A(\alpha)$ in (5.11) is of order $\omega_0 (V_0/\hbar\omega_0)^{\frac{1}{2}}$ in the weak-damping limit and of order $\alpha^{7/2}\omega_0$ in the strong-damping limit.† It should be possible to extend these results to different forms of $V(q)$ and/or different forms of damping should it become of experimental interest (cf. ref. 42).

The main qualitative lesson is that while (normal) dissipation always tends to suppress quantum tunnelling, it need not do so to the point of unobservability provided the damping is not too strong.

A rather different approach to the calculation of the tunnelling rate is taken by Zwerger[42], who has also, in very recent work,[51] considered the problem of finite temperature. Starting from a Lagrangian which is a simple generalization of (5.1) and distinguishing explicitly, as was not done in the method just described, between 'elastic' processes in which the environmental oscillators are left in their groundstate and 'inelastic' ones in which they are excited, he shows that to lowest order in the damping the elastic rate is identical to expression (5.11) (with A_0 given its weak-damping value) while the inelastic rate is zero. For higher damping he concludes that the expression for the tunneling rate which follows from (5.5) – (5.8), i.e. expression (5.12), is actually the *elastic* tunnelling rate ‡ and that there are other contributions from inelastic processes, which enter first as a term of order $\alpha^{3/2}$. In ref. 51 he uses a somewhat different method, which is based on the crucial assumption that 'the probability of a dissipative process

† Some of these results have been independently obtained on the basis of Eqs. (5.5)–(5.9) by Golub and Iordatin[49] and by Chakravarty and Chang[50]. Cf. also ref. 42. The strong-damping behaviour quoted in ref. 1, $A(\alpha) \sim \alpha^{9/2}$, is incorrect (an Erratum is currently in preparation).

‡ The present author disagrees with this conclusion.

with energy loss E [is the same as] that of a bare process at the correspondingly reduced energy'. In this way he obtains (in the notation of the present paper) a correction ΔB to the bounce exponent for a state of energy E of the form

$$\Delta B(E:\alpha) = A(E,T)\,\eta(\Delta q)^2 \tag{5.13}$$

where the dimensionless coefficient A does not depend on α. (See ref. 51, Eq. (20).) In particular. for states with energies E much less than the barrier height (e.g., the groundstate) in the quadratic-plus-cubic potential he finds

$$A(E \to 0,T) = \frac{2\pi}{15}\left(1 - \left(\frac{T}{T^*}\right)^4\right) \tag{5.14}$$

where $T^* \equiv \hbar\omega_0/2\pi$ is the temperature above which classical thermal nucleation dominates (cf. ref. 27). This result is expected [51] to be valid for fairly small α. Note that the suppression of the zero-temperature tunnelling rate in the weak-damping limit given by (5.14) is very slightly less than that found by the functional-integral method and in ref. 42 (cf. above): this is interpreted [51] as due to the fact that inelastic processes do in fact contribute to the exponent even to order α, although their contribution is very small. (See however last footnote.) The prediction (5.14) should be experimentally testable, since it will dominate the temperature-dependence of the total escape probability at temperatures $T \ll T^*$ (the effect of population of states other than the groundstate decreases exponentially in this limit).

Let us turn now to the special case of the charge-density-wave (CDW) system. The original work of Bardeen [18] took no account of dissipation, but in a subsequent paper [20] he modified the theory in a phenomenological way by introducing a 'correlation length' L across which the electric field must be applied to be effective in accelerating the CDW. This idea leads to a threshold field $E_T = \varepsilon_g/e^*L$ below which there is essentially no tunnelling. Using an intuitive analogy with superconductivity theory, he argued that the length L should be twice the electron mean free path l. In very recent work Wonneberger [52] has looked at this problem by taking into account the CDW-phonon interaction. Starting from a (nonfactorizable) generalization of the Lagrangian (5.1) and making physically reasonable approximations, he derives a correction to the effective action which is the generalization to finite bounce time(cf. ref. 1, section 4) of Eqs. (5.7) and (5.9). Using a representation of Zener tunnelling in real space, he then derives a correction ΔB to the WKB exponent which in the limit of weak damping is just $\eta\xi^2(E)/4$, where $\xi(E) \equiv \varepsilon_g/eE$. Actually, the weak-damping limit does not apply in the physical situation, and by considering the details of the pinning and dissipation mechanism Wonneberger finally concludes that in the case of impurity pinning Bardeen's length L is just $(2m^*/M^*)^{\frac{1}{2}}\,\xi_0$ where $\xi_0 \equiv 2\hbar v_F/\pi\varepsilon_g$. It would be interesting to see whether any of these results could be reproduced using the method used above for the nondissipative CDW system.

One interesting related paper not so far mentioned is that of Koch *et al.*[53] who calculate the behaviour of a Josephson junction using a quantum Langevin equation; that is, they assume (in our language) that the variable $q(t)$ obeys the equation

$$M\ddot{q} + \eta\dot{q} + \partial V/\partial q = F(t) \ , \tag{5.15}$$

where the average of $F(t)$ is zero and the correlations are given by the *quantum* expression

$$\int \langle F(t)\,F(t')\rangle\, e^{-i\omega(t-t')}\, d(t-t') = \eta\omega \coth \beta\hbar\omega/2 \ . \qquad (\beta \equiv 1/kT) \tag{5.16}$$

In the limit where all the frequencies of interest are small compared to $kT/\hbar$ this reduces to the classical expression $\langle F(t)\,F(t')\rangle = 2kT\eta\delta(t-t')$, but it gives finite fluctuations also for $\hbar\omega \geqslant kT$. Although in ref. 53 Koch *et al.* were primarily interested in the mixing-down of quantum noise in the classical accessible region of motion, it is also possible[54] to use their method to compute numerically quantum tunnelling probabilities. The question then arises: Is the quantum Langevin approach reliable for this kind of purpose? It is known to give the correct answer for the damped quantum harmonic oscillator provided the system-environment interaction is linear, and Schmid[55] has recently argued that it should be applicable in the region of semiclassical motion (in the classically allowed region); but does it give correct results in the cases of interest to us in this paper? I believe that in *general* it cannot, for the following reason. If we compare the interaction term in the Lagrangian (5.1) (with the constraint (5.2) with the more general nonfactorizable coupling $\sum_\alpha f_\alpha(q)\,x_\alpha$ satisfying the constraint (5.3), we see that both will by construction give rise to the same force correlations and therefore the same quantum Langevin equation. Yet we know (see ref. 1, section 4) that they will give rise to quite different quantitative results for the quantum tunnelling behaviour. Thus I believe that in the general case the quantum Langevin approach cannot give a complete description. Whether it does so once we impose the condition that the system-environment interaction is linear (which is the physically relevant situation for the RSJ model) is however an open question. For further remarks on this question see ref. 56.

Let me now turn more briefly to the question of the effect of dissipation on quantum *coherence*. This question is currently the subject of intense theoretical interest, and any remarks I make are almost certain to be out of date by the time this paper is published, so I shall just make a few general comments. This question was discussed qualitatively, from the point of view of the quantum theory of measurement, in ref. 2, section 5. Although this treatment is, I believe, qualitatively correct, it suffers from two inadequacies. First, it was not sufficiently emphasized that the efficiency of the 'observation by the environment' invoked there may itself be frequency-dependent, and that the relative effects on tunnelling and coherence depend on the nature of the relative spectrum† (the $J(\omega)$ of Eq. (5.2), or a related quantity). Secondly, the *reactive* effect of the environment, which in general tends to suppress tunnelling rates, was ignored. The two effects tend to act in opposite directions, so it is not *prima facie* clear whether the conclusions of ref.2 err on the side of optimism or of pessismism. Recently Chakravarty[57] has studied the problem of quantum coherence in a SQUID, describing the dissipation by the effective action (5.10) derived for an ideal oxide tunnel junction (a calculation using the formulae (5.7) — (5.9) appropriate to the RSJ model would presumably give very similar results). Recognizing the analogy with the Kondo and long-range Ising problems, he was able to show *inter alia* (transposing to the notation used in this paper) that there is a critical value η_c of the dissipative coefficient η, which is generally of the order of $\hbar/(\Delta q)^2$ where Δq is the distance under the barrier, such that for $\eta > \eta_c$ the system will spontaneously break the reflection symmetry and become localized in a single well at $T = 0$. A similar conclusion was reached for the weak-damping limit ($\alpha \equiv \eta/M\omega_0 \ll 1$) by Bray and Moore[58], who further concluded that for $\eta > \eta_c$ the real-time correlation function of the system at low temperatures would not oscillate as in the undamped case but would relax exponentially with a time constant which tends to infinity as $T^{-\beta}$ where $\beta \equiv 2\left(\frac{\eta}{\eta_c} - 1\right)$ (see ref. 58, Eqs. (14) and (21)). It is an urgent task, in the light of possible experimental searches for macroscopic quantum coherence, to extend these results into the region $\eta < \eta_c$ and in particular to the region $\eta/\eta_c \sim \hbar\omega_0/2\pi V_0 \ll 1$, where preliminary investigations indicate that the behaviour may change qualitatively from the undamped case. A related question, that of the effect of dissipation on photo-assisted quantum tunnelling, has been very recently investigated by Chakravarty and Kivelson[30] using a truncation approximation (cf. ref. 43); they conclude that the absorption lineshape should be very sensitive to η and should change qualitatively at

---

† In particular, the statement (ref. 2, p.97) that for ohmic dissipation the decay rate of the 'inversion' oscillation is $1/RC$ is incorrect; it is generally much smaller than this.

the point $\eta = \eta_c$. Although many features of these problems are not yet well understood, it is clear that the case of ohmic dissipation ($J(\omega) = \eta\omega$) is a very special ('marginal') one and that the quantity η_c has a fundamental significance.†

6. The Experimental Situation‡

Apart from the experiments on charge density waves in one-dimensional compounds (for a review of which see ref. 19) and one experiment on a ring laser,[22] all experiments known to the author which have been interpreted in terms of macroscopic quantum tunnelling (and/or coherence) have been performed on SQUID's or CBJ's. In the earliest such experiment Fulton and Dunkleberger[61] measured the lifetime of the zero-voltage state in a CBJ as a function of bias current. They saw some indications that quantum tunnelling might be playing a role, but could not be definite about this as, even ignoring the effect of dissipation, their experiments only went to the edge of the region where tunnelling might be expected to dominate. Thereafter no experiments seem to have been done until 1980, when den Boer and de Bruyn Ouboter[13] reported direct measurements of flux transitions in a niobium point-contact SQUID ring as the externally applied flux was varied. From the fact that the transition rate, which was strongly temperature-dependent at high temperature, flattened off to a constant value for T in the region of 1K, plus the observed absence of hysteresis in this regime, they concluded that tunnelling might be the operative mechanism at these temperatures. This would have required a junction capacitance of the order of 10^{-15} F or less; see above for a discussion of this point. Shortly thereafter Prance *et al.*[14] reported measurements of the tank circuit current-voltage characteristics of an rf SQUID operated in the standard of rf mode (cf. ref. 7); They observed a curious periodic fine structure of the characteristic which they interpreted as evidence for macroscopic quantum effects. (The theory used by Prance *et al.*[14] to interpret their data is formulated in the language of macroscopic quantum coherence, but it is not clear that anything more than macroscopic quantum tunnelling is required to produce their 'theoretical' I-V curves. Cf. ref. 62.) It should be noted that if we use the values of the SQUID parameters quoted in ref. 14, then according to the standard WKB

---

† For some recent developments see refs. 59,60, and references cited therein.

‡ There is unfortunately no space here to discuss the experiment of Koch *et al.*,[53] which although not on tunnelling is relevant to the general question of the quantum mechanics of a macroscopic variable. On this, see ref. 67.

calculation† (even neglecting dissipation) appreciable tunnelling in the experimental situation would have been obtained only if the capacitance C of the junction was 10^{-18} F or smaller.

Further experiments on CBJ's along the lines of those of Fulton and Dunkleberger[61] were performed by Voss and Webb[63] and by Jackel *et al.*[64] The latter group measured the width σ of the distribution in current of the onset of the finite-voltage state in Pb—Pb(In) junctions with critical currents of a few hundred μA and capacitances which they estimated at $5\text{-}10 \times 10^{-15}$ F. With these parameters the crossover between a temperature-dependent σ due to thermal nucleation and a constant low-temperature value due to quantum tunnelling is predicted by the simple WKB theory without damping to occur at a temperature of about 8K (cf. ref. 17). In fact, the data of Jackel *et al.*[64] do show a flattening-off of the curve of $\sigma(T)$, but only below about 2.2K. (The lowest temperature reached was ~ 1.5K.) They concluded that their data were consistent with the theory given in ref. 38. Voss and Webb[63] made similar measurements on all $-Nb$ junctions with critical currents of the order of 100 nA—1 μA and capacitances which they estimated[65] at ≈ 0.1 pF with an upper limit of 0.15 pF; they were able to go to much lower temperatures (~ 3 mK). For a junction with very weak dampening ($\beta_c \equiv 2\pi I_c R^2 C/\phi_0 \sim 5000$) they found that their data, both on the lifetime as a function of current and on the widths $\sigma(T)$, were in fairly good agreement with the thermal predictions at high T but flattened off at low temperatures to a value which (taking the value $C = 0.1$ pF for the capacitance) is very close to the value predicted by WKB theory without damping. For a second junction with non-negligible damping ($\beta_c \sim 50$) they found, again, fairly good agreement with the thermal formula at high T and a flattening-off at low T: however, the nearly constant value reached at low T is not only below the undamped WKB prediction but also substantially below the value which would follow from the theory of ref. 38 if one inserts in the formulae a value of C of 0.1 pF and the measured (dc) resistance and uses the weak-damping value of the constant A_0 in Eq. (5.12). Voss and Webb suggest that the discrepancy could be remedied either by treating A_0 as an adjustable parameter and taking it equal to about 4.5 times the weak-damping value, or by taking R to be a high-frequency resistance which may be lower than the measured d.c. value. In the light of the rigorous limits set on A_0 by more recent work (ref. 1, section 5) the first suggestion seems untenable within the general framework

† The calculation actually performed in ref. 14, which obtains appreciable tunnelling for a much larger value of C, appears not to be equivalent to the WKB theory or to any other standard method.

of the theory of ref. 38, but the second is not implausible. It should be observed that whether or not the data of refs. 63 and 64 give support to the theory of ref. 38, they certainly seem to refute fairly conclusively any suggestion (cf., e.g., ref. 66) † that dissipation in a system such as the one investigated should increase the tunnelling rate above the undamped WKB prediction.

Finally, in a recent experiment designed specifically to look for the effects of dissipation on tunnelling, de Bruyn Ouboter and Bol [16] carried out a further set of measurements on a *Nb* point contact SQUID similar to those of ref. 13, but with the difference that they could insert in the SQUID cavity a rod of some material which would be strongly absorbing in the frequency range relevant to tunnelling (in practice, either copper or the microwave absorber Eccosorb). They observed that the effect of inserting the rod was to decrease the transition rate between the (nearly degenerate) wells, but that both with and without the rod the observed rate was many orders of magnitude *higher* than that predicted by the theory of ref. 38 (in contrast to the above results on current-biassed junctions). However, as they point out, quantitative comparison is difficult both because the frequency is sufficiently high that the simple RSJ model may break down (cf. above) and, in the case of the experiments with the rod in, because it contributes to the effective self-inductance of the ring as well as to the dissipation. Any extrinsic noise not described by the (measured) bath temperature would of course also complicate the interpretation, as in all the experiments mentioned.

7. Conclusion

From the above it is evident that there is plenty of scope for more work, both theoretical and experimental, in the area of macroscopic quantum tunnelling and related phenomena. On the experimental side, one particularly desirable development (from the point of view of the theorist!) would be the ability to do tunnelling (and possibly coherence) experiments in a Josephson system in which one could simultaneously determine both the capacitance and the resistance *at the frequencies appropriate to tunnelling* by experiments interpretable in classical terms, e.g., by detecting the junction plasma resonance and measuring its frequency and linewidth. On the theoretical front, the subject of tunnelling in a dissipative system at finite temperatures urgently needs more work, and so does the question of the effect of dissipation on macroscopic quantum coherence, especially in the (theoretically

---

† Ref.66 reaches this conclusion explicitly for a SQUID rather than a CBJ.

difficult) case of ohmic dissipation. Finally, it is highly desirable to try to analyze further just what it is that success (or failure) in observing these phenomena is telling us about the foundations of quantum mechanics.

After this paper was completed I became aware of the recent review paper[67] by K.K. Likharev on this topic. (I am grateful to Dr Likharev for sending me a copy of this.) This review, which concentrates on the experimental aspects, may be profitably read in conjunction with the present paper.

I am grateful to the members of the Cornell Quantum Tunnelling Seminar, in particular Vinay Ambegaoker, Sudip Chakravarty, Jim Sethna, and David Waxman, for many useful discussions on this subject. I am also grateful to John Bardeen, John Tucker, Leigh Sneddon and Alfred Zewadowski for very helpful discussions on the charge-density-wave problem, and to R. de Bruyn Ouboter and W. Zwerger for helpful correspondence.

References

1. Caldeira, A.O. and Leggett, A.J., Ann. Phys. (NY), **149**, 374 (1983).
2. Leggett, A.J., Prog. Theor. Phys., Suppl. **69**, 80 (1980).
3. d'Espagnat, B., *Conceptual Foundations of Quantum Mechanics*, 2nd ed. (Benjamin, New York, 1976), pt. iv.
4. Schrödinger, E., Die Naturwissenschaften, **23**, 807 (1935).
5. Bohr, N, *Essays 1958-62 on Atomic Physics and Human Knowledge* (Interscience Publishers, New York, 1963).
6. Scalapino, D.J. In *Tunnelling Phenomena in Solids*, (eds. E. Burstein and S. Lundqvist)(Plenum Press, New York, 1969) p.477.
7. Lounasmaa, O.V., *Experimental Principles and Methods below 1 K*, (Academic Press, New York, 1974), chap. 7.
8. Anderson, P.W., In *Lectures on the Many-Body Problem*, (ed. E. Caianello) (Academic Press, New York, 1964), p.113.
9. Einstein, A., Podolsky, B. and Rosen, N., Phys. Rev., **47**, 777 (1935).
10. Bell, J.S. Physics, **1**, 195 (1964).
11. Clauser, J.F. and Horne, M.A., Phys. Rev., **D10**, 526 (1974).
12. Anderson, P.W., Revs. Mod. Phys., **38**, 298 (1966).
13. den Boer, W. and de Bruyn Ouboter, R., Physics B & C, **98**, 185 (1980).
14. Prance, R.J. *et al.*, Nature, **289**, 542 (1981)
15. McCumber, D.E., J. Appl. Phys., **39**, 3113 (1968).
16. de Bruyn Ouboter, R. and Bol, D., Physica, **112B**, 15 (1982).
17. Leggett, A.J., *Proc. 6th International Conference on Noise in Physical Systems*, Gaithersburg, MD., 1981, (NBS Special Publication 614), p.355.
18. Bardeen, J., Phys. Rev. Lett., **42**, 1498 (1979).
19. Grüner, G., Comments on Solid State Physics (GB), **10**, 183 (1983).
20. Bardeen, J., Phys. Rev. Lett., **45**, 1978 (1980).
21. Sneddon, L., Cross, M.C. and Fisher, D.S. Phys. Rev. Lett., **49**, 292 (1982).
22. Lett, P., Christian, W., Singh Surendra and Mandel, L., Phys. Rev. Lett., **47**, 1892 (1981).
23. Kramers, H.A. Physica, **7**, 284 (1940).
24. Caldeira, A.O., D.Phil. Thesis, University of Sussex, 1980 (unpublished)
25. Likharev, K.K., Physica B & C, **108**, 1079 (1981).

26. Waxman, D. (unpublished)
27. Affleck, I.K. Phys. Rev. Lett. **46**, 388 (1981).
28. Kurkijärvi, J., J. Low Temp. Phys., **45**, 37 (1981).
29. Chakravarty, S. (unpublished).
30. Chakravarty, S. and Kivelson, S., Phys. Rev. Lett. **50**, 1811 (1983); and Erratum, ibid, **51**, 1109 (1983).
31. Barone, A. and Paterno, G. *Physics and Applications of the Josephson Effect*, (Wiley-Interscience, New York, 1982).
32. Widom, A. and Clark, T.D., preprint: Phys. Rev. B. (in press)
33. Leggett, A.J. and Takagi, S., Ann. Phys. (NY) **106**, 79 (1977).
34. Caldeira, A.O. and Leggett, A.J., Phys. Rev. Lett., **48**, 1571 (1982).
35. Bailin, D. and Love, A., J. Phys., **A13**, L271 (1980).
36. Leggett, A.J., J. de Physique Colloque, **39**, C6-1264 (1978).
37. Dekker, H., Physics Reports, **80**(1), 1 (1981).
38. Caldeira, A.O. and Leggett, A.J., Phys. Rev. Lett., **46**, 211 (1981).
39. Feynman, R.P. and Vernon, F.L., jr., Ann. Phys. (NY), **24**, 118 (1963).
40. Leggett, A.J., Phys. Rev. B. (in press).
41. Ambegaokar, V., Eckern, U. and Schön, Phys. Rev. Lett., **48**, 1745 (1982).
42. Zwerger, W., Z. Phys., **B47**, 129 (1982).
43. Sethna, J.P., Phys. Rev., **B24**, 698 (1981).
44. Sethna, J.P., Phys. Rev., **B25**, 5050 (1981).
45. Feynman, R.P. and Hibbs, A.L., *Quantum Mechanics and Path Integrals*, (McGraw-Hill, New York, 1965).
46. Langer, J.S., Ann. Phys. (NY), **41**, 108 (1967).
47. Stone, M., Phys. Letters, **67B**, 186 (1977).
48. Callan, C.G. and Coleman, S., Phys. Rev., **D16**, 1762 (1977).
49. Golub, A.A. and Iordatin, V.P., Pis'ma v Zh. Eksp. Teor. Fiz., **36**, 184 (1982) (translation: JETP Lett. **36**, 227 (1982)).
50. Chakravarty, S. and Chang., L.-D., (to be published).
51. Zwerger, W., preprint.
52. Wonneberger, W., Z. Phys., **B50**, 23 (1983).
53. Koch, R., van Harlingen, D.J. and Clarke, J., Phys. Rev. Lett., **45**, 2132 (1980).
54. Koch, R. (private communication).
55. Schmid, A., J. Low Temp. Phys., **49**, 609 (1982).
56. Kurkijärvi, J., Phys. Lett., **88A**, 241 (1982).
57. Chakravarty, S., Phys. Rev. Lett., **49**, 681 (1982).
58. Bray, A.J. and Moore, M.A., Phys. Rev. Lett., **49**, 1545 (1982).
59. Chakravarty, S. and Leggett, A.J., Phys. Rev. Lett., (in press).
60. Behrman, E.C., Jongeward, G.A. and Wolynes, P.G., J. Chem. Phys., 15 December 1983 (in press).
61. Fulton, T.A. and Dunkleberger, Bull. Am. Phys. Soc., **19**, 205 (1975).
62. Leggett, A.J., University of Sussex internal memo., 1981 (unpublished).
63. Voss, R.F. and Webb, R.A., Phys. Rev. Lett., **47**, 265 (1981).
64. Jackel, L.D., *et al*. Phys. Rev. Lett., **47**, 697 (1981).
65. Voss, R.F. and Webb, R.A., Physica, **B108**, 1307 (1981).
66. Widom, A., Megaloudis, G., Sacco, J.E., and Clark, T.D., Nuovo Cimento, **61B**, 112 (1981).
67. Likharev, K.K., Usp. Fiz. Nauk, **139**, 170 (1983) (translation: Soviet Physics Uspekhi, **26**, 87 (1983)).

Unitary Transformation Methods in Intense Fields Atomic Physics

C. A. S. Lima* and L. C. M. Miranda**
*Instituto de Fisica, Universidade Estadual de Campinas
13100 Campinas, SP, Brazil
**Instituto de Estudos Avancados, Centro Técnico Aeroespacial,
12200 S. J. Campos, SP, Brazil

ABSTRACT

The unitary transformation method in dealing with bound states systems under strong electromagnetic fields has been considered in recent literature as a rather promising one. The present work reviews some aspects of this work with specific regard to atomic systems.

1. Introduction

In dealing with the interaction of an atomic or molecular system with radiation one has to face two different, yet complementary aspects. On the one side there are the system's structural changes as reflected on the shifts, line broadening mixing or splitting of the system's field-free energy levels. On the other side, one has to consider the field induced transitions, i.e., the ways by which the field affects the occurrence of dynamical processes within the system, the most outstanding example being probably that of the transitions to the continuum states. Of course, at high intensities, one expects these two aspects to be strongly coupled to one another. Both aspects, isolated or together have been the subject of several investigations using a variety of approaches, e.g., those which, by way of example, will be considered in the present short review.[1-6]

The current availability of adequately powerful short pulsed lasers ranging from the infrared to the ultraviolet[7] is making it possible to carry on detailed investigations of such effects that are bound to disclose many new features of the atom/molecule-radiation interaction. Indeed, a vigorous upsurge of experimental papers, in the past few years, on the subject has been clearly observed. A concurrent flux of theoretical papers has accompanied, sometimes preceded, this experimental effort. Among those, quite a few have been devoted to propose or extend model approaches to the problem. A number of excellent reviews [8-16] covering the specific subjects of atom/molecule-radiation interaction and related topics have come to light in the past few years. In one of the recent ones the models developed to 1978 have been considered in some detail.[8] The main purpose of the present short review is to single out amongst the forementioned works those exploring one of the most promising general theoretical framework to approach the problem, namely, that of the unitary transformations.[1-6,9] Thus, a choice of representation models, under this approach, will be considered in some detail; specifically, we shall focus our attention on the momentum translation approximation [3], the rotating frame approach [4], and the space translation approximation.[2,5,6] Among the earliest attempts to produce a consistent approach to strong field ionization problems, the pioneering work of Keldysh [1] stands out as a landmark. Despite its self-recognized limitations, it inspired quite a few works and still receives widespread citation in the area of strong field physics in both solids and gases. Thus, both for historical recognition and for having a summary of it readily available for eventually contrasting its features with those of the unitary transformation approach, we begin in the section to follow, by discussing the Keldysh model in some detail.

2. Oscillating Electronic Orbitals — Keldysh's Approach

One approach to the questions posed in the introduction uses the so-called oscillating electron orbitals. This approach was originally introduced to the literature of strong field physics by Keldysh's pioneering 1964 paper[1]. Keldysh tackles the problem of field induced ionization of atoms within the semiquantal framework.[8] His approach was directed to produce an expression for the ionization probability of atoms under an intense laser field whose frequency was lower than that equivalent to the ionization potential (he also aptly converted his findings to apply them to crystals as well), under the assumption that the radiation electric field affects maximally the continuum states and the degenerate excited bound states. A special feature of his results was that with a single expression he obtains, under the appropriate limits, both the tunnel-effect auto-ionization probability and the multiphoton absorption. These limits are reached, respectively, by requiring

$\gamma_k \gg 1$ or $\gamma_k \ll 1$. The parameter $\gamma_k = \omega/(eE_0/\sqrt{2mI_0})$, which became to be known as the Keldysh parameter, depends on the atomic ionization potential I_0, the oscillating field strength E_0 and frequency ω, and the electronic charge (e) and mass (m). It gives a measure of the adiabaticity of the field variation, namely, when $\gamma_k \ll 1$ the field appears to be almost static, so that the electron experiences ordinary tunnelling. In the opposite limit, $\gamma_k \gg 1$, the time dependence of the external radiation field is essential[1,10]; γ_k can be seen to be the ratio between the field frequency and the field-induced tunnelling frequency.

The main steps in Keldysh procedure will now be followed. In an oscillating field $\boldsymbol{E} = \boldsymbol{E}_0 \cos\omega t$, e.g., that of an electromagnetic wave in the dipole approximation, the free electron wavefunction can be written as

$$\psi(\boldsymbol{r},t) = \exp\left\{\frac{i}{\hbar}\left[\boldsymbol{q}(\boldsymbol{p},t)\cdot\boldsymbol{r} - \frac{1}{2m}\int_0^t q^2(\boldsymbol{p},\tau)\,d\tau\right]\right\} \tag{1}$$

where $\boldsymbol{q}$ is the electron effective momentum

$$\boldsymbol{q}(\boldsymbol{p},t) = \boldsymbol{p} + \frac{e\boldsymbol{E}_0}{\omega}\sin\omega t\ . \tag{2}$$

If one considers an electric dipole transition between a field-free initial (electronic ground state in H-atom) state and a field-free final (free electron with momentum $\boldsymbol{p} = \hbar\boldsymbol{k}$) state, the relevant matrix element is

$$\begin{aligned} V_0(\boldsymbol{p}) &= \int \frac{d^3r}{(\pi a_0^3)^{\frac{1}{2}}}\, e^{i\boldsymbol{k}\cdot\boldsymbol{r}}\, e\boldsymbol{E}\cdot\boldsymbol{r}\; e^{-r/a_0} \\ &= 8(\pi a_0^3)^{\frac{1}{2}}\, e\hbar\boldsymbol{E}\cdot\boldsymbol{\nabla}_p(1 + p^2a_0^2/\hbar^2)^{-2}\ . \end{aligned} \tag{3}$$

where a_0 is the Bohr radius. Now the corresponding matrix element when the final state is reached in the presence of the electromagnetic field is given as above with the change $\boldsymbol{p} \to \boldsymbol{q}(\boldsymbol{p},t)$. One can now use this matrix element in combination with the usual quantum procedure (Fermi's rule) to write down the unit time probability of a direct transition from the ground state to the continuum spectrum[1]

$$\begin{aligned} W_0 = \frac{1}{\hbar^2}\lim_{t\to\infty} \mathrm{Re}\int \frac{d^3p}{(2\pi\hbar)^3}\int_0^t dt'\cos\omega t\cos\omega t'\, V_0[\boldsymbol{q}(\boldsymbol{p},t)]\; V_0[q(\boldsymbol{p},\boldsymbol{t}')] \times \\ \times \exp\left\{\frac{i}{\hbar}\int_t^{t'}\left[I_0 + \frac{1}{2m}q^2(\boldsymbol{p},\tau)\right]d\tau\right\} \end{aligned} \tag{4}$$

or equivalently

$$W = \frac{2\pi}{\hbar}\int \frac{d^3p}{(2\pi\hbar)^3}\sum_{s=-\infty}^{+\infty} |L_s(\boldsymbol{p})|^2\,\delta\left(I + \frac{p^2}{2m} + \frac{e^2E^2}{4m^2} - s\hbar\omega\right) \tag{5}$$

where use was made of the Fourier expansion

$$V_0(\mathbf{q}(\mathbf{p},t)) \exp\left\{\frac{i}{\hbar}\int_0^t \left[I_0 + \frac{q^2}{2m}\right] d\tau\right\}$$

$$= \sum_{s=-\infty}^{-\infty} L_s(\mathbf{p}) \exp\left\{\frac{i}{\hbar}\left(I_0 + \frac{p^2}{2m} + \frac{e^2E_0^2}{4m\omega^2} - s\hbar\omega\right) t\right\} \tag{6}$$

with

$$L_s(\mathbf{p}) = \frac{1}{2\pi} \oint_c du\, V_0\left[\mathbf{q}(\mathbf{p},u)\right] \exp\left\{\frac{1}{\hbar\omega}\int_0^u \left[I_0 + \frac{1}{2m} q^2(\mathbf{p},u)\right] \frac{dv}{(1-v^2)^{\frac{1}{2}}}\right\} \tag{7}$$

The contour C in Eq. (7) encloses the interval (-1 , 1). The rapidly oscillating character of the argument of the exponential in the integrand of $L(\mathbf{p})$ makes suitable its evaluation by the saddle point method as a sum of contributions from each saddle point u_s. These are to be found as usual from

$$q^2(\mathbf{p},u_s) + 2m\, I_0 = 0\ . \tag{8}$$

Then

$$L(\mathbf{p}) = \sum_s L_s(\mathbf{p}) \tag{9}$$

with

$$L_s(\mathbf{p}) = 2(\pi a_0^3)^{\frac{1}{2}} \frac{I_0}{eE_0 a_0} \frac{\hbar\omega}{(1-u_s^2)^{\frac{1}{2}}} \exp\left\{\frac{1}{\hbar\omega}\int_0^{u_s} \left[I_0 + \frac{q^2}{2m}(\mathbf{p},v)\right] \frac{dv}{(1-v^2)^{\frac{1}{2}}}\right\} \tag{10}$$

being the contribution to $L(\mathbf{p})$ arising from the saddle point u_s. As a final step in this procedure observe that while the u_s do depend on $\mathbf{p}$ it is only from small p ($p^2 \ll 2mI_0$) that one gets contributions to the ionization probability. Thus, in $L_s(\mathbf{p})$, one can safely set $p^2 = 0$ in the pre-exponential factor, expand the exponential and keep terms up to p^2 only. Then, upon carrying the indicated integration one gets:

$$W_0 = \left(\frac{2I_0}{\hbar}\right)^{\frac{1}{2}} \omega\left(\frac{\gamma}{\sqrt{1+\gamma^2}}\right)^{\frac{3}{2}} S(\gamma, I_0/\hbar\omega) \exp\left\{-\frac{2\tilde{I}_0}{\hbar\omega}\left(\text{arc sin}\,\gamma - \gamma\frac{\sqrt{1+\gamma^2}}{1+2\gamma^2}\right)\right\} \tag{11}$$

where γ is the Keldysh parameter, and $S(\gamma, \tilde{I}_0/\hbar\omega)$ is a slowly varying factor (as compared to an exponential) depending on both γ and on the ionization potential I_0. In fact, the effective threshold energy of absorption $\tilde{I}_0$ exceeds I_0 by $e^2E_0/4m\omega^2$, i.e., by the value of the average electron oscillation energy in the field of the wave. In other words $\tilde{I}_0 = I_0(1+1/2\gamma^2)$. As to the specific form of S one has [1], letting $y = [x+1] - x + n$,

$$S(\gamma,x) = \sum_{n=0}^{\infty} \Phi\left[\left(2\frac{\gamma y}{\sqrt{1+\gamma^2}}\right)^{\frac{1}{2}}\right] \exp\left\{-2y\left(\text{arc sinh}\,\gamma - \sqrt{\frac{\gamma^2}{1+\gamma^2}}\right)\right. \tag{12}$$

Here $[x]$ means the integer part of x, and

$$\Phi(z) = \int_0^z dy \, \exp\,(y^2 - z^2)\,. \tag{13}$$

The limiting cases referred to before can now be worked out. For the case where $\gamma \ll 1$ (low frequencies and very strong fields) the value of $S(\gamma,x)$ is mainly determined by large values of $n \sim \gamma^{-1}$, in which case $S = \sqrt{3\pi}/4\gamma^2$ (ref. 1) and one gets:

$$W_0 = \left(\frac{3\pi e E_0}{8\hbar}\sqrt{\frac{I_0}{m}}\right)^{\frac{1}{2}} \exp\left\{-\frac{4\sqrt{2m I_0^3}}{3e\hbar E_0}\left(1 - \frac{m\omega^2 I_0}{5e^2 E_0^2}\right)\right\}. \tag{14}$$

Note that as $\omega \to 0$ the exponential in Eq. (14) goes exactly into that describing the well known process of tunnel ionization of atoms in a d.c. electric field. However, the same is not true for the pre-exponential factor. The observed discrepancy arises from the neglect of the residual Coulomb field in the final state interactions. In the opposite limit of $\gamma \gg 1$ (high frequencies and not too strong fields) one obtains the ionization probability with the simultaneous absorption of several photons:

$$W_0 = A\omega\left(\frac{I_0}{\hbar\omega}\right)^{\frac{3}{2}}\left(\frac{e^2 E_0^2}{8m\omega^2 I_0}\right)^{\tilde{I}_0/\hbar\omega + 1} \phi\left[\left(2\left\{\frac{\tilde{I}_0}{\hbar\omega} + 1\right\} - \frac{2I_0}{\hbar\omega}\right)^{\frac{1}{2}}\right] e^{u} \tag{15}$$

where

$$u = 2\left[\frac{I_0}{\hbar\omega} + 1\right] - \frac{I_0}{\hbar\omega}\left(1 + \frac{e^2 E_0^2}{2m\omega^2 I_0}\right). \tag{16}$$

Again this lacks the inclusion of the ion-electron final state interaction. If this correction is included both Keldysh formulae are directly applicable to atoms beyond hydrogen. This is seen from the fact that in the Keldysh approach the action of an electromagnetic field is considered upon the free electron field state and not on the atomic ground state. Accordingly, to apply it to an atom other than hydrogen it suffices to recalculate the matrix elements $V_0(\mathbf{q})$ appropriately. The forementioned residual Coulomb interaction in the final state was briefly considered to by Keldysh [1] himself, who simply quoted that an overall factor of $\gamma I_0/\hbar\omega(1+\gamma^2)^{\frac{1}{2}}$ crudely accounts for it and must be included if theoretical and experimental probabilities are to be compared. Semiquantal calculations including this interaction have been carried out disclosing the importance of the corrections. However, since our aim in reviewing Keldysh's paper was more at confronting approaches rather than discussing details we shall refrain from pursuing this refinement here. For details refer to refs. 8, 17–20.

We shall now consider the case when the transition ground state continuum does not proceed directly but rather through an intermediate step, namely, a field-modified atomic excited state (an oscillating bound state)

$$\psi_s(\boldsymbol{r},t) = \phi_s(\boldsymbol{r}) \exp\left\{\frac{i}{\hbar}\left(I_s t - \frac{e\sigma_s}{\omega} E_0 \sin \omega t\right)\right\} \tag{17}$$

where I_s and $e\sigma_s$ stand, respectively, for the ionization potential and dipole moment of the s-th state. By following a procedure similar to the direct transition case to describe each of the assumed steps (field-free ground state → oscillating bound orbital → free oscillating electron) with the appropriately calculated dipole matrix elements, one obtains ω as a sum of partial probabilities W_s that the ionization proceeds first through a bound-bound transition into the orbital ψ_s and then by a bound-continuum transition. The result is[1]

$$W_s = \frac{\left|J_{n_s-1}(\rho_s) + J_{n_s} + (\rho_s)\right|^2 \left|V_{0s}\right|^2}{4(I_0 - I_s - n_s \hbar\omega)} p_s \tag{18}$$

where $p_s \equiv W_0^{\text{corr.}}(I_0 \to I_s)$ is the probability for the direct process from the s-th excited state, the ionization potential for which is I_s. Now, the expression for $W_0^{\text{corr.}}$ is given by the corrected general expression given before where one should use the substitution $I_0 \to I_s$. In Eq. (18), J_k is a Bessel function, $\rho_s = e\sigma_s E_0/\omega$, and n_s stands for the integer closest to $(I_0 - I_s)/\hbar\omega$. In the calculation of W_s the assumption was made that the contribution arising from the situation closest to a multiphoton resonance absorption between the initial and intermediate levels (namely $I_0 - I_s$ is closer to an integer multiple of $\hbar\omega$) far outweighs all the others and was, accordingly, the only one kept. Keldysh's final expression for W_s incorporates two additional changes: the inclusion in the energy denominator of both the 2nd order Stark shift of the atomic levels and the attenuation of the s-th level due to its ionization and spontaneous decay. Specifically one has

$$W = \frac{1}{4} \sum_s \left|J_{n_s-1}(\rho_s) + J_{n_s+1}(\rho_s)\right| \frac{|V_{0s}|^2 \, p_s}{(I_0 - I_s + \frac{1}{2}\alpha_s e^2 E^2 - n_s \hbar\omega)^2 + \hbar^2 (\omega_s + \Gamma_s)^2} \tag{19}$$

where $\hbar\omega_s = I_0 - I_s$ and Γ_s is the s-th-level width.

2. The Momentum Translation Method

In a consolidated form extending previous reports[21,22], the momentum translation method (MTM) was introduced[3,23,24] as an alternative to the treatment of bound-state problems in semiclassical electrodynamics via perturbation theory which was deemed inadequate to deal with the then emerging intense field atomic and molecular physics. Inserted within the various unitary transformation approaches transforming strong field problems into a more manageable representation and allowing for a clearer interpretation of the physics involved, the MTM was soon to be the target of varied criticisms [31-35] the strongest perhaps being the assertion by Cohen-Tannoudji *et al* [31] that it is no more than a gauge transformation and hence devoid of physical content. This point, as well as the other criticisms [32-35] relating to the appropriateness of its use and to the accuracy of the calculations based on the MTM, were contested and explained by its author [30] but they had already undoubtedly influenced against its use. Recently, in a thorough evaluation of the MTM, Reiss [29] sets the appropriate grounds on which one may take advantages of simple analytical forms, breadth of applicability, and simplicity as one goes to higher order multiphoton processes.

The MTM starting point is the single particle Schrödinger equation

$$i\hbar\dot{\psi}(\boldsymbol{r},t) = \left\{\frac{1}{2m}\left(\frac{i}{\hbar}\nabla - \frac{e\boldsymbol{A}}{c}(t)\right)^2 + V(\boldsymbol{r})\right\}\psi(\boldsymbol{r},t) \tag{20}$$

where the additional effect of an external electromagnetic field has been included. By applying a unitary transformation U to ψ:

$$\tilde{\phi} = U\psi\ , \quad \text{with} \quad U = \exp(-i\,e\,\boldsymbol{A}\cdot\boldsymbol{r}/\hbar c)\ , \tag{21}$$

one has for $\tilde{\phi}(\boldsymbol{r},t)$:

$$i\hbar\dot{\tilde{\phi}}(\boldsymbol{r},t) = \left\{-\frac{\hbar^2}{2m}\nabla^2 + V(\boldsymbol{r}) - e\,\boldsymbol{r}\cdot\boldsymbol{E}(t)\right\}\tilde{\phi}(\boldsymbol{r},t)\ , \tag{22}$$

where $E(t)$ is the electromagnetic wave electric field in the dipole approximation. If one further requires $|e\,\boldsymbol{r}\cdot\boldsymbol{E}| \ll \varepsilon$, where ε is any characteristic energy or energy difference in the problem, then $\tilde{\phi} \to \phi(\boldsymbol{r},t)$, the field-free wave function of the system. Thus, the complete solution can be written as

$$\psi(\boldsymbol{r},t) = e^{ie\boldsymbol{A}(t)\cdot\boldsymbol{r}/\hbar c}\ \phi(\boldsymbol{r},t)\ , \tag{23}$$

where

$$i\hbar\tilde{\phi} = \left\{-\frac{\hbar^2}{2m}\nabla^2 + V(\boldsymbol{r})\right\}\phi\ . \tag{24}$$

For a linearly polarized plane wave in the dipole approximation $\boldsymbol{A}(t) = \boldsymbol{a}\cos\omega t$. Therefore, if we take the Bohr radius a_0 as a representative value of $|\boldsymbol{r}|$ then $|e\boldsymbol{r}\cdot\boldsymbol{E}| \ll \varepsilon$ reads $e\,a_0 a\omega \ll \varepsilon$, which is to be taken as the validity condition for the MTM with, of course, the $\omega a_0 \ll c$ condition already implied by the dipole approximation for the wave field. A formal solution to the (complete) $\tilde{\phi}$ equation can be used to write ψ as [3]

$$\psi = \frac{\exp(ie\boldsymbol{A}\cdot\boldsymbol{r}/\hbar c)}{(\tilde{\phi},\tilde{\phi})}\left[\phi + \sum_n \phi_n\left(-\frac{i}{\hbar}\int_{-\infty}^{t} dt'\,(\phi_n - e\boldsymbol{r}\cdot E\,\tilde{\phi})\right)\right] \tag{25}$$

where the summation is to be carried over the complete orthonormal set of field-free states $\{\phi_n\}$. Dropping the summation altogether produces the previously mentioned

$$\psi(\boldsymbol{r},t) = e^{ieA(t)\cdot\boldsymbol{r}/\hbar c}\ \phi(\boldsymbol{r},t)$$

approximate solution which became known as the 'momentum translation approximation' (MTA) wave function.

The MTA wavefunction can be generated from at least two different approaches, e.g., from the path-integral technique [36], and from time-dependent perturbation theory [29]. The conditions for the validity of the MTA given above are seen to be the main assumptions needed to extract from these two formalisms the MTA wavefunction. It should also be remarked that the MTA has been shown to be obtainable as a special case of some more elaborate formalisms in electrodynamics. For instance, one can use a unitary transformation with the generating function $S = \int d^3r\ \boldsymbol{p}\cdot\boldsymbol{A}$ where $\boldsymbol{p}$ is the polarization operator [37, 38]. The non-relativistic limit of the relativistic formulation of the electrodynamics problem yields the MTA, since $\boldsymbol{p}$ is then limited to the electric-dipole contributions. Another approach, based on the Magnus expansion, solves the equation of motion for the time-evolution operator [38]. The MTA is shown to be the leading term of this approach.

An important remark is made by Reiss [27-29] in regard to the criticism mentioned above that the unitary transformation U leading to the MTA is a gauge transformation of the potentials from Coulomb gauge to the electric-field gauge. He claims that this transformation was used in the MTM to accomplish a unitary transformation of the wave function within the Coulomb gauge. In support of that he calls attention to the fact that the MTA wavefunction is, under the unitary transformation approach, an approximate solution in the Coulomb gauge to the full equation

$$i\hbar\dot{\psi} = (H_0 + H')\ \psi \tag{26}$$

where

$$H_0 + H' = \frac{1}{2m}\left(\frac{\hbar}{i}\boldsymbol{\nabla} - \frac{e\boldsymbol{A}}{c}(t)\right)^2 + V(\boldsymbol{r}) \tag{27}$$

with the field-free Hamiltonian being

$$H_0 = -\frac{\hbar^2}{2m}\nabla^2 + V(\boldsymbol{r}) \tag{28}$$

and the bound-system — electromagnetic field interaction given by

$$H' = -\frac{e\boldsymbol{A}(t)}{mc}\cdot\tilde{p} + \frac{e^2\boldsymbol{A}^2(t)}{2mc^2}, \tag{29}$$

in exactly the same sense as it is a solution under both the path-integral approach and the time-dependent perturbation theory with H' as the perturbing Coulomb-gauge interaction Hamiltonian. Clearly no gauge transformation is implied in any of these cases. Another source of major criticism [31, 32] is the lack of consistent improvement over the MTA result when additional terms in the MTM series are considered. Failure to recognize that cancellations do occur among terms beyond the basic MTA term is at the root of the difficulty [29]. Accordingly, the MTM cannot be viewed as a successive approximation method. The MTA term has a physical meaning as a momentum-translated result but none of the others, individually, is physically significamt. It is the *complete* sum that could be viewed as a correction, one that would then bring the MTA into the full, exact result. Adding only one or more supplementary terms to the MTA produces a hybrid of MTM and perturbation theory which has no physical or numerical meaning [29].

The MTA as it stands now may well get new recognition among practitioners in the field. It may help to bring things into proper perspective to say that the MTA does provide a very straightforward method to obtain wave functions and transition probabilities for multiphoton transitions, for a large variety of one-field and two-fields problems. A few examples are bound-bound transitions in hydrogen without intermediate near resonances [30], and two-photon detachment rates of negative ions [39]. Overall agreement within a factor of 2 to 10 accuracy in calculated transition probabilities using MTA seems to be the case in general, maintained so even in high photon multiplicity processes. The overall behaviour of polarization effects in photoionization [25] and polarization ratios in particular, [25,40] have been studied with MTA. It can also be said that the MTA wave function automatically predicts replica states [29] though the simple MTA does not contain a level shift. Yet, a calculation of level shift in a momentum translation framework can be shown to produce the same result as the corresponding perturbation calculation [29]. A well known and undisputed difficulty with the MTM is that the use of the MTA wave function in the context of field-induced transitions when intermediate near-resonant states are present can be seriously in error, as has been found to be the case with both bound-bound [29,41] and bound-free [33,42-47] transitions with intermediate near resonances possible.

We should add that it can be shown, on rather general grounds, that the MTA describes the bound states in strong radiation field problems as an appropriate limit of the exact wavefunction, namely, the case when the field oscillation period is much smaller than the system (field-free) smallest characteristic time. This will be demonstrated in the last section of this report.

3. ROTATING FRAME TRANSFORMATION

Another interesting approach specially suited to deal with the problem of bound states in a circularly polarized plane wave field has been introduced by Manakov and Rappoport[4]. The main ansatz takes advantage of the considerable simplifications introduced when it is appropriate to apply a transformation to a rotating coordinate frame. This is the case[4,17,48,49] when we have a system having spherical symmetry in the field of a circularly polarized wave. Under this assumption it is possible to eliminate the time-dependence by having the coordinate frame rotating at the wave frequency, thus reducing the problem to a stationary one. In order to make it clear that the method affords a non-perturbative treatment of the wave-system interaction, an application was made in[4] to the problem of electrons moving in an attractive three-dimensional delta-function potential. The overall procedure goes as follows. For electrons interacting with a right-handed circularly polarized wave propagating along the z-axis, the corresponding interaction Hamiltonian, within the usual dipole approximation, can be written as

$$V(\boldsymbol{r},t) = eFr \sin\Theta \cos(\omega t - \phi) \tag{30}$$

so that the Schrödinger equation reads:

$$i\hbar\dot{\psi}(\boldsymbol{r},t) = \left\{H + V(\boldsymbol{r},t)\right\} \psi(\boldsymbol{r},t) \tag{31}$$

where H_0 is the field-free Hamiltonian.

Performing the unitary transformation

$$\psi(\boldsymbol{r},t) = \exp(-i\omega\hat{L}_z t/\hbar)\, \phi(\boldsymbol{r},t) \tag{32}$$

Eq. (31) reduces to an equivalent form[4]

$$(\hat{Q}(\boldsymbol{r}) - E_n)\, \phi_n(\boldsymbol{r}) = 0 \,, \tag{33}$$

where $\tilde{Q}(\boldsymbol{r})$ is the quasi-energy operator given by

$$\hat{Q}(\boldsymbol{r}) = H_0 - \omega\tilde{L}_z + eFx \,. \tag{34}$$

It can be seen that $\hat{Q}(\boldsymbol{r})$ is the equivalent stationary problem Hamiltonian associated with the system as seen from a rotating frame of reference. In the foregoing Eqs. (30)—(34), ω is the frequency of the electromagnetic wave, which in the rotating frame has an electric field of strength F directed along the x-axis, and E_n are the quasi-energies.

For the specific problem of the delta potential considered by Manakov and Rappoport[4], Eq. (33) thus reduces to

$$\left[-\frac{\hbar^2}{2m}\nabla^2 - \omega\tilde{L}_z + eFx - E\right] \delta(\boldsymbol{r})\, \phi(\boldsymbol{r}) \tag{35}$$

with the boundary condition[4,50]

$$\phi(\boldsymbol{r}) \xrightarrow[(r\to 0)]{} (1/r - 1/b_0)/4\pi \,, \quad E_0 = -\hbar^2/2mb_0^2 \tag{36}$$

where b_0 is the scattering length, and E_0 is the level energy in the absence of the radiation field. Note that, in the rotating frame, the wave function ϕ of our problem coincides with the Green's function of a free particle in a constant (d.c.) electric field, which is readily available, namely[4,51]:

$$\phi(\boldsymbol{r}) = \frac{1}{4\pi}\left(\frac{m}{2\pi i\hbar}\right)^{\frac{1}{2}} \int_0^\infty \frac{dt}{t^{\frac{1}{2}}} \exp\frac{i}{\hbar}\left\{Et + \frac{mr^2}{2t} + \frac{eFy}{\omega t}\left(\frac{\sin\omega t}{\omega t} - 1\right) - \right.$$

$$\left. - \frac{2eFx}{\omega^2 t}\sin^2\left(\frac{\omega t}{2}\right) - \frac{e^2F^2}{2m\omega^2}\left(t - \frac{4}{\omega^2 t}\sin^2\left(\frac{\omega t}{2}\right)\right)\right\} . \tag{37}$$

By separating that part of Eq. (37) which diverges at $r = 0$, and then imposing the boundary condition given by Eq. (36), the equation for the complex energy E of the quasi-stationary state is found. One gets[4]:

$$\left(\frac{E}{E_0}\right)^{\frac{1}{2}} = 1 - \frac{1}{(4\pi i)^{\frac{1}{2}}} \int_0^\infty \frac{dt}{t^{\frac{3}{2}}} \exp\left(-i\,\frac{E}{E_0}\,t\right)\left\{1 - \exp\left[-\frac{it}{\gamma_{\mathbf{k}}^2}\left(1 - \frac{\sin^2\delta t}{\delta^2 t^2}\right)\right]\right\} \tag{38}$$

where $\gamma_{\mathbf{k}}$ is the Keldysh parameter, $I = -E_0$ is the ionization potential, and $2\delta = \hbar\omega/I$ is a quantity related to the number of photons k needed to ionize the bound level in accordance with the energy conservation law

$$k_0 = \left\{I/\hbar\omega + 1\right\} .$$

The results of the numerical evaluation of Eq. (38) to get the energy shift and the ionization rate (width) are shown in Figs. 1 and 2. In their calculation Manakov and Rapoport[4] have introduced a characteristic field strength F_0 given by

$$F_0 = 8m^2 I^3/e\hbar \tag{39}$$

and a characteristic frequency ω_0 equal to

$$\omega_0 = I/\hbar . \tag{40}$$

It follows from these results that[4,8], for the energy shift, the deviation from the second order perturbation theory result seems to occur for $\omega < 0.5\,\omega_0$ and $F > 0.1\,F_0$. Thus, the deviations from perturbation theory are large primary when F/F_0 is not small. On the other hand, the deviations from the power law dependence $F^{2\mathbf{k}_0}$ of non-resonant perturbation theory are large when $(k_0\gamma_{\mathbf{k}})^2$ is not small. This latter result is an important one, as the transition region had been believed to be near $\gamma_{\mathbf{k}} \approx 1$, whereas now it seems shifted to $\gamma_{\mathbf{k}} \approx k_0^{-1}$, in agreement with the Nd:glass laser multiphoton ionization results of Lompre *et al.*[52]

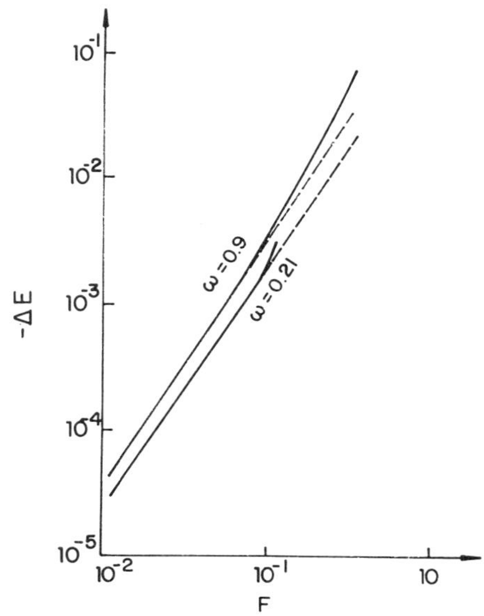

Fig. 1 Bound-level energy shift, after Manakov and Rapoport.[4] The strength of the circularly polarized field is F and the frequency is ω.

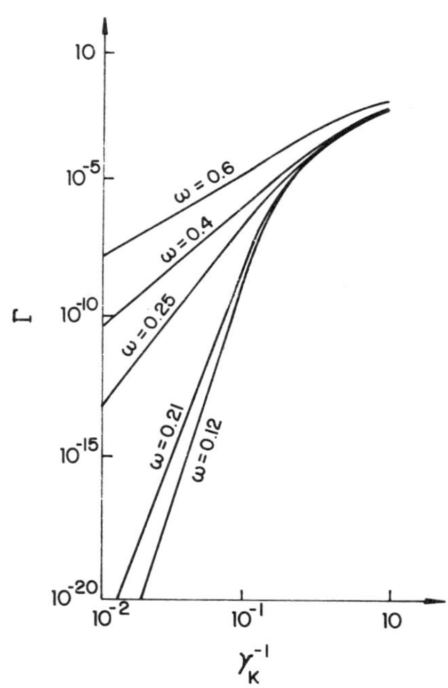

Fig. 2 Ionization rate as a function of the inverse Keldysh parameter γ_k and the frequency ω of a circularly polarized field, after Manakov and Rapoport.[4]

4. Space Translation Method

The space translation method (STM) was introduced by Henneberger [2] as an alternative to the time-dependent perturbation theory description of the atomic states in the presence of intense laser fields. The original idea was to find an effective interaction potential determined by the time average of the combined effects of the Coulomb and the laser field. The atomic states corresponding to this effective potential would then represent an improvement over the unperturbed states used in conventional perturbation theory. Basically, the STM consists in using a time-dependent unitary transformation such that, in the transformed Schrödinger equation for the atomic electron, the nucleus appears to be oscillating with the frequency of the radiation field with amplitude given by the classical displacement of a free electron in a radiation field. The unitary transformation used by Henneberger is essentially the same as that proposed by Kramers [53] to remove divergences in non-relativistic quantum electrodynamics in the dipole approximation. For a detailed discussion on the use of unitary transformations in quantum electrodynamics we refer to the work of Power [9], in which Kramer's transformations and its implications are fully discussed.

Let us consider, as an example, the case of a hydrogen atom in a classical electromagnetic field. The electronic states are governed by the Hamiltonian

$$H = \frac{1}{2m}\left(\boldsymbol{p} + \frac{e}{c}\,\boldsymbol{A}(t)\right)^2 - \frac{e^2}{|\boldsymbol{r}|}\ , \tag{41}$$

where $\boldsymbol{A}(t)$ is the vector potential describing the electromagnetic field in the dipole approximation. Applying the unitary transformation

$$U = \exp\left(i\boldsymbol{\delta}(t)\cdot\boldsymbol{p}/\hbar\right)\exp\left(i\eta(t)/\hbar\right)\ , \tag{42}$$

where

$$\boldsymbol{\delta}(t) = -\frac{e}{mc}\int^{t} dt'\,\boldsymbol{A}(t') \quad \text{and} \quad \eta(t) = -\frac{e^2}{2mc^2}\int^{t} dt'\,A^2(t') \tag{43}$$

the Hamiltonian H changes into $\tilde{H}$, namely

$$H \rightarrow \tilde{H} = U^{+}\left\{-i\hbar\frac{\partial}{\partial t} + H\right\}U \quad \frac{p^2}{2m} - \frac{e^2}{|\boldsymbol{r}-\boldsymbol{\delta}(t)|}\ . \tag{44}$$

It is Henneberger's contention [2,54] that by working with the time average of $\tilde{H}$, namely,

$$\langle\tilde{H}\rangle = \frac{p^2}{2m} - \frac{e^2}{2\pi^2}\int d^3k\,\frac{e^{i\boldsymbol{k}_0\cdot\boldsymbol{r}}}{k^2}\,J_0(\boldsymbol{k}\cdot\boldsymbol{a}) \tag{45}$$

the corresponding eigenstates become intensity-dependent and represent an improvement over the usual perturbation theory. In Eq. (45) $\boldsymbol{a}$ is the amplitude of the electron oscillation in the radiation field obtained from the expression for $\boldsymbol{\delta}(t)$. Although it will be discussed in greater detail at the end of this section, we shall briefly mention here that Eq. (45) is valid in the high-frequency limit, where the laser period is shorter than the characteristic field-free bound electron oscillation time.

Since its original proposal, the STM has been used for studying a number of problems. These include the energy-level structure [5,54-60], multiphoton ionization [6], laser-assisted atomic collisions [61-69], and so on. Nevertheless, some criticisms[55,56,70] have been raised regarding the calculations of the ionization potential shifts[54-58] using the time-independent transformed Hamiltonian (c.f. Eq. (45)). These criticisms were essentially motivated by the fact that the energy shift calculated with the time-independent effective potential of Eq. (45) alone does not account for the full shift. This point, however, was later clarified by Lambropoulos *et al.*[71] who showed that the proper use of the transformed Hamiltonian for calculating the energy shifts demands considerations beyond the d.c. term arising from the transformed Hamiltonian. That is, to get the correct ionization potential shift one should also include, in second-order perturbation theory, the contribution of the first time-dependent term of the transformed Coulomb potential of Eq. (44). When this procedure is adopted, it can be shown [71] that the STM leads to the same predictions as conventional perturbation theory. This comparison between the STM and the conventional perturbation theory is, of course, valid only in the case where the electromagnetic field is not so intense that its interaction energy with the atom is greater than the binding energy.

To assess further the full implications of the STM one can alternatively try to produce a suitable parametrization of the time-averaged dressed potential so as to be able, by an appropriate series expansion, to accumulate in a reasonably small number of terms most of the dynamical effects of the field upon the system. Such a program has been carried out by the present authors[5,58,59] by resorting to the following expansion of the dressed potential (for a circularly polarized electromagnetic wave):

$$V(\boldsymbol{r}-\boldsymbol{\delta}(t)) = -\frac{e^2}{(r^2+a^2)^{\frac{1}{2}}}\left[1-\frac{2\boldsymbol{r}\cdot\boldsymbol{\delta}(t)}{r^2+a^2}\right]^{-\frac{1}{2}} = V_{\mathbf{dc}} + V_{\mathbf{ac}} \tag{46}$$

where $a = |\boldsymbol{\delta}(t)| = eA/mc\omega$, and

$$V_{\mathbf{dc}} = \frac{e^2}{(r^2+a^2)^{\frac{1}{2}}}\left\{1+\frac{3}{4}\frac{r^2a^2\sin^2\theta}{(r^2+a^2)^2}+\ldots\right\} = V_0 + V_2 + \ldots$$

$$= -\frac{e^2}{(r^2+a^2)^{\frac{1}{2}}}\sum_{\mu=0}^{M}\frac{(4\mu)!}{2^{4\mu}(2\mu)!(\mu!)^2}\left(\frac{ra}{r^2+a^2}\right)^{2\mu}\sin^{2\mu}\theta \tag{47}$$

$$V_{\mathbf{ac}} = -\frac{e^2}{(r^2+a^2)^{\frac{1}{2}}}\left\{\frac{\boldsymbol{r}\cdot\boldsymbol{\delta}(t)}{r^2+a^2} - \frac{3}{2}\,\frac{r^2a^2\sin^2\theta\cos[2(\omega t-\phi)]}{(r^2+a^2)^2} + \right.$$
$$\left. + \frac{3.5}{4!}\left(\frac{\boldsymbol{r}\cdot\boldsymbol{\delta}(t)}{r^2+a^2}\right)^3 + \ldots\right\} . \tag{48}$$

Here, the upper limit M in the sum over μ representing $V_{\mathbf{dc}}$ indicates the order up to which one wants to describe the d.c. potential. The laser-induced distortion of the binding potential is contained in the d.c. potential. On the other hand, the a.c. terms provide the coupling among the dressed states thereby inducing transitions. They also contribute to the energy shifts of the dressed states of $V_{\mathbf{dc}}$ through second (or higher) order dynamical effects.

The eigenstates and eigenvalues of the Hamiltonian corresponding to an electron in $V_{\mathbf{dc}}$ of Eq. (47) have been calculated by Landgraf *et al.*[59]. In their calculations[59] these authors have approximated the d.c. potential by the first term of Eq. (47), namely, $V_{\mathbf{dc}} = -e^2(r^2+a^2)^{-\frac{1}{2}}$, and solved the corresponding radial Schrödinger equation numerically using the Numerov method. The influence of the higher-order terms of the expression of $V_{\mathbf{dc}}$ on the zeroth-order energy spectrum so obtained have been shown by Lima and Miranda[58] to be small, and can be taken into account, if needed, via perturbation theory. These higher-order terms, with the additional energy-shift contribution from the $V_{\mathbf{ac}}$ terms, give the total correction to the zeroth-order spectrum. The dependence of the energy spectrum on the laser field strength as obtained by Landgraf *et al.* is shown in Fig. 3. Here the parameter λ is defined as the amplitude of the electron oscillation in the radiation field in units of the Bohr radius, namely, $\lambda = a/a_0$; it measures the laser field strength and is related to the laser frequency ω and intensity I (in W/cm^2) by $\lambda = 6.5\times10^{24}\,\omega^{-2}\,I^{\frac{1}{2}}$. The results of Fig. 3 tell us that the effect of the laser field is to weaken the Coulomb binding to the nucleus as had previously been noticed by several authors[54,57,5].

To conclude a few words regarding the criteria of validity are in order. The STM is valid provided two conditions are met. The first one is that the dipole approximation obtains. As discussed in refs. 5, 58-60, 70, this condition sets an upper limit intensity $I_{\mathbf{u}}$ for the method to be valid. The second and actually the most relevant criterion has in fact been overlooked in one of our previous papers[59]. It refers to the condition which must be met for the time-average potential (i.e., $V_{\mathbf{dc}}$ in Eq. (47) and in Eq. (45)) to be valid. If the period of oscillation of the laser field is much larger than the characteristic time τ of the electron in the bound state (i.e., $\tau = \hbar/E_{\mathbf{b}}$, where $E_{\mathbf{b}}$ is the field-free bound state energy), then the electron

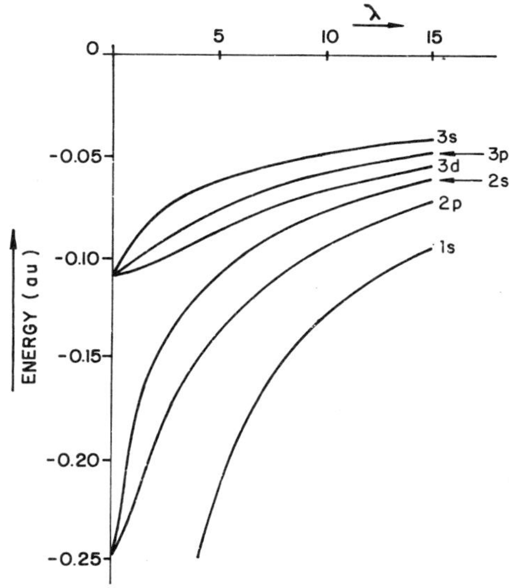

Fig. 3 Dependence of the six lowest states' energies (in *Ry*) of a hydrogen atom on the laser field strength (λ). The parameter λ is related to the laser intensity I (in W/cm^2) by $\lambda = 6.5 \times 10^{24}\, \omega^{-2}\, I^{\frac{1}{2}}$.

can reach a steady state in a time $\Delta t \gg \tau$ but still short enough compared to ω^{-1} so that $V(\boldsymbol{r}-\boldsymbol{\delta}(t)) = V(\boldsymbol{r}-\boldsymbol{\delta}(t+t))$. In other words, for $\omega\tau \ll 1$, the electron is too fast to see the potential distortion induced by the laser field. Accordingly, in this low-frequency (or adiabatic limit one may ignore the laser dressing of the binding potential, so that the solutions to Eq.(44) have the same eigenvalues as the laser-free system; i.e., the electron-radiation field interaction can be taken as a perturbation to the field-free system. Physically, this regime of a slowly moving binding potential is characterized by the electron motion being dominated by the undistorted potential. In contrast, in the high-frequency limit, where $\omega\tau \gg 1$, the electron motion is dominated by the oscillation of the binding potential in the laser field and, consequently, sees a laser-dressed potential (i.e., $V_{\mathbf{dc}}$ of Eq. (47) or Eq. (45)). In this case the energy eigenstates become parametrically dependent upon the laser intensity. We note that this case of high-frequency i.e., $\omega\tau \gg 1$, is thus applicable to lasers having ionizing frequencies ($\hbar\omega \gg E_b$) and not to infrared lasers as implied in ref. 59. Furthermore, the adiabaticity condition, as given by $\omega\tau \ll 1$, does not involve the laser field strength as in the case of the Keldysh model. In other words, the low- and high-frequency regimes are determined by the laser period and the period of oscillation of the electron in the field-free bound states.

To appreciate fully the above reasoning regarding the frequency criterion let us consider the one-dimensional harmonic oscillator in a laser field. This problem is a good example for discussion purposes since it is an exactly soluble model. The Schrödinger equation for this problem reads

$$i\hbar\dot{\psi}(x,t) = \left\{\frac{1}{2m}\left(\hat{p} + \frac{e}{c}\,A(t)\right)^2 + V(x)\right\}\psi(x,t) \tag{49}$$

where $a(t) = A_0\cos\omega t$, and $V(x) = \frac{1}{2}m\omega_0^2 x^2$. Following the STM procedure, we perform the transformation:

$$\psi(x,t) = e^{i\delta(t)\hat{p}/\hbar}\, e^{i\eta(t)/\hbar}\,\phi(x,t) \tag{50}$$

where

$$\begin{aligned}\delta(t) &= -a\sin\omega t \quad ; \quad a = eA_0/mc\omega \\ \eta(t) &= -\frac{e^2A_0^2}{4m\omega^2}\left(t + \frac{\sin 2\omega t}{2\omega}\right).\end{aligned} \tag{51}$$

One gets

$$i\hbar\dot{\phi}(x,t) = \left\{\frac{\hat{p}^2}{2m} + V(x-\delta(t))\right\}\phi(x,t)\,. \tag{52}$$

The STM consists in replacing $V(x-\delta(t))$ by the time-average potential $V_{\mathbf{dc}}$

$$\begin{aligned}V_{\mathbf{dc}} &= \langle V(x-\delta(t))\rangle \\ &= \tfrac{1}{2}m\omega_0^2x^2 + \tfrac{1}{4}m\omega_0^2a^2\end{aligned} \tag{53}$$

so that the solution to Eq. (52) is readily written in terms of the field-free harmonic oscillator wave function $\chi_{\mathbf{n}}$ as

$$\phi_{\mathbf{n}}(x,t) = \chi_{\mathbf{n}}(x)\,e^{-iE_{\mathbf{n}}t/\hbar}\,e^{-im\omega_0^2a^2t/4\hbar}\,. \tag{54}$$

One then gets for the STM solution:

$$\psi_{\mathbf{n}}(x,t) = e^{i\delta(t)\cdot\hat{p}/\hbar}\,\chi_{\mathbf{n}}(x)\,e^{-iE_{\mathbf{n}}t/\hbar}\,e^{-im\omega_0^2a^2t/4\hbar}\,e^{-i\eta(t)/\hbar}\,. \tag{55}$$

Now, the exact solution to Eq. (49) can be written as [72-74]:

$$\psi_{\mathbf{n}}(x,t) = e^{i\beta(t)x/\hbar}\,e^{i\alpha(t)\hat{p}/\hbar}\,e^{i\gamma(t)/\hbar}\,\chi_{\mathbf{n}}(x)\,e^{-iE_{\mathbf{n}}t/\hbar} \tag{56}$$

where

$$\begin{aligned}\alpha(t) &= \frac{eA_0}{mc\omega}\,\frac{\omega^2}{\omega_0^2-\omega^2}\,\sin\omega t \\ \beta(t) &= \frac{eA_0}{c}\,\frac{\omega_0^2}{\omega_0^2-\omega^2}\,\cos\omega t \\ \dot{\gamma}(t) &= \tfrac{1}{2}m\omega_0^2\alpha^2(t) - \frac{1}{2m}\left(\beta + \frac{eA(t)}{c}\right)^2.\end{aligned} \tag{57}$$

In the high-frequency limit, namely, for $\omega \gg \omega_0$ it follows from Eq. (57) that

$$\alpha(t) \to \delta(t) \;;\quad \beta(t) \sim 0(\omega_0^2/\omega^2) \;;\quad \gamma(t) \to -\tfrac{1}{4} m\omega_0^2 a^2 t + \eta(t) \;. \tag{58}$$

Thus, for $\omega \gg \omega_0$, upon substituting Eq.(58) into Eq. (56) it reduces to the STM solution given by Eq. (55). In contrast, in the low-frequency limit, where $\omega \ll \omega_0$, one has from Eq. (57):

$$\begin{aligned} a(t) &\sim 0(\omega^2/\omega_0^2) \quad ; \quad \beta(t) \to -\frac{eA(t)}{c} \\ \gamma(t) &\to \tfrac{1}{4}\,\frac{e^2E_0^2}{m\omega_0^2}\left(t - \frac{\sin 2\omega t}{2\omega}\right) \;;\quad E_0 = \frac{\omega A_0}{c} \;. \end{aligned} \tag{59}$$

I.e., in the low-frequency limit Eq. (56) reduces to

$$\psi_n(x,t) = e^{-ieA(t)x/\hbar c}\,\chi_n(x)\,e^{-i(E_n+\Delta E_n)t/\hbar} \tag{60}$$

where $\Delta E_n = e^2E^2/4m\omega_0^2$ is the second order d.c. Stark shift. Equation (60) tells us that in the low-frequency limit the wave function is essentially described by the momentum-translation approximation with the energy eigenvalues being shifted by the second-order d.c. Stark shift. This is physically consistent with the fact that if $A(t)$ is slowly varying with time the main effect of the radiation field is to shift the particle momentum by $eA(t)/c$, as originally suggested by Reiss[3] for justifying his method. In contrast, in the high-frequency limit the solution corresponds to that of a harmonic oscillator oscillating around the $\delta(t)$-shifted centre of force as described by the STM.

References

1. Keldysh, L.V., Zh. Eksp. Teor. Fiz., **47**, 1945 (1964); [Sov. Phys. - JETP, **20**, 1307 (1965)].
2. Henneberger, W.C., Phys. Rev. Lett., **21**, 838 (1968).
3. Reiss, H.R., Phys. Rev. **A1**, 803 (1970).
4. Manakov, N.L. and Rapoport, L.P., Zh. Eksp. Teor, Fiz., **69**, 842 (1975). [Sov. Phys. - JETP, **42**, 430 (1976)].
5. Lima, C.A.S. and Miranda, L.C.M., Phys. Rev., **A23**, 3335 (1981).
6. Faisal, F.M.H., J. Phys., **B6**, L89 (1973).
7. Hutchinson, M.H.R., Appl. Opt., **19**, A237 (1980); Lax, B. and Cohn, D.R. Appl. Phys. Lett., **23**, 363 (1973).
8. Bayfield, J.E., Phys. Reports, **51**, 317 (1979).
9. Power, E.A., In *Multiphoton Processes*, (eds. P. Lambropoulos and J. Eberly) (Wiley, 1978) p.11.
10. Stenholm, S., Contemp. Phys., **20**, 37 (1979); Phys. Reports, **6**, 1 (1973).
11. Lamb, F.K. and Haar, D.ter, Phys. Reports, **2**, 253 (1971).
12. Delone, N.B., Zon, B.A. , Krainov, V.P. and Khodovoi, V.A., USP. Fiz. Nauk, **120** (1976)³; [Sov. Phys. - USP, **19**, 711 (1976)].
13. Zeldovich, Ya.B., USP, Fiz. Nauk, **110**, 139 (1973); [Sov. Phys. - USP, **16**, 427 (1974)].
14. Samson, J.A.R., Phys. Reports, **28**, 303 (1976).

15. Letokov, V.S. and Makarov, A.A., USP Fiz. Nauk, **134**, 45 (1981); [Sov. Phys. - USP, **24**, 366 (1981)].
16. Lambropoulos, P., *Advances in Atomic and Molecular Physics*, **12**, 87 (Academic Press, New York).
17. Bunkin, F.V. and Prokhorov, A.M., Zh. Eksp. Teor. Fiz., **46**, 1090 (1964). [Sov. Phys. - JETP, **19**, 739 (1964)].
18. Perelomov, A.M., Popov, V.S. and Terentev, M.V., Zh. Eksp. Teor. Fiz., **50**, 1393 (1969); [Sov. Phys. - JETP, **23**, 924 (1966)].
19. Nikishov, A.I. and Ritus, V.I., Zh. Eksp. Teor. Fiz., **50**, 255 (1966). [Sov. Phys. - JETP, **23**, 168 (1966)].
20. Perelomov, A.M. and Popov. V.S., Zh. Eksp. Teor. Fiz. **52**, 514 (1967). [Sov. Phys. - JETP. **25**. 336 (1967)].
21. Reiss, H.R., Bull. Am. Phys. Soc., **12**, 109 (1967).
22. Reiss, H.R., Bull. Am. Phys. Soc., **13**, 80 (1968).
23. Reiss, H.R., Phys. Rev. Lett., **25**, 1149 (1970).
24. Reiss, H.R., Phys. Rev., **D4**, 3533 (1971).
25. Reiss, H.R., Phys. Rev. Lett., **29**, 1129 (1972).
26. Reiss, H.R., Phys. Rev., **D11**, 388 (1975).
27. Reiss, H.R., Phys. Rev., **A19**, 1140 (1979).
28. Reiss, H.R., Phys. Rev., **A22**, 770 (1980).
29. Reiss, H.R., Phys. Rev., **A23**, 3019 (1981).
30. Rogers, J.E., Witt, De. R.E. and Reiss, H.R., Phys. Rev., **A7**, 21 (1973).
31. Cohen-Tannoudji, C., Dupont-Roc, J., Fabré, C. and Grynberg, G. Phys. Rev., **A8**, 2747 (1973).
32. Decoster, A., Phys. Rev., **A9**, 1446 (1974).
33. Rachman, A. and Laplanche, G., J. Phys., **B8**, 826 (1975).
34. Tewari, S. and Tewari, S., Phys. Lett., **50A**, 369 (1974).
35. Zaretskii, D.F. and Krainov, V.P., Zh. Eksp. Teor. Fiz., **66**, 517 (1974). [Sov. Phys. - JETP, **39**, 257 (1974)].
36. Babiker, M., Proc. Roy. Soc., London, **A342**, 113 (1975).
37. Babiker, M., Phys. Rev., **A12**, 1911 (1975).
38. Campos, D. and Krüger, H., Z. Phys., **A274**, 23 (1975).
39. Faisal, F.H.M., J. Phys., **B5**, L258, (1972).
40. Jones, H.D. (unpublished; see also ref. 29)
41. Tareja, R.K. and Haque, S.N., Phys. Lett., **48A**, 231 (1974).
42. DeWitt, R.N., J. Phys., **B6**, 803 (1973).
43. Mohan, M. and Tareja, R.K., J. Phys., **B6**, 809 (1973).
44. Parzynski, R., Phys. Lett., **43A**, 171 (1973).
45. Vitushkin, L.F., Opt. Spectrosc., **40**, 619 (1976).
46. Jdouen, M., Laplanche, G. and Rachman, A., Phys. Lett., **43A**, 101 (1973).
47. DeWitt, R.N., Ph.D. Dissertation, The American University, Washington, (1973) (unpublished).
48. Manakov, N.L., Ovsyannikov, V.D. and Rapoport, L.P., Zh. Eksp. Teor. Fiz., **70**, 1697 (1076) [Sov. Phys. - JETP, **43**, 885 (1976)]
49. Berson, I.J., J. Phys., **B8**, 3078 (1975).
50. Demkov, Yu.N. and Drukarev, G.F., Zh. Eksp. Teor. Fiz., **47**, 918 (1964); [Sov. Phys. - JETP, **20**, 614 (1965)].
51. Drukarev, G.F. and Monozon, B.S., Zh. Eksp. Teor. Fiz., **61**, 956 (1971). [Sov. Phys. - JETP, **34**, 509 (1972)].
52. Lompre, L.A., Mainfray, G., Manus, C., Repoux, S. and Tehbout, J. Phys. Rev. Lett., **36**, 949 (1976); Phys. Rev., **A15**, 1604 (1977).
53. Kramers, H.A., In *Les Particules Elementaires*, Proc. of the Eighth Solvay Conference, (ed. R. Stoops) (Wiley, New York, 1950).
54. Choi, C.K., Henneberger, W.C. and Sanders, F.C., Phys. Rev., **A9**, 1895 (1974).
55. O'Connel, R.F., Phys. Rev., **A12**, 1132 (1975).
56. Ford, W. and O'Connel, R.F., Phys. Rev., **A13**, 1281 (1976).
57. Gersten J.I. and Mittleman, M.H., J. Phys., **B9**, 2561 (1976).

58. Lima, C.A.S. and Miranda, L.C.M., Phys. Lett., **86A**, 367 (1981).
59. Landgraff, T.C., Leite, J.R., Almeida, N.S., Miranda, L.C.M. and Lima, C.A.S.. Phys. Lett., **92A**, 131 (1982).
60. Almeida, N.S. and Miranda, L.C.M., Opt. Commun., **42**, 241 (1982).
61. Brandi, H.S., Koiller, B., Lins de Barros, H.G.P., Miranda, L.C.M. and Castro, J.J., Phys. Rev., **A17**, 1900 (1978); Phys. Rev., **A25**, 2429 (1982).
62. Brandi, H.S., Koiller. B., Lins de Barros, H.G.P. and Miranda, L.C.M., J. Chem. Phys., **69**, 1069 (1978).
63. Henneberger, W.H., Phys. Rev., **A16**, 1379 (1977).
64. Choi, C.K., Mian, S.N., Henneberger, W.C. and Shatas, R.A., Phys. Rev., **A12**, 2635 (1975).
65. Mohan, M. and Chand, P., Phys. Lett., **63A**, 245 (1979).
66. Mohan, M., Phys. Lett., **50A**, 283 (1974).
67. Faisal, F.H.M., J. Phys., **B6**, 1313 (1973).
68. Pundir, R.S. and Mathur, K.C., J. Phys., **B15**, L471 (1982).
69. Ferrante, G. and Leone, C., Phys. Rev., **A26**, 3101 (1983).
70. Ehlotzky, F., Opt. Commun., **40**, 135 (1981).
71. Lambropoulos, P., Power E.A. and Thirnnamachandan, T., Phys. Rev., **A14**, 1910 (1976).
72. Haar, D.ter, In *Problems in Quantum Mechanics*, (Academic Press, New York, 1975).
73. Galvão, R.M.O. and Miranda, L.C.M., Am. J. Phys. (in press).
74. Seely, J.F., Am. J. Phys., **42**, 326 (1974).

Densely Packed Magnetic Insulator Glasses

M. E. Lines
Bell Laboratories, Murray Hill, New Jersey 07974

1. Introduction

Solids in which the atomic arrangement is not spatially periodic are said to be amorphous. Many are prepared from the liquid phase by rapid cooling (or quenching) while some require preparation by other techniques, such as vapour deposition. The term glass (or vitreous material) is sometimes reserved for the former, although a distinction between amorphous and vitreous materials is not universally accepted and the terms are often used interchangeably.

In understanding the structure of crystalline materials, solid state physics leans heavily on the mathematical simplifications which result from lattice periodicity. A periodic lattice (even one with a complicated unit-cell) is an extremely special organization of atoms, and all such structures together occupy only an infinitesimal volume in the coordinate space available to solids in general. This, combined with the fact that the absence of periodicity invalidates most of the mathematical tools available for the detailed interpretation of experimental data from conventional structural probes, emphasizes the difficulties encountered in trying to understand atomic arrangements in glasses.

Amorphous materials are expected to embrace a wide variety of structures depending on the nature of the atoms concerned and on the method of preparation. Some glasses form very readily, even on comparatively slow cooling, while others require rapid quenching. Still others cannot be quenched from a melt at all and must be prepared by sputtering, or vapour, electro-, and chemical deposition techniques. Chemically they range from strongly covalent glasses, like vitreous silica and related tetrahedral systems of silicon and germanium, through the more complex chalcogenide glasses, to amorphous metals, alloys, and ionic materials.

In spite of this variety, it is believed that all such structures can be understood within the same basic conceptual framework; that of an irregular topology of linked atoms or molecules. This implies that models involving micro-crystallinity can now be excluded as a description of truly amorphous structure. The basic differences between the different types of amorphous structure involve the range over which a distorted, but locally recognizable, 'structure' exists.[1] In covalent glasses, with strongly oriented bonds and low coordination, nearest-neighbour coordinations are dominantly determined by the chemistry of molecular and inter-molecular bonding (which is similar in both the crystalline and amorphous phases) and a recognizable structure is undoubtedly maintained over at least a few molecular distances. In amorphous metals and alloys the situation is less clear. Directional bonds are certainly weaker here, and concepts involving the dense random packing of spherical atoms (which neglect directional bonding completely) have been used with limited success in this area.[2] Nevertheless evidence is now mounting that some locally recognizable (if badly distorted) structure, generally related to crystalline structure with the same or similar chemical composition, is present in most amorphous metals and alloys as well, at least at the level of nearest neighbour bonds.[3-5] It is in the close-packed amorphous ionic materials, where, perhaps, we should anticipate that the concept of random packing might be most physically realistic. There is already some evidence that this may be true, at least in some circumstances, and it is the recent progress in understanding the amorphous ionic insulators which will be the concern of the present chapter. This understanding has been aided very considerably by the use of magnetism as a probe of the local amorphous structure. This is a somewhat unconventional approach but one which is potentially powerful in the context of the amorphous ionic magnetic materials.

X-ray, electron, and neutron diffraction techniques have been the traditional basic tools for understanding atomic arrangements in both crystalline and non-crystalline materials. In the latter context they measure only a radial distribution (or atomic pair-correlation) function which records the probability of finding an atom at a given distance from another.[6,7] Since this distribution function is both a statistical average and a 'one-dimensional' projection of a three-dimensional structure, its information is limited. Much local structural detail is inevitably smeared out or lost, and quite often a variety of quite different structural models may be consistent with the observed pair-correlation function. In addition, raw data must be Fourier transformed to yield the spatial correlations, and this often leads to series truncation errors which can easily produce spurious details in the range of intermediate order.

Another valuable probe of amorphous structure is Mössbauer spectrometry. Unlike diffraction techniques the Mössbauer resonance directly probes the real-space local environment of the Mössbauer atom via the interaction of the nuclear quadrupole of the Mössbauer level with the crystal field gradient at the relevant nuclear site.[8] In principle this can provide information concerning directional bonding, and nicely complements the information concerning packing which results from diffraction methods. The particular value of amorphous magnetism in the Mössbauer context is that the onset of any magnetic ordering (or spin freezing) modifies the Mössbauer spectra (via Zeeman splitting) in a manner which provides additional local structural information, sometimes of a very detailed kind.

In this article we shall introduce the reader to a new class of amorphous materials, the densely packed magnetic insulator glasses. We shall discuss their magnetic properties and, in particular, report on recent structural investigations, using Mössbauer spectrometry involving ^{57}Fe, for a few of them. Although diffraction data, leading to radial distribution functions, have not yet been reported for these materials, it has been possible, through the combined use of Möddbauer spectrometry and magnetism, to establish that no vestige of distorted crystalline environment is retained, at least for some ionic amorphous magnets containing the formally spherical ($3d^5$; half-filled shell) ferric ion. Work on ferrous materials has as yet barely begun. The concept of random packing therefore seems to be a promising approach for these materials, although the packing is at best only random from an orientational point of view (implying an essential absence of directional bonding). We do find that Coulomb forces play an important role in the radial distribution, preventing highly ionized small cations (such as the ferric ions themselves) from coming into close proximity.

Amorphous ionic magnetic materials are a sub-class of amorphous ionic materials, and are the most recent new class of non-crystalline materials to receive attention in the literature. The term 'ionic' is a qualitative one, of course, and in the present context we shall use it to include those non-metals in which Madelung energies are thought to provide the dominant part of the binding energy, and in which bond-hybridization is consequently small. We are therefore aiming to look as those amorphous insulators for which directional bonding characteristics should be smallest. In practice these are primarily fluorides, chlorides, and bromides, but include some oxides. In particular, they contain none of the elements which are normally thought of as efficient glass-formers (e.g., Si, B, P, C), since the latter are all disposed to form strongly directional hybridized bonds. As a consequence they often have to be prepared as films by vapour deposition, although a few have been obtained by rapid quenching from the melt. To date the magnetic members of this new amorphous class have primarily been those involving the Fe^{3+} ferric ion. Some examples (where we symbolize the amorphous phase by a prefix a-) are a-FeF_3, a-$KFeF_4$, a-$NaFeF_4$, a-$RbFeF_4$, a-K_2FeF_5, a-K_3FeF_6, a-$Y_3Fe_5O_{12}$, a-$Gd_3Fe_5O_{12}$ and a-Fe_2O_3. Among the amorphous ferrous ionic insulators which have been prepared (but for which far less structural investigation has yet been carried out) are a-FeF_2, a-$FeCl_2$ and a-$FeBr_2$.

At low temperatures most of these ionic magnetic insulator glasses undergo a spin freezing to a magnetic state in which each individual spin is essentially fixed in direction. In most cases this magnetically 'ordered' phase is of the 'spin-glass' type, with magnetic moments frozen in a fixed, but random, orientational pattern. In the present context, for which the magnetic ions are densely concentrated, the term 'speromagnetism' or 'mictomagnetism' is also used.[9] There is a claim in the literature [10] that a-FeF_2 is an amorphous ferromagnet but, if confirmed, it would seem as yet to be unique in this category of amorphous ionic ferromagnets.

The present article will be broadly set out as follows. In section 2 we shall discuss the materials themselves and their bulk magnetic behaviour. Section 3 introduces the techniques for computer modelling using a random packing of spheres and highlighting the role played by the repulsive Coulomb forces. Section 4 discusses the use of Mössbauer quadrupole spectroscopy in ^{57}Fe as a measure of local crystal field environment in the amorphous context, and probes the structural interpretation in terms of the predictions from the computer model. Section 5 and 6 discuss the splitting of the paramegnetic quadrupole spectrum by the hyperfine magnetic field, which is induced by spin freezing, and sets out the details of the manner in which line positions, shapes, and widths, can all contribute valuable information concerning the structure of these glassy materials.

2. Materials and Bulk Magnetism

Of the amorphous close-packed magnetic insulators yet prepared, only amorphous yttrium iron garnet[11] ($Y_3Fe_5O_{12}$; a-YIG) and amorphous gadolinium iron garnet[12] ($Gd_3Fe_5O_{12}$; a-GIG) have been obtained by quenching from the melt, and they can therefore be termed glasses in the traditional sense. Crystalline yttrium (and rare earth) iron garnets are all ferrimagnetic, with high Curie temperatures ($T_c \approx 560$ K) which are determined essentially by a strong Fe-O-Fe antiferromagnetic superexchange and are closely equal for all. This invariance of T_c is dictated by the approximate invariance of the Fe-O-Fe bond angle in the garnet crystal structure, and is a feature which will not survive in the amorphous phase if random packing concepts are realistic. The crystalline iron garnets therefore all possess a spontaneous bulk magnetism at room temperature.

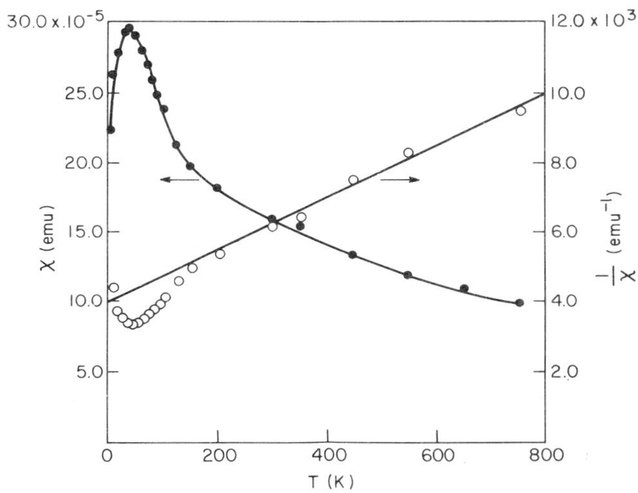

Fig. 1 Magnetic susceptibility χ (solid circles) and its inverse (open circles) as a function of temperature for amorphous yttrium iron garnet ($Y_3Fe_5O_{12}$) showing a broad magnetic spin-freezing anomaly near 40 K. From ref. 11.

In contrast, the corresponding amorphous materials are paramagnetic at room temperature and, indeed, retain this magnetic disorder to much lower temperatures. Typically they undergo a spin freezing, at temperatues well below the nitrogen range, to a spin-glass state which possesses no spontaneous bulk magnetic moment. The amorphous magnetic susceptibility χ (and its reciprocal) for a-YIG are shown in Fig. 1. The susceptibility exhibits a Curie-Weiss temperature dependence at $T \geqslant 200$ K with a negative Weiss-constant (i.e. extrapolated intercept of the reciprocal susceptibility on the temperature axis) of $\Theta \approx -500$ K. This indicates the presence of strongly antiferromagnetic interactions in the glass in spite of a severely depressed magnetic ordering temperature. From Fig. 1 the susceptibility is seen to pass through a broad peak at about 40 K. This is indicative of the spin freezing although accurate measurements in very small magnetic fields have not yet been reported for the amorphous garnets. These would be required in order to examine the precise nature of the susceptibility anomaly at the peak; spin-glass transitions typically show a sharp cusp anomaly [13] in the zero field limit which is rapidly broadened by increasing applied fields. Also from Fig. 1 we see that below 200 K the susceptibility deviates from Curie-Weiss form in an opposite direction from that observed for simple crystalline antiferromagnets. This sign of Curie-Weiss deviation seems to be common for amorphous magnetic insulator susceptibilities and is qualitatively that expected if some magnetically isolated (Curie-law) ferric ions or small isolated (superparamagnetic) clusters are contained in the amorphous matrix [14], although it has yet to be established that such is, in fact, the actual physical origin of the effect.

Another method of preparation of amorphous densely-packed magnetic insulators which seems to produce a good stoichiometric sample is the depositing of films by evaporation of the respective crystalline materials. Film thicknesses of several mils are readily obtained by the procedure and the method has been particularly useful in the preparation of amorphous fluorides (which by present techniques will not quench from the melt). Materials prepared in this manner include [15-16] a-FeF_3, a-$KFeF_4$, and a-$NaFeF_4$, and all are stable as amorphous films to temperatures well above room temperature.

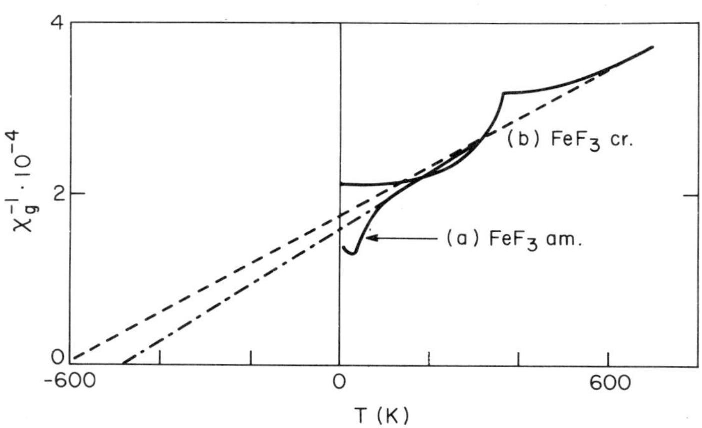

Fig. 2 Reciprocal magnetic susceptibility as a function of temperature for the crystalline and amorphous forms of FeF_3, showing their respective extrapolations to negative Curie-Weiss intercepts. From Ref. 15.

Crystalline ferric fluoride (FeF_3) is a weak ferromagnet (i.e. canted antiferromagnet) at room temperature with a Néel temperature of $T_N = 363$K, and a Curie-Weiss constant of $\Theta = -610$ K. Amorphous FeF_3, on the other hand, exhibits a magnetic susceptibility qualitatively similar to that of a-YIG with a Curie-Weiss form at high temperatures ($\Theta = -486$ K) and a rounded maximum near 30 K. In Fig. 2 we show the reciprocal susceptibility for both the crystalline (c-FeF_3) and (a-FeF_3) forms. Note that the deviations from Curie-Weiss forms are of opposite signs for the two forms on approach to the respective magnetic ordering temperatures.

Crystalline FeF_3 is basically a simple cubic antiferromagnet, with magnetic nearest neighbour coordination number $z = 6$ and magnetic interactions symmetrically distributed in all three dimensions. Crystalline $KFeF_4$ and $NaFeF_4$, though not crystallographically identical, are like virtually all the $AFeF_4$ alkali-metal (A) iron tetrafluorides, being layered materials having a bonding topology which leads to basically two-dimensional antiferromagnetism[17,18] with $z = 4$. The effects of a change in (magnetic) dimensionality on magnetic susceptibility are well-known and are quite profound.[19] In particular, Θ/T_N ratios are typically higher (≈ 2-3) in layered magnets than are corresponding ratios (≈ 1-2) in three dimensional ones.

On passage to the amorphous state, all three of these fluorides maintain a large, though slightly reduced, Weiss constant indicating that strong antiferromagnetic exchange is still present. On the other hand, the magnetic ordering temperatures are drastically reduced (by a factor of 10 or more). The magnetic susceptibilities for a-$NaFeF_4$ and a-$KFeF_4$ are shown in Fig. 3. Although qualitatively similar to that of Fig. 2 for a-FeF_3, the minimum in reciprocal susceptibility (which presumably marks the onset of spin freezing) is progressively shifted to lower temperatures as we progress along the series a-FeF_3, a-$NaFeF_4$, a-$KFeF_4$, and is below 4.2 K (if it exists at all) for the last. It seems likely that iron-anion-iron exchange pathways (or magnetic connectivity) decrease both as the relative concentration of iron decreases (e.g. within a series like a-FeF_3, a-$KFeF_4$, a-K_2FeF_5, a-K_3FeF_6) and as the size of the alkaline cation increases within a particular isochemical series like a-$NaFeF_4$, a-$KFeF_4$, a-$RbFeF_4$, a-$CsFeF_4$, although experimental details are as yet scant. For example, the crystalline materials A_2FeF_5 (where A is an alkali metal cation) are all low T_N linear chain antiferromagnets.[20] It seems likely that when randomized in the amorphous context, they will have an iron concentration below the percolation limit and will not form an ordered (i.e. spin-glass) phase at any temperature. This should be equally true for A_3FeF_6 materials which are so magnetically dilute that they do not order magnetically even in the crystalline phase.

In a-FeF_3 the magnetization is observed to increase from zero as a linear function of applied magnetic field at all temperatures,[15] confirming the absence of spontaneous magnetization in the magnetically ordered phase.

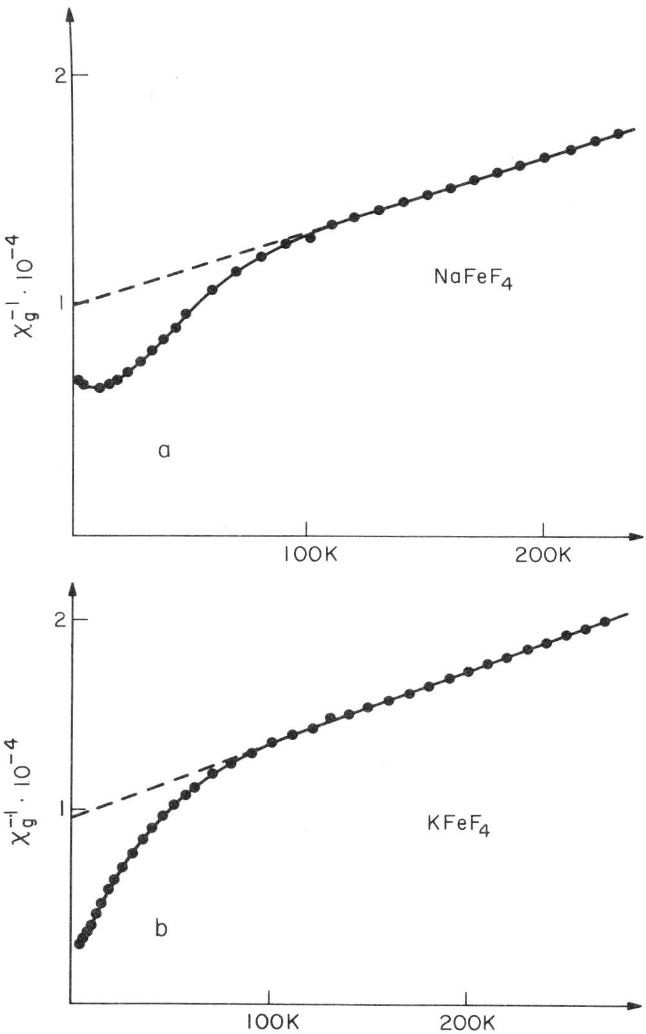

Fig. 3 Reciprocal magnetic susceptibility as a function of temperature for amorphous $NaFeF_4$ and amorphous $KFeF_4$. From Ref. 16.

In a-$NaFeF_4$, a considerable sensitivity of magnetic susceptibility to applied field has been noted near and below the susceptibility maximum.[16] In low fields the maximum sharpens and a tendency towards a cusp-shape is observed. This is very similar to the observations in crystalline insulator spin-glasses.[13] In addition, a saturation of magnetization in fields of order 50 kOe has been achieved in a-$NaFeF_4$ at 4.2 K,[16] a behaviour which is again qualitatively in accord with expectations for a concentrated magnetic spin-glass or 'mictomagnet'.

The clearest examples to date of a classic spin-glass susceptibility cusp in a magnetic insulator glass have been obtained for a-$PbMnFeF_7$ and a-Pb_2MnFeF_9 (Fig. 4), although since Pb^{2+} is not a formally spherical ion, these lead-containing materials may not be such good examples of spherical random packed structures. These measurements[21] were made using a very small low-frequency a/c field. The presence of a d/c field causes a rounding of the cusp in a manner familiar from spin-glass observations in dilute magnetic

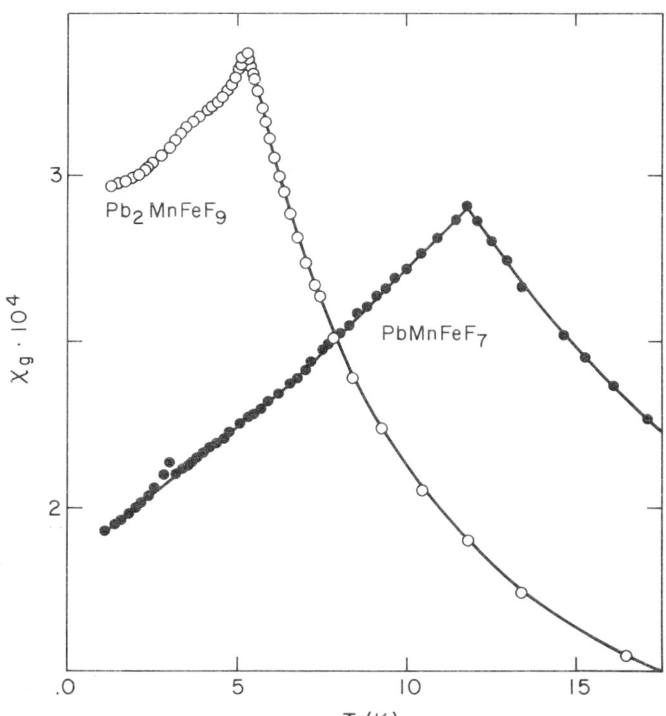

Fig. 4 The temperature dependence of magnetic susceptibility taken in a very small low frequency field for amorphous Pb_2MnFeF_9 and amorphous $PbMnFeF_7$, each showing the sharp cusp anomaly expected for spin glasses at the onset of magnetic order (or freezing). From Ref. 21.

amorphous alloys such as Au(Fe) and Cu(Mn). The fundamental difference, from a physical standpoint, is that the magnetic interactions in amorphous insulators are essentially of nearest neighbour superexchange origin, while those in amorphous alloys are long range itinerant-electron-carried interactions between widely spaced localized spins.

Amorphous ferrous (Fe^{2+}) materials like a-FeF_2, a-$FeCl_2$ and A-$FeBr_2$ have been prepared by deposition from a molecular beam onto a substrate kept at liquid helium temperatures.[22-24] They are, in general, more difficult to prepare than their ferric counterparts, and I am aware of magnetic susceptibility measurements only for the flouride, although all appear to undergo some spin-freezing process at about 21 K as deduced from Mössbauer spectra. The amorphous chloride and bromide are less stable than the fluoride and begin to crystallize on heating well below room temperature. The amorphous fluoride, however, is stable at room temperature and magnetic susceptibility measurements from room down to 21K have been reported for it by Litterst.[10] Very interestingly, they follow a Curie-Weiss law throughout this range with a *ferromagnetic* Weiss constant $\Theta = +(22 \pm 1)$K, (Fig. 5). The diverging form of the susceptibility near 21K suggests the onset of amorphous ferromagnetism. Below this temperature magnetic hysteresis and remanence are observed. The remanence (i.e. spontaneous magnetization as an applied field H is reduced to zero) is of order 1.5 μ_B per spin, or a little less than half the fully saturated value of about $4\mu_B$ which is observed at high fields.

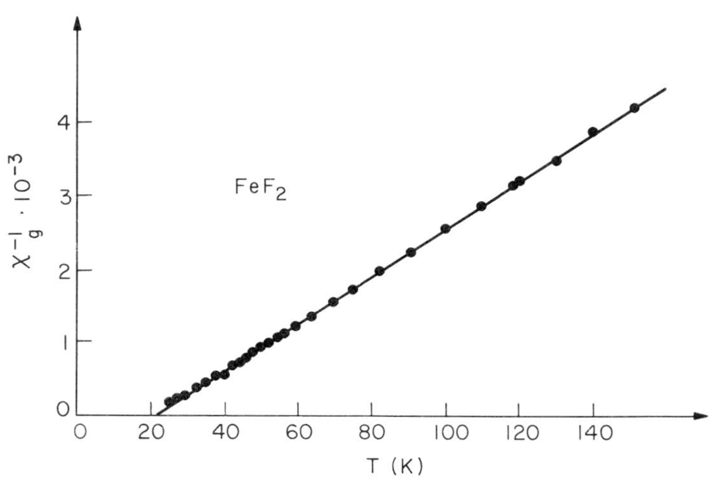

Fig. 5 Reciprocal magnetic susceptibility as a function of temperature for amorphous ferrous fluoride, showing a divergence of magnetic response of a Curie-Weiss form near 21K. From Ref. 10.

3. Computer Models and Dense Random Packing

The idea of modelling disorder by densely packing together a large number of spheres was explored originally in connection with the organization of simple liquids.[25-27] Later on, randomly packed aggregates of equal spheres were also proposed as possible models for the then hypothetical elemental metallic glass.[28] The theory was that such a dense amorphous assembly of atoms which lacked strongly angular-dependent bonds should possess a structure which, in lowest order at least, was dominated simply by the repulsive core of each 'spherical' constituent atom coupled with a random packing which maximized the density. The simplest model of all would use 'hard spheres' (i.e. a step function repulsive core potential) recognizing that the final hard-sphere aggregate could always be computer-relaxed at a later stage (using a more realistic core potential) so as to locally minimize the total energy.

The first hard sphere models were made by physically packing together spheres in containers and shaking or kneading them to maximize the packing density. These packing techniques resulted in non-crystalline assemblies consistent with infinite extension. In due course, simple algorithms were proposed to simulate these physical assemblies of hard spheres by using computers from the outset. Most of these use sequential building methods[29] starting from a central 'seed cluster'. In either case the final hard-sphere positions could be recorded and a computer relaxation procedure performed.[30]

The models were initially used to interpret X-ray, electron, and neutron diffraction experiments on amorphous metals and alloys. These diffraction observations measure primarily a radial distribution function (RDF) which records atomic pair correlation, which is the probability of locating one atom at a particular distance from another. The extension of the model to alloys is accomplished by randomly packing spheres of more than one size,[31] and early comparisons with experiment were encouraging. However, since the RDF's are only one-dimensional projections of a three dimensional structure, the information they carry is limited and many of the structural details are inevitably smeared out or lost.

More recent work[32] using pulsed neutron sources, EXAFS (extended X-ray absorption fine structure), X-ray anomalous scattering techniques, and magnetic probes, has now clearly eliminated the possibility of a dense random mixture of atoms being present in most amorphous alloys. Since amorphous elemental metals are extremely difficult to prepare, and generally recrystallize at low temperatures, this could well exclude the dense random packing (DRP) computer aggregate as a viable model for the majority of stable amorphous metals. Although various modifications of the dense random packing principle can be adopted, such as preventing certain close contacts from arising in amorphous alloys,[33] it now appears that the introduction of directionally oriented bonds may be necessary in most cases.[34] The most recent indications are that the local structure of amorphous alloys retains a distorted form of the closest (compositional) crystalline structure.[3,35,36] It therefore appears that, in the context of amorphous alloys, more complex models which retain a structural local unit[5,37] may be required for a more quantitative understanding, although some work with DRP assemblies still continues using more dramatic relaxation schemes.[4]

Physically, perhaps the closest amorphous manifestation of orientational random packing might be expected to occur in close packed ionic insulators; particularly those containing formally spherical ions (i.e. filled or half-filled shells) and for which binding energy is dominated by the electrostatic Madelung term. To test this hypothesis efforts have been made to computer model certain ferric fluorides and oxides of the kind discussed in section 2. In the rest of this section we shall describe some of these efforts in detail.

The ferric ion Fe^{3+} ($3d^5$) is a small highly charged formally spherical ion with a radius $R(Fe) \approx 0.56 - 0.6$ Å (depending on bonding coordination in crystals) which is less than one half of that of anions like F^-($R(F) \approx 1.36$ Å) and O^{2-} ($R(O) \approx 1.40$ Å). In the context of amorphous insulators, one essential additional feature must be added to the DRP models used for alloys — it is the presence of strongly repulsive Coulomb forces between the triply ionized cations. These prevent the small ferric ions from approaching each other any closer than is absolutely ncessary to maintain concentration. In the computer building algorithm it is therefore essential to include a feature which recognizes the importance of these forces and we refer to such models by the acronym RPHS (random packing of hard spheres). As the simplest example of such a scheme we shall describe below the computer building algorithm for the binary amorphous ionic insulator a-FeF_3.[38,39]

From the structure of crystalline FeF_3(c—FeF_3), and using the accepted ionic radius for F^- of $R(F) = 1.36$ Å, we find that a ferric ion radius of $R(Fe) = 0.56$ Å satisfies the average Fe-F bond length measurement. Using two types of spheres with radii R(F) and R(Fe) we first form a seed cluster of two F spheres and one Fe sphere all in contact in an *xy*-plane. Other spheres are then added sequentially subject to a criterion that no two ferric ions shall touch, or even come closer than an initially arbitrary centre-to-centre distance $d > 2R(Fe)$. Starting with the seed, the computer calculates the possible positions (or pockets) for a new fluorine sphere which contacts all $N = 3$ spheres of the seed (the first pockets must be anion pockets since they involve contact with a cation). Thus, the seed pocket calculation results in M = 2 new sites for an additional anion. Labelling all substrate ions and pockets by their formal valence charges (F = -1, Fe = 3) the following building procedure is now followed for general N and M.

Using a 'global' criterion,[29] we first choose the pocket nearest an arbitrarily chosen 'centre' or origin of the seed cluster, and fill it as long as it does not violate the d-spacing criterion. If it does, we strike out this pocket and proceed to the next global pocket until one is found which does not violate the d-criterion. After filling this pocket, we strike out all pockets which overlap the newly added (N+1)th sphere, and then compute newly created pockets involving this sphere. If a newly created pocket involves contact with a cation, then it is designated an anion pocket. If not, the pocket-type is chosen which tends towards establishing electrical neutrality (i.e. an anion if the sum of the valence charges on the substrate is positive; a cation if it is zero or negative). After calculating all the new pockets we strike out those which overlap the existing substrate and then redefine N and M to count the new substrate and pocket sets and repeat the entire procedure.

In modelling ternary systems[40] such as a-YIG, with two trivalent cation species, an extra building rule is required for the choice of cation type when a cation pocket is called for. The simplest such rule is always to add that type of cation pocket which will bring the cluster composition closer to its stoichiometric crystalline value when added to the substrate.

To this point the distance d has remained unspecified. If chosen too large, the number of pockets M goes to zero at some stage of the building process, and the procedure terminates. This is because no cation pockets can be found, or be created, which are more distant than d from the existing substrate cations. The largest value of d for which the building program will propagate therefore, in some sense, minimizes the local cationic Coulomb energy within the building algorithm, and it is this value which is used in constructing the final random-packed assembly. For FeF_3, this d-value turns out to be close to 2.7 Å. Using fast computer facilities, aggregates of up to 3000 spheres can be built up in about a half-hour of computing time.

Because of the global criterion the basic clusters are spherical. However, since R(Fe) is so much smaller than R(F), occasional iron pockets are created and immediately filled as much as 1.8 Å inside the spherical 'surface' defined by the centres of the outermost anions. This tends to make the surface region a little iron deficient. Plotting the radius of the N-sphere cluster aggregate as a function of 1/N as the cluster grows, one can extrapolate to $1/N \to 0$ to obtain the reciprocal atomic density $(12.8 \pm 0.15)Å^3$ per atom, which compares with the crystalline value of $13.00\,Å^3$. The model therefore suggests that a-FeF_3 should be a little more densely packed than c-FeF_3, with a packing fraction (i.e. fraction of space filled) equal to 0.632 as against 0.622 in the crystal). No experimental information is yet available on this comparison.

From the computer cluster one can immediately calculate the pair correlation functions $g(r)$ for the iron-iron, iron-fluorine, and fluorine-fluorine pairs. The pair correlation function $g_{\alpha\beta}(r)$ measures the average density of ions of type β at a distance r from ions of type α and, since $g_{\alpha\beta}$ is directly proportional to $g_{\beta\alpha}$ when $\alpha \neq \beta$, only the three correlation functions $g_{FF}(r)$, $g_{FeF}(r)$ and $g_{FeFe}(r)$ are independent. They are shown as histograms in Figs. 6-8. Certain features of these correlations for comparatively short range r can be identified with obvious hard-sphere geometric configurations (see the figure captions) and would be expected to be less sharp in a relaxed (or 'soft'-sphere) approximation. Such a relaxation procedure has not yet been performed for the ionic glasses. The amplitude of the Fe-F 'contact-

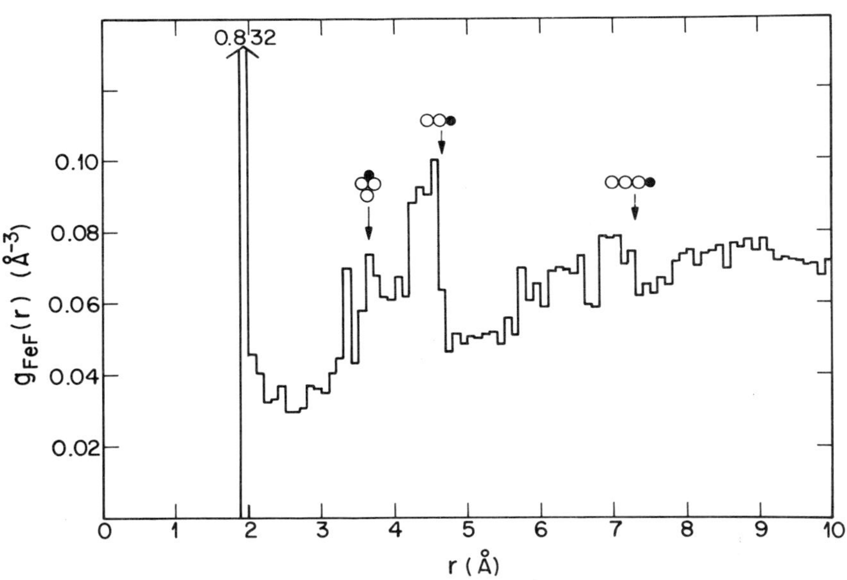

Fig. 6 The pair correlation function $g_{FeF}(r)$ which measures the density of fluorine ions at a distance r from an iron ion in a-FeF_3, as determined directly from the computer sample. Shown in this and the two following figures are r-values corresponding to simple planar geometric configurations of the constituent hard spheres as illustrated, where open circles represent fluorine ions and closed circles iron ions. From Ref. 38.

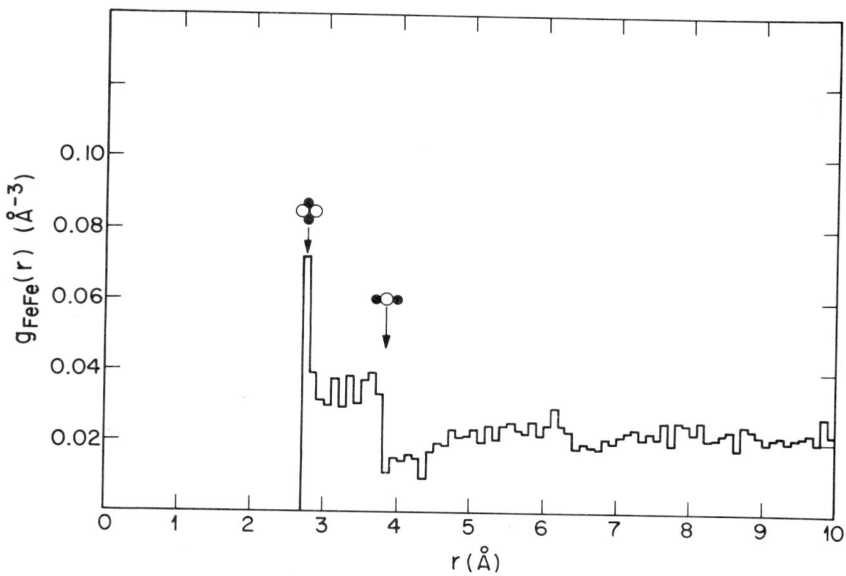

Fig. 7 As Fig. 6 but for the correlations $g_{FeFe}(r)$ of iron-iron density

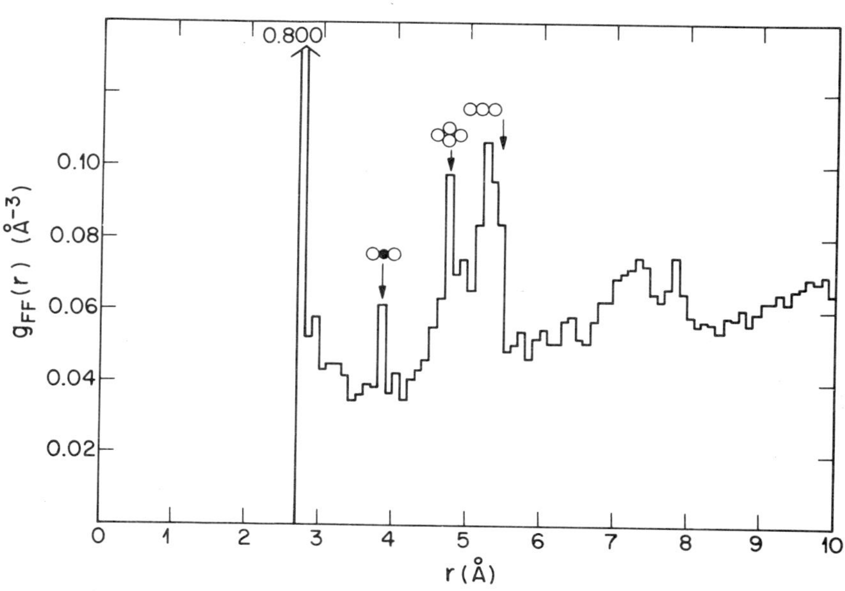

Fig. 8 As Fig. 6 but for the correlations $g_{FF}(r)$ of fluorine-fluorine density

spike' at a radial distance of 2.9 - 3.0 Å in Fig. 6 establishes that approximately 4 anions actually 'touch' the average ferric cation. It is interesting to note that the ratio R(Fe)/R(F) = 0.56/1.36 = 0.412 is extremely close to that value ($\sqrt{2}-1$ = 0.414) which would enable 6 fluorines to be perfectly packed around one ferric ion, although such a dense packing cannot propagate in RPHS FeF_3 in three dimensions. At this writing, no X-ray or neutron measurements of correlation functions $g(r)$ have been reported for any amorphous ionic insulators and the major contact with experiment has come from Mössbauer experiments using the ^{57}Fe Mössbauer nucleus. The rest of this article will set out some rather detailed studies of local amorphous environment which can be conducted in this manner.

4. Electric Field Gradients and Mössbauer Quadrupole Spectra

One useful probe of local structure in amorphous materials containing Mössbauer nuclei is the Mössbauer quadrupole spectrum. In a-FeF_3, for example, a simple two-line spectrum is expected for the ^{57}Fe quadrupole resonance and is observed above the spin-freezing temperature (30 K). In ^{57}Fe the ground nuclear level has a nuclear spin quantum number $I = 1/2$ and the excited Mössbauer level $I = 3/2$. The latter is therefore perturbed by non-zero electric field gradients (EFG) at the iron sites. For an orbitally non-degenerate ($L = 0$) ion like Fe^{3+}, this EFG results dominantly from the charge distribution on neighbouring ions and may be decomposed into contributions from monopoles (i.e. point charges), dipoles, etc.

In a RPHS amorphous model there exists, not surprisingly, a broad distribution of EFG reflecting the different possible environments. What is not so self-evident, however, is why this EFG distribution should not be centered about zero. It is, however, a topological property of a random assembly of one (or more) types of sphere that the probability of finding a site with high local symmetry (that is, EFG → 0) itself approaches zero.[41] A histogram of the EFG distribution can readily be obtained from the computer assembly in a point charge approximation (dipole and higher order contributions are, unlike the situation in crystalline analogues, generally small because of the random orientation of all vector and tensor properties in the amorphous context). The first contact between experiment and the RPHS model can therefore be made via the measured quadrupole spectrum and a point charge EFG computation.

Using the formal valence charges $+3e$ and $-1e$ for Fe^{3+} and F^- respectively (where $-e$ is the electronic charge) we first calculate the EFG components V_{xx}, V_{yy}, V_{zz} along the principal EFG axes at each iron site. A general representation[8] is obtained in terms of only two quadrupole parameters (since, from Laplace's equation, $V_{xx}+V_{yy}+V_{zz}=0$ at each site). They are the component $q = eV_{zz}$ of largest magnitude V_{zz}, and an asymmetry parameter $\eta=(V_{xx}-V_{yy})/V_{zz}$. The parameter q may be positive or negative while η, since x,y,z, are defined such that $|V_{zz}|>|V_{yy}|>|V_{xx}|$ is always positive with $0<\eta<1$. In the context of iron-containing ionic glasses the difference between a positive and a negative q is readily pictured. Since the ferric cations are surrounded by contact and quasicontact anions, the negative q results (to the extent that the EFG is dominated by these anions) when these near-neighbours (nn) tend towards a clustering near the opposite poles of a ferric ion diameter, while a positive q results when they tend towards a clustering around a ferric equatorial plane. These two types of configuration have different characters and there is no *a priori* reason for anticipating identical q-distributions for $q>0$ and $q<0$.

In Figs. 9a and 9b we show the a-FeF_3 model EFG distributions $p_+(q)$ and $p_-(q)$ for positive and negative q respectively.[39] Although the shapes are different, the number of sites with $q>0$ (n_+) is, in fact, very closely equal to the number (n_-) with $q<0$. In detail one finds

$$n_+/n_- = 0.99 \pm 0.03 \; , \quad (\text{a-FeF}_3) \tag{1}$$

showing that the two types of anion configuration do indeed occur with closely equal probabilities in a-FeF_3. Of the two distributions p_+ and p_-, both peak at about the same $|q|$-value although $p_+(q)$ is slightly broader, and these two properties appear to be common to random packed models of other ionic ferric glasses, although the ratio n_+/n_- and the actual value of $|q|$ at the peaks do differ in different materials. In a model of a-YIG,[40] for example, we find much wider q-distributions (due primarily to the larger point charges on the oxygen anions) and a value

$$n_+/n_- = 1.19 \pm 0.05 \; , \quad (\text{a-YIG}). \tag{2}$$

The difference in n_+/n_- between (1) and (2) is caused by the greater concentration of 3+ cations (Y and Fe) in a-YIG which reduces the minimum cation spacing d markedly from its value in a-FeF_3. In a-YIG, cation near-neighbour contributions to q are no longer small and the two most probable configurations, with one or two (well-separated) nn cations at distance d from the probe ferric nucleus, both contribute positive amounts to q which bias the $p(q)$-distribution to positive values.

With all this modelling, it is now appropriate to search for experimental probes which will test the viability of the RPHS model for ionic glasses. As a first step we appeal to the quadrupole splitting itself. The quadrupole

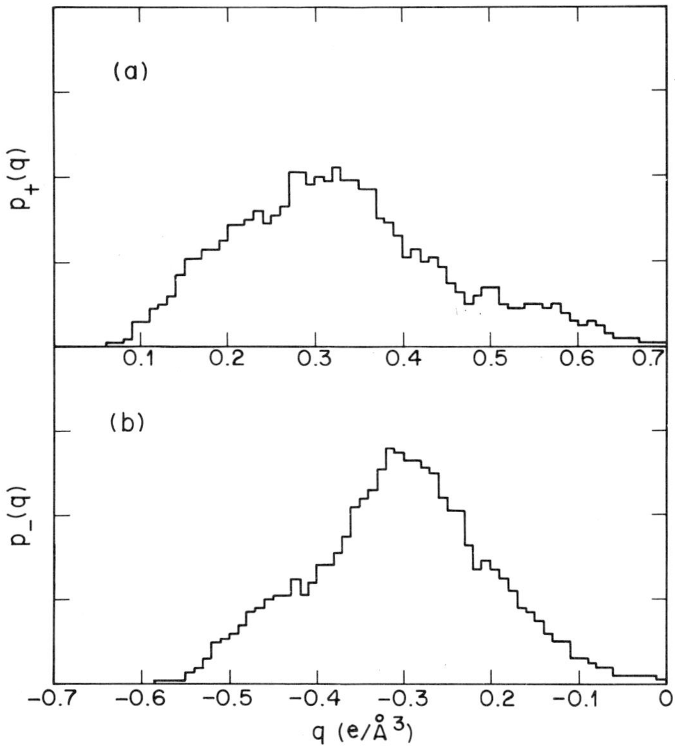

Fig. 9 The distribution $p(q)$ of principal EFG component $q = eV_{zz}$ for a point-charge calculation in a random packed hard sphere (RPHS) model of a-FeF_3. Fig. 9a shows the distribution $p_+(q)$ for positive EFG $q>0$ and Fig. 9b the equivalent distribution $p_-(q)$ for negative EFG $q<0$. From Ref. 39.

Hamiltonian for the Mössbauer $I = 3/2$ level of ^{57}Fe

$$\mathcal{H}_Q = q'\left[3I_z^2 - 15/4 + \eta(I_x^2 - I_y^2)\right], \tag{3}$$

in which $q' = e^2Qq/12$ is an energy measure of q, and Q is the nuclear quadrupole moment. Its eigenvalues are

$$E_Q = \pm\, 3q'(1+\eta^2/3)^{\frac{1}{2}}. \tag{4}$$

Since the factor $(1+\eta^2/3)^{\frac{1}{2}}$ is never larger than 1.16 at any site $(0 < \eta < 1)$ the distribution E_Q is essentially a measure of the distribution $p(|q|) = p_+(q) + p_-(q)$. Although insensitive to the sign of q it will produce a broad single line if the EFG distribution is centred at $q = 0$ and a broad doublet if, as the RPHS simulations suggest, $p(|q|)$ is peaked at a non-zero $|q|$-value.

Experimentally, a broad quadrupole-doublet is observed in all cases. Since each constituent transition making up the complete spectrum has, for ^{57}Fe, a natural Lorentzian lineshape of width (full width at half maximum) $w \approx 0.2$ mm/sec (where 1 mm/sec in Doppler shift velocity units corresponds to an energy of 7.69×10^{-20} ergs), the quadrupole doublet lineshape $f(z)$ can in general be expressed as a distribution $p(E_Q)$ of Lorentzian lines in the form

$$f(z) = h \int_{-\infty}^{\infty} (w/2)^2 \frac{p(E_Q)}{(w/2)^2 + (z-y)^2} dy , \tag{5}$$

where h is an amplitude parameter.

The quadrupolar energy distribution $p(E_Q)$, which since $(1+\eta^2/3)^{\frac{1}{2}}$ is never bigger than 1.16 is essentially proportional to $p(|q|)$, can therefore be experimentally determined by least squares fitting the expression (5) to the measured quadrupole spectrum. For example, for the amorphous iron fluorides, an excellent fit can be obtained using the asymmetric Gaussian[42] distribution

$$\begin{aligned} p_1(E-E_m) &= \exp\left\{-\left[(E-E_m)/aE_m\right]^2\right\} , \quad 0<E<E_m , \\ p_2(E-E_m) &= \exp\left\{-\left[(E-E_m)/bE_m\right]^2\right\} , \quad E_m<E<\infty , \end{aligned} \tag{6}$$

in which E_m marks the peak of the distribution, and $a \approx 0.6$, $b \approx 1.2$. The respective quadrupolar fits for a-FeF_3, a-$KFeF_4$, and a-$RbFeF_4$ are shown in Figs. 10a, b, c and the corresponding distribution function $p(E) = p(|E_Q|)$ in Fig. 11. In the latter figure we compare (6) with another analytic form

$$p(E) = h(E/\sigma)^4 \exp(-E^2/2\sigma^2) , \tag{7}$$

in which h and σ are amplitude and width parameters respectively.

The form (7), unlike (6) which is heuristic, has some physical justification and is analytically derivable[41] from a model consisting of randomly positioned spheres constrained to a spherical shell around the site of interest. We therefore term it the 'shell model' approximation and, since a ferric ion in a RPHS ionic aggregate is (to a good approximation) surrounded by a shell of nearest neighbour anions, we should expect it to describe the computer estimate for $p(|q|) = p_+(q) + p_-(q)$ for iron sites quite well, and such is found to be the case. The 'shell' curve in Fig. 11 therefore represents fairly accurately the distribution of EFG magnitudes predicted by the computer model of section 3. It is seen to be a little narrower than the experimentally observed distribution (6) and a little less asymmetric. The width discrepancy may well be due to a distribution of isomer shifts (centres of the individual Lorentz doublets (4)) which has so far been neglected and

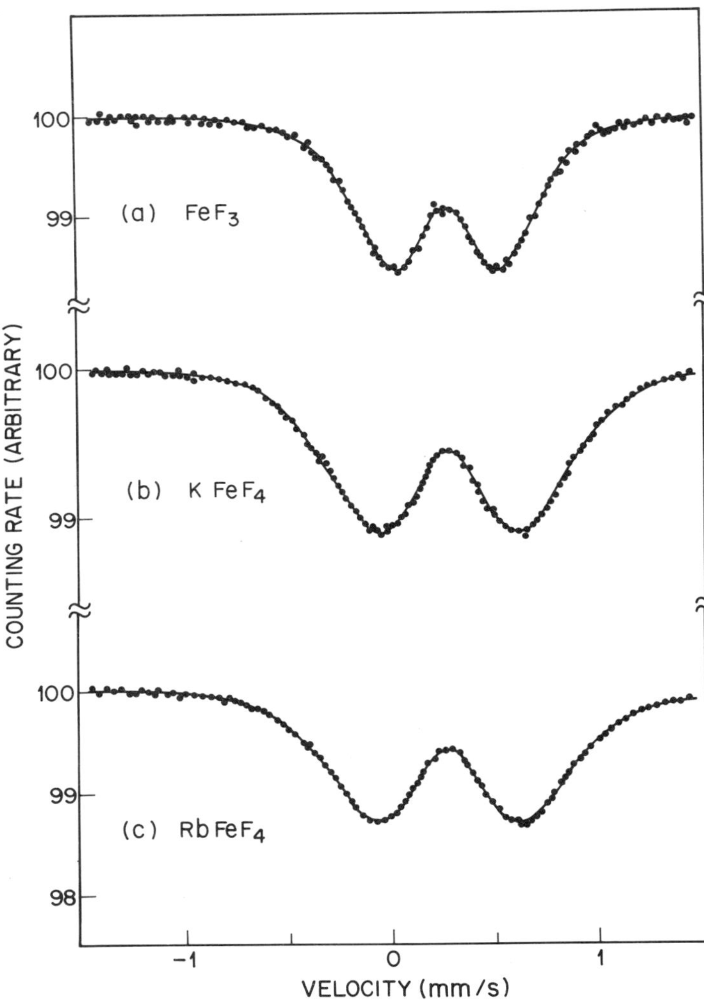

Fig. 10 Room temperature paramagnetic Mössbauer spectra for the amorphous fluorides a-FeF_3, a-$KFeF_4$ and a-$RbFeF_4$. The solid lines represent nonlinear least-squares fits to the data using an asymmetric distribution of EFG as described in the text. From Ref. 42.

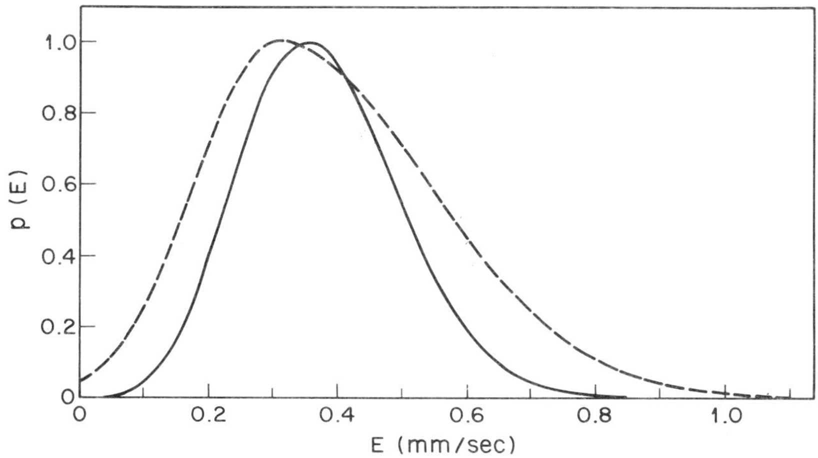

Fig. 11 Comparison of the 'best-fit' shell-distribution function (solid curve) and the 'best-fit' asymmetric Gaussian (dashed curve) for the EFG distribution function in the amorphous fluorides. The energy scale corresponds to a-$KFeF_4$ and a-$RbFeF_4$ and should be reduced by 25% for a-FeF_3. From Ref. 42.

which measures different *s*-electron densities at individual iron nuclear sites. Although details of this isomer distribution in glasses are not known, the spread of such shifts in different ferric oxide crystalline environments is sufficient to account for the width discrepancy. Ferric fluorides all seem to crystallize with iron in the same basically octahedral fluorine coordination so that the sensitivity of isomer shift to (specifically) fluoride environmental changes is not known.

What is most impressive about the Mössbauer paramagnetic doublet findings is that the magnitudes of the peak-to-peak quadrupole splittings in the amorphous phases of the various amorphous ferric insulators bear no relationship whatsoever to their respective crystalline counterparts. Thus, while c-FeF_3 has an immeasurably small quadrupole splitting, those for c-$KFeF_4$ and c-$RbFeF_4$ are about double their amorphous equivalents.[15,42]

The RPHS point charge EFG computation is in quantitative accord with the magnitude of the quadrupole splitting of Fig. 10a if the point charges are 70% of their formal values (a seemingly realistic number). RPHS models for a-$KFeF_4$ and a-$RbFeF_4$ have not yet been devised and the problem of the treatment of large monovalent cations within the building algorithm not yet confronted.

5. Mössbauer Zeeman Spectra: Theory

To this point in the Mössbauer study of amorphous materials, no use has been made of magnetic phenomena. In materials like a-YIG and a-FeF_3 there is, as we have already discussed, a spin freezing transition at low temperatures. Below the onset of spin freezing, the 'frozen' magnetic moment at each iron site creates a spontaneous local magnetic field (called the hyperfine field) at each respective iron nucleus. This magnetic field splits the ground G $(I = 1/2)$ and excited Mössbauer $E\,(I = 3/2)$ nuclear levels of ^{57}Fe and therefore severely modifies the Mössbauer spectrum from its paramagnetic doublet form.

The low temperature (or Zeeman) spectrum consists of six lines, and can be understood by examining the manner in which a combined hyperfine field H and EFG q' split the nuclear levels of G and E. Consider an arbitrary iron nuclear site for which the EFG has *local* principal axes x, y, z, while the local direction of the field H is denoted by Z. If we define polar angles θ and ϕ to measure the direction if H in the xyz coordinate system, then the total nuclear magnetic Hamiltonian for this particular nucleus can be expressed in a coordinate system XYZ, where

$$\begin{bmatrix} X \\ Y \\ Z \end{bmatrix} = \begin{bmatrix} \cos\theta\,\cos\phi & \cos\theta\,\sin\phi & -\sin\theta \\ -\sin\phi & \cos\phi & 0 \\ \sin\theta\,\cos\phi & \sin\theta\,\sin\phi & \cos\theta \end{bmatrix} \begin{bmatrix} x \\ y \\ z \end{bmatrix} \tag{8}$$

for which the magnetic field energy (which normally dominates for saturated spins) is diagonal. In the XYZ system the quadrupole Hamiltonian $\mathcal{H}_Q$ for the Mössbauer $I = 3/2$ level transforms to the matrix form

$$\mathcal{H}_Q = \begin{bmatrix} u & e & f & 0 \\ e^* & -u & 0 & f \\ f^* & 0 & -u & -e \\ 0 & f^* & -e^* & u \end{bmatrix} \tag{9}$$

in which

$$u = (3/2)\,[3c^2 - 1 + \eta s^2 \cos(2\phi)]\,q', \tag{10}$$

$$e = \sqrt{3}\,[-3sc + \eta sc\cos(2\phi) + i\eta s\sin(2\phi)]\,q', \tag{11}$$

$$f = [\sqrt{3}/2][3s^2 + \eta[1 + c^2]\cos(2\phi) + 2i\eta c\sin(2\phi)]\,q', \tag{12}$$

where $c = \cos\theta$, $s = \sin\theta$, and * indicates a complex conjugate.

To second order in perturbation theory, the Hamiltonian

$$\mathcal{H} = -g\mu_N I_Z H + \mathcal{H}_Q \tag{13}$$

for the quadrupole perturbed Zeeman levels of the $I = 3/2$ state has eigenvalues

$$E_1^E = -(3/2)\,h_E + u - (e^* e + f^* f/2)/h_E\,,\quad E_3^E = +(1/2)h_E - u - (e^* e - f^* f/2)/h_E\,,$$
$$E_2^E = -(1/2)\,h_E - u + (e^* e - f^* f/2)/h_E\,,\quad E_4^E = +(3/2)h_E + u + (e^* e + f^* f/2)/h_E\,, \tag{14}$$

while for the ground state G the exact eigenvalues are

$$E_1^G = -(1/2)\,h_G\,,\quad E_2^G = +(1/2)\,h_G\,, \tag{15}$$

where $h_G = g_G \mu_N H$, $h_E = g_E \mu_N H$ ($\mu_N = 5.05 \times 10^{-24}$ erg/gauss is the nuclear magneton) and $g_G = +0.1805$ and $g_E = -0.1030$ are the relevant nuclear g-factors.

Magnetic dipole transition intensities between these levels are readily calculated using the 'golden rule' and show that only 6 of the 8 possible ground to excited level transitions are allowed. They are, in order of increasing energy, the lines

$$\begin{aligned} L_1 &= \delta - g_1 \mu_N H + u - \alpha_+\,, & L_4 &= \delta + g_3 \mu_N H - u + \alpha_-\,, \\ L_2 &= \delta - g_2 \mu_N H - u + \alpha_-\,, & L_5 &= \delta + g_2 \mu_N H - u - \alpha_-\,, \\ L_3 &= \delta - g_3 \mu_N H - u - \alpha_-\,, & L_6 &= \delta + g_1 \mu_N H + u + \alpha_+\,, \end{aligned} \tag{16}$$

where δ is the isomer shift (that is the energy separation of the 'centres of gravity' of the E and G levels),

$$\alpha_\pm = (e^* e \pm f^* f/2)\,|g_E|\,\mu_N H\,, \tag{17}$$

and

$$g_1 = (g_G - 3g_E)/2 = 0.2448\,,\quad g_2 = (g_G - g_E)/2 = 0.1418\,,\quad g_3 = (g_G + g_E)/2 = 0.0388\,. \tag{18}$$

In Mössbauer spectrometry the spectra are sampled by varying energy via the Doppler effect between γ-ray source and sample. The energies are therefore usually quoted in velocity units (mm/sec) and, as mentioned earlier, can in the present context be transformed to regular energy units by using the conversion 1 mm/sec = 7.69×10^{-20} ergs. The Zeeman lines (16) in amorphous samples generally consist of rather broad lines since all of the quantities δ, H, u, and $\alpha_\pm$ are expected to vary from site to site. A typical example[43], for a-YIG at 4.2 K, is shown in Fig. 12a.

The equations (16) are therefore potentially a source of a great deal of information in the iron-containing ionic glasses. Firstly, the line positions themselves (that is to say their centroids, or centres of 'gravity') provide measures of $\langle\delta\rangle$, $\langle H\rangle$, $\langle u\rangle$ and $\langle\alpha_\pm\rangle$, where the angular brackets denote averages over all sites. One common experimental finding for virtually all amorphous materials yet analyzed in this context is that $\langle u\rangle$ is extremely close to zero. Ruling out accidental correlation effects over a wide range of materials, this presumably implies the independence of (θ,ϕ) and q', and therefore, from Eq. (10), either that θ and ϕ are random variables, with

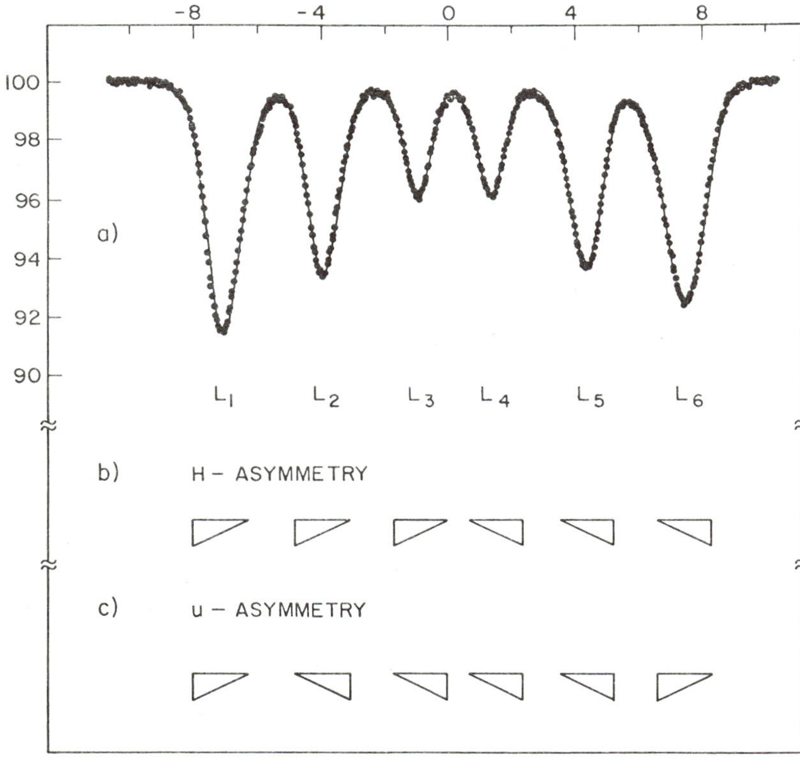

Fig. 12 (a) The experimental Mössbauer Zeeman spectrum (solid circles) of ^{57}Fe in speromagnetically ordered amorphous YIG at 4.2K least-squares computer fitted (continuous curve) to six independent symmetric distributions of natural width Lorentzian lines. The velocity scale of the abscissa is mm/sec. (b) A symbolic depiction of the relative sign of the line asymmetries resulting from an asymmetry in the distribution of hyperfine field (H) about its mean. (c) A symbolic depiction of the sign of the line asymmetries resulting from a preponderence of iron sites with positive field gradients $q>0$. From Refs.43 and 46.

$$\langle 3\cos^2\theta - 1 + \eta \sin^2\theta \cos 2\phi \rangle = 0 , \tag{19}$$

or that there are an equal number of Mössbauer sites with positive and negative EFG, i.e.

$$\langle q' \rangle = 0 . \tag{20}$$

A choice between these two possibilities can be made by observing the mean second order shifts $\langle \alpha_{\pm} \rangle$ in Eq. (16). If (θ, ϕ) are random variables in the amorphous environment then, from Eq. (17), via (11) and (12), the ratio $\langle \alpha_{+} \rangle / \langle \alpha_{-} \rangle = 3$ so that the second order shift pattern should be of symmetry +3, −1, +1, −1, +1, −3. If an approximate 3 to 1 ratio is not observed in these amplitudes then Eq. (20) must hold.

In addition to the mean line positions, a great deal of further information can be obtained from the same Zeeman spectrum by measuring the root mean square (RMS) full linewidths $2W_i = 2\langle (L_i - \langle L_i \rangle)^2 \rangle^{\frac{1}{2}}$. Directly from Eq. (16) we can derive two sets of relationships, one for the 'symmetrized linewidths' of the three line pairs (L_1, L_6), (L_2, L_5) and (L_3, L_4) as follows:

$$\begin{aligned}
(1/2)(W_1^2 + W_6^2) &= a^2 + g_1^2 b^2 + 2\langle \Delta u \Delta \delta \rangle + 2g_1 \mu_N \langle \Delta H \Delta \alpha_+ \rangle , \\
(1/2)(W_2^2 + W_5^2) &= a^2 + g_2^2 b^2 - 2\langle \Delta u \Delta \delta \rangle - 2g_2 \mu_N \langle \Delta H \Delta \alpha_- \rangle , \\
(1/2)(W_3^2 + W_4^2) &= a^2 + g_3^2 b^2 - 2\langle \Delta u \Delta \delta \rangle + 2g_3 \mu_N \langle \Delta H \Delta \alpha_- \rangle ,
\end{aligned} \tag{21}$$

and the other for pair linewidth differences

$$\begin{aligned}
(1/4)(W_6^2 - W_1^2) &= g_1 \mu_N \langle \Delta H (\Delta \delta + \Delta u) \rangle + \langle \Delta \alpha_+ (\Delta \delta + \Delta u) \rangle , \\
(1/4)(W_5^2 - W_2^2) &= g_2 \mu_N \langle \Delta H (\Delta \delta - \Delta u) \rangle - \langle \Delta \alpha_- (\Delta \delta - \Delta u) \rangle , \\
(1/4)(W_4^2 - W_3^2) &= g_3 \mu_N \langle \Delta H (\Delta \delta - \Delta u) \rangle + \langle \Delta \alpha_- (\Delta \delta - \Delta u) \rangle ,
\end{aligned} \tag{22}$$

where $\Delta x = x - \langle x \rangle$ $(x = H, \delta, u, \alpha_{\pm})$,

$$a^2 = \langle (\Delta u)^2 \rangle + \langle (\Delta \delta)^2 \rangle , \tag{23}$$

$$b^2 = \mu_N^2 \langle (\Delta H)^2 \rangle , \tag{24}$$

and we have neglected $\langle (\Delta \alpha_{\pm})^2 \rangle$ since such higher order terms are nearly always negligibly small in the present context.

The 'symmetrized linewidths' of Eq. (21) are usually dominated by the a^2 and b^2 terms, and therefore provide a measure of the isomer shift fluctuations $\langle (\Delta \delta)^2 \rangle$ and the hyperfine field fluctuations $\langle (\Delta H)^2 \rangle$, the quadrupole fluctuations $\langle (\Delta u)^2 \rangle$ already being known from the paramagnetic doublet spectrum or from the second order line shifts if (as seems normally to be the case in the ferric materials) θ and ϕ are approximately random variables. In such a case, for example, from Eqs. (10) and (4)

$$\langle (\Delta u)^2 \rangle = (9/5) \langle q'^2 (1 + \eta^2/3) \rangle = \langle E_Q^2 \rangle / 5 . \tag{25}$$

Equations (22) imply that the Zeeman spectrum as a whole is generally not symmetric about its center of 'gravity' $\langle \delta \rangle$. A linewidth analysis using these equations is, as we shall demonstrate in the next section, often sufficient to allow for a direct estimate of hyperfine field, isomer shift, and quadrupole correlations. Since the isomer shift is a measure of covalency, these provide an insight into the manner in which local site symmetry affects both hyperfine field and covalency.

There is still more information to be gleaned from the Zeeman spectrum. The six individual Zeeman lines L_i themselves are in general asymmetric about their own centres of 'gravity' and an analysis of lineshape can provide even more structural detail. The normal situation, since $g_1 \gg g_3$, is for the outer lines L_1 and L_6 to be dominated by the hyperfine field term $g_1 \mu_N H$ in Eqs. (16) and the inner lines L_3 and L_4 by the isomer shift or quadrupole variable u. A lineshape analysis of the outside lines therefore provides a direct measure of the hyperfine field distribution probability p_H. In some cases, like a-YIG, it happens that the inside lines are primarily influenced by quadrupole energy fluctuations so that the lineshapes of L_3 and L_4 provide a measure of quadrupole induced asymmetry. The latter is particularly important since it provides an estimate of the *sign* distribution of EFG, a property which the paramagnetic spectrum was unable to determine, but which is of great value in testing the viability of the RPHS computer model of section 3. In the following section we shall detail how these analyses are actually carried out in the specific case of a-YIG.

6. Magnetism as a Probe of Amorphous Structure

Let us now look at a particular example of the sort of information which can be obtained by using magnetism, via the Mössbauer Zeeman effect, in an amorphous ionic material. Of the amorphous ferric materials yet prepared, amorphous yttrium iron garnet ($Y_3Fe_5O_{12}$) possesses the largest quadrupole peak-to-peak splitting (≈ 1.1 mm/sec) of the paramagnetic doublet spectrum[44] (Fig. 13). Its shape is well described by the EFG histogram computed (in a point charge approximation) from a 2555-sphere RPHS assembly of spheres of three different sizes R(Fe) = 0.54 Å, R(Y) = 1.00 Å, and R(O) = 1.40 Å. The point charge calculation, using formal valence charges of +3, +3, −2, respectively for Fe^{3+}, Y^{3+}, and O^{2-} ions, also gives a good approximate estimate for the magnitude of the splitting, the larger value than for the fluorides of Fig. 10 being primarily due to the larger formal charge (−2) for the oxygen anions surrounding the Mössbauer nuclear sites than (−1) for the fluorines. Within the model, the Coulomb forces keep all trivalent cations (Y and Fe) at least 2.4 Å apart.

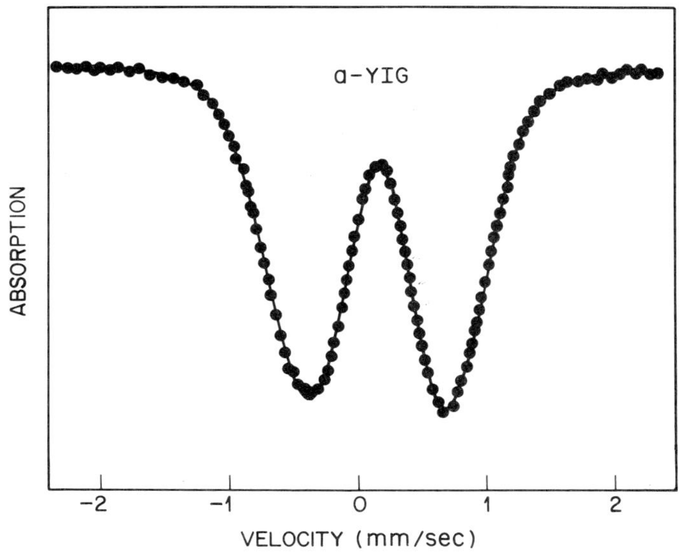

Fig. 13 The Mössbauer paramagnetic doublet spectrum for a-YIG at room temperature. The solid curve is a heuristic fit to the data using an asymmetric Gaussian-like distribution of natural-width Lorentzian lines. From Ref. 44.

A particularly interesting point in this EFG calculation is that it shows the number of sites n_+ with positive EFG q' to exceed those with n_- with negative EFG by close to 20%. In crystalline YIG, on the other hand, there are only two crystallographically distinct iron sites (one quasi-octahedral and one quasi-tetrahedral) and both possess a *negative* EFG of substantial magnitude.[45] It follows that an estimate of the ratio n_+/n_- from the Mössbauer Zeeman lineshapes should therefore be close to zero if the amorphous structure retains a distorted crystalline form on a local scale, but should be a little larger than 1 if the random packing concept is valid. This very marked distinction between the two models enables a clear cut decision to be made on the character of local structure for this case.

Fitting the $T = 4.2K$ Mössbauer Zeeman spectrum heuristically with six independent symmetric Gaussian distributions of natural-width Lorentzian lines gives an excellent 'zeroth order' fit to the experimental data as can be seen from Fig. 12a. Nevertheless, there are significant residuals R_i (measured data minus best least-squares-fitted symmetric lines) for each of the six lines when plotted on an expanded scale[46] (Fig. 14). Each line contains a small but unmistakeable asymmetry and, as is evident from the figure, this asymmetry is of opposite sign for lines 1 and 6, and also for lines 2

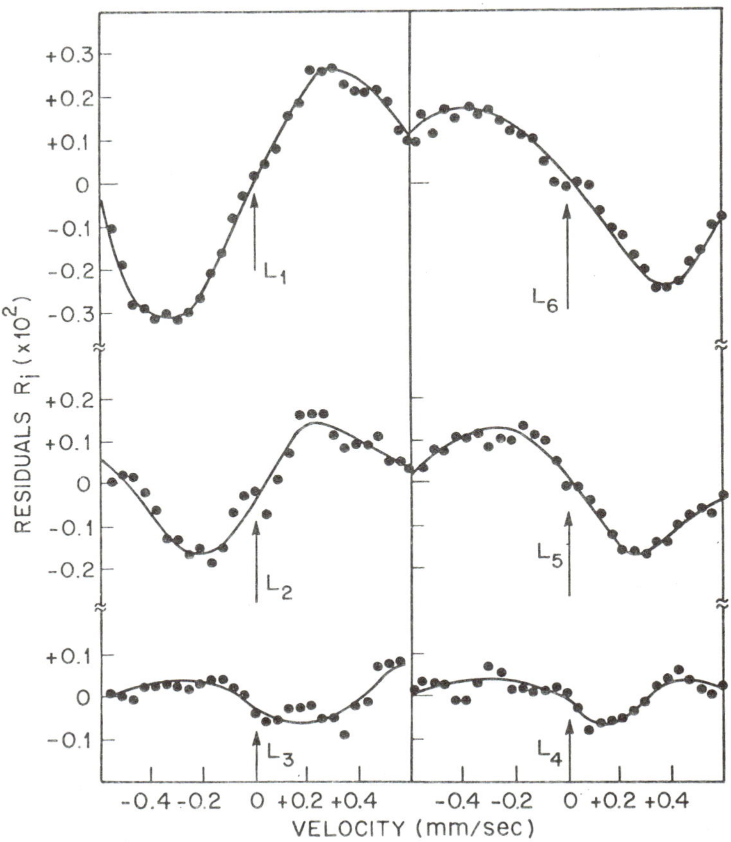

Fig. 14 The residual R_i for each of the six Mössbauer Zeeman Lines L_i (i = 1-6) measuring the difference between the data points and the 'best-fit' computer generated curve of Fig. 12a for a-YIG at 4.2 K. The continuous curves are merely guides to the eye. The arrows labelled L_1 to L_6 denote the peaks of the Zeeman lines, and the velocity scale for each line is zeroed at its peak value. From Ref. 46.

and 5, but is of the same sign for lines 3 and 4. From Eqs. (16), these asymmetries can have origins in the probability distribution functions for isomer shift δ, hyperfine field H, or quadrupole parameter u (the second order terms $\alpha_{\pm}$ are too small to be of significance in this context). It is evident, from Eqs. (16), that the isomer shift distribution would produce contributions of the same sign for all lines, while the H and u-induced asymmetry contributions would have the relative signs for the six lines as shown in Figs. 12b and 12c. Only the field induced asymmetry can produce the opposite sign pattern observed in Fig. 14 for the pairs (L_1 , L_6) and (L_2 , L_5). The inside line asymmetry (L_3 , L_4) might, until we have obtained further knowledge, arise from either u or δ. In pursuit of this further knowledge we first look at the information available from an analysis of line positions and line widths, and then follow up with a discussion of line shapes.

Line Positions

From a measure of the mean line positions $\langle L_i \rangle$ we first obtain a measure of mean isomer shift $\langle \delta \rangle$, and find a value which is typical for ferric ions in oxygen coordination. This immediately settles any question concerning the valence state of the iron in the amorphous structure; it is exclusively Fe^{3+}. Also from $\langle L_i \rangle$ we find a mean hyperfine field (at 4.2K)

$$\langle H \rangle = 29.58\,\text{mm/sec} = 450.4\,\text{kOe}\ , \qquad (26)$$

a mean u-value

$$\langle u \rangle = -0.002 \pm 0.002\,\text{mm/sec}, \qquad (27)$$

indistinguishable from zero, and second order quadrupole shifts

$$\langle \alpha_+ \rangle = 0.066 \pm 0.003\,\text{mm/sec}\ , \quad \langle \alpha_- \rangle = 0.026 \pm 0.003\,\text{mm/sec}. \qquad (28)$$

From the findings $\langle u \rangle = 0$ and $\langle \alpha_+ \rangle / \langle \alpha_- \rangle \approx 3$ we conclude that the angles θ , ϕ (which measure the hyperfine field direction with respect to the local principal EFG axes at each site) are essentially randomly distributed throughout the material. Since each local hyperfine field direction is essentially that of the frozen electronic spin on its site, this implies that local spin directions are largely independent of the local crystal field orientation and sign. This result is perhaps not unexpected for amorphous ferric ions since crystal field anistropy is known to be very small for the orbitally quenched ($3d^5$) ferric ion, and anisotropy (even in ferric *crystals*) is usually dominated by the long range inter-spin dipole-dipole forces.

Using the fact that angles θ, ϕ are random variables, to an excellent approximation, the average values of $\langle \alpha_+ \rangle$ and $\langle \alpha_- \rangle$ can be calculated from Eqs. (17), (11), (12), and (4) in the form

$$\langle \alpha_+ \rangle \approx (3/5)\,\langle E_Q^2 \rangle / |g_E| \mu_N \langle H \rangle, \quad \langle \alpha_- \rangle \approx (1/5)\,\langle E_Q^2 \rangle / |g_E| \mu_N \langle H \rangle, \qquad (29)$$

so that the second order shift pattern, leading to Eqs. (28), determines the mean square quadrupole splitting $\langle E_Q^2 \rangle \approx 0.360\ (\text{mm/sec})^2$ at 4.2 K, and consequently also determines

$$\langle u^2 \rangle = \langle E_Q^2 \rangle / 5 = 0.072\ (\text{mm/sec})^2\ . \qquad (30)$$

The value for $\langle E_Q^2 \rangle$ at 4.2K is extremely close to that $\langle E_Q^2 \rangle = 0.383\ (\text{mm/sec})^2$

calculated directly from the room temperature QS spectrum of Fig. 13, and indicates that the q'-distribution is largely independent of temperature.

Symmetrized Linewidths

Looking at the mean-square full linewidths $2W_i$ for the three different pairs of Zeeman lines (L_1,L_6), (L_2,L_5), and (L_3,L_4) we find that

$$\begin{aligned} &\text{(i)} \quad (1/2)(W_1^2+W_6^2) = 0.3082(\text{mm/sec})^2 , \\ &\text{(ii)} \quad (1/2)(W_2^2+W_5^2) = 0.2218(\text{mm/sec})^2 , \\ &\text{(iii)} \quad (1/2)(W_3^2+W_4^2) = 0.1527(\text{mm/sec})^2 . \end{aligned} \tag{31}$$

These widths $2W_i$ are RMS widths after the natural Lorentz broadening has already been subtracted out, so that they should be directly explicable in terms of Eqs. (21). We first attempt a fit neglecting inter-parameter correlations altogether. This 'zeroth order' calculation gaves a least squares best fit to the data of Eqs. (31) when $a = 0.396$ mm/sec and $b = 1.068$ mm/sec. The resulting errors (data minus the best-fit) are respectively -0.004, $+0.0125$, and -0.0085 (all in $(\text{mm/sec})^2$) for Eqs. (i), (ii) and (iii) respectively.

Since the maximum possible value of correlations $\langle xy \rangle$ between any *fully*-correlated pair of variables x and y, for which $\langle x \rangle = \langle y \rangle = 0$, is given by $|\langle xy \rangle_f| = \langle x^2 \rangle^{\frac{1}{2}} \langle y^2 \rangle^{\frac{1}{2}}$, we are now able to estimate the largest possible contributions from the correlation terms $\langle \Delta u \, \Delta \delta \rangle$ and $\langle \Delta H \, \Delta \alpha_{\pm} \rangle$ in (21) as follows

$$\begin{aligned} &|2g_1 \mu_N \langle \Delta H \Delta \alpha_+ \rangle_f| = 0.018, \quad |2g_2 \mu_N \langle \Delta H \Delta \alpha_- \rangle_f| = 0.019 , \\ &|2g_3 \mu_N \langle \Delta H \Delta \alpha_- \rangle_f| = 0.005, \quad |2 \langle \Delta \delta \Delta u \rangle_f| = 0.160 , \end{aligned} \tag{32}$$

all in $(\text{mm/sec})^2$ where the subscript f denotes full correlations. Since the last is a full order of magnitude larger than the others, it is highly probable that the dominant correlation effects, as far as symmetrized linewidths are concerned, come from $\langle \Delta\delta \, \Delta u \rangle$. With this assumption, Eqs. (21) can again be fitted to the experimental values of Eqs. (31) when we find $a = 0.341$ mm/sec, $b = 1.927$ mm/sec, and $\langle \Delta\delta \Delta u \rangle = -0.0154$ $(\text{mm/sec})^2$ leading, via Eqs. (23) and (24), to estimates

$$\langle (\Delta\delta)^2 \rangle = 0.044 \ (\text{mm/sec})^2 \tag{33}$$

and

$$\mu_N \langle (\Delta H)^2 \rangle^{\frac{1}{2}} = 1.927 \text{ mm/sec} = 29.3 \text{ kOe} . \tag{34}$$

The RMS isomer fluctuation $\langle (\Delta\delta)^2 \rangle^{\frac{1}{2}} \approx 0.21$ mm/sec is a little larger than the variation of isomer shifts ($\sim \pm 0.15$ mm/sec) commonly found for Fe^{3+} in different crystalline coordinations.[47] For comparison, the two different ferric sites in crystalline YIG have isomer shifts which differ by 0.22 mm/sec.[45] Note also that the correlation value of -0.0154 (mm/sec)^2 for $\langle \Delta\delta \Delta u \rangle$ represents correlations between isomer shift and u-fluctuations of about -20% of their maximum possible value.

Lineshape Analysis

Using the above estimate for the RMS extent of isomer shift, hyperfine field and u-fluctuations (*viz.* 0.044, 3.71, and 0.072 (mm/sec)^2 respectively) we can now compare the various primary fluctuation contributions to linewidth in the different line-pairs as follows:

	$g_i^2\langle(\Delta H)^2\rangle$	$\langle(\Delta u)^2\rangle$	$\langle(\Delta\delta)^2\rangle$
(L_1, L_6)	0.223	0.072	0.044
(L_2, L_5)	0.074	0.072	0.044
(L_3, L_4)	0.006	0.072	0.044

It is clear that the outside lines (L_1, L_6) are dominated by hyperfine field fluctuations, immediately explaining (Fig. 12b) the approximate antisymmetry between R_1 and R_6 in Fig. 14. The inside lines (L_3, L_4) on the other hand are primarily influenced by $\langle(\Delta u)^2\rangle$ which leads us to anticipate also the symmetric relationship (Fig. 12c) between R_3 and R_4 in Fig. 14. Since field and quadrupole fluctuations contribute about equally to lines L_2 and L_5, the approximate antisymmetry of R_2 and R_5 indicates that the *degree* of asymmetry of the distribution function p_H of field must be considerably larger than the degree of asymmetry for the quadrupole u-distribution function p_u.

The asymmetry of p_H is therefore directly deducible as the function

$$p_H^a(\Delta E) = (1/2)[R_6(\Delta E) + R_1(-\Delta E)] \tag{35}$$

which is shown in Fig. 15a, normed with respect to its symmetric Gaussian component. Its degree of asymmetry is about $\pm 3\%$. Adding to its symmetric Gaussian component, we finally arrive at the total hyperfine distribution line-*shape* $p_H(\Delta E)$ shown in Fig, 15b. In like manner the approximate asymmetry for the u-distribution can be measured by

$$p_u^a(\Delta E) = (1/2)[R_3(-\Delta E) + R_4(-\Delta E)]\,. \tag{36}$$

This function is shown in Fig. 16a, again normalized with respect to its symmetric Gaussian component. Its degree of asymmetry is about $\pm 1\%$.

The u-asymmetry is particularly important since it can be shown to be a direct measure of the proportion of positive to negative EFG sites. Using

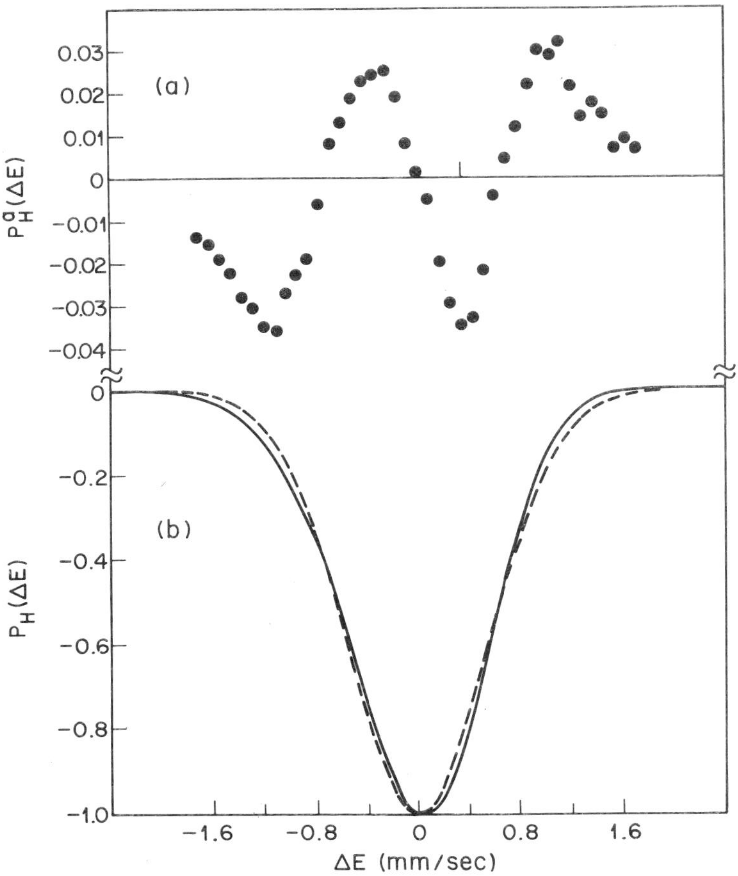

Fig. 15 (a) the asymmetric component of hyperfine-generated lineshape $p^a_H(\Delta E)$ calculated by averaging the L_6 and energy reversed L_1 residuals of Fig. 14 for a-YIG at 4.2K. The ordinate scale has been normalized to the peak amplitude of the averaged symmetric L_6 and L_1 line components and is therefore a physically relevant measure of hyperfine asymmetry in a-YIG. (b) The total hyperfine distribution lineshape (full curve) obtained by adding the symmetric Gaussian (dashed curve) and asymmetric $p^a_H(\Delta E)$ components.

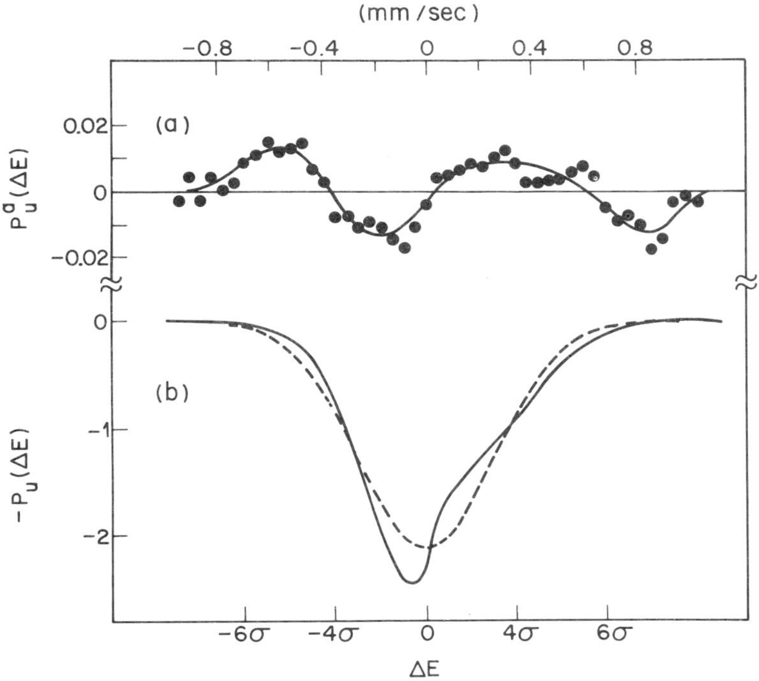

Fig. 16 (a) The asymmetric component of u-distribution $p_u^a(\Delta E)$, as calculated from Eq. (36) using the L_3 and L_4 residuals of Fig. 14. The ordinate scale is normalized to the peak amplitude of the averaged L_3 and L_4 line components and is therefore a physically relevant measure of the degree of EFG-induced line asymmetry in a-YIG. The full curve is merely a guide to the eye. (b) Full curve: the EFG-generated lineshape $p_u(\Delta E)$ as computed numerically from the shell model EFG distribution via Eq. (37) for the case in which all iron sites have a positive principal EFG component $q > 0$. Dashed curve: a Gaussian line of equal mean position, area, and RMS width to the full curve. The scale of the abscissa is a measure of the width of the EFG distribution via Eq. (7). The ordinate scale is arbitrary. From Ref. 46.

the full shell-model distribution function[41] $p(q',\eta)$ as a fair analytic approximation of both the measured and the computer generated distribution for EFG, the lineshape $p_u(\Delta E)$ can be calculated by integrating the delta function $\delta(\Delta E - u)$ over all angles θ and ϕ and over the complete variable ranges of q' and η as follows:

$$p_u(\Delta E) = \iiiint p(q',\eta)\, \delta(\Delta E - u) \sin\theta \, d\theta \, d\phi \, dq' \, d\mu . \qquad (37)$$

The result depends on the sign distribution n_+/n_- of positive to negative q' sites. It is necessarily symmetric if $n_+ = n_-$ and is maximally asymmetric if

$n_+ = 0$ or $n_- = 0$. For $n_- = 0$ (all sites $q' > 0$) the numerically integrated lineshape of Eq. (37) is shown in Fig. 16b and has an asymmetric component of the same sign and approximate shape as the experimentally determined a-YIG asymmetry shown in Fig. 16a, but with a fifteen-fold greater amplitude ($\approx \pm 15\%$). The relative amplitude 1:15 of the asymmetric components in Fig. 16a and 16b are therefore a direct measure of $(n_+ - n_-)/(n_+ + n_-)$ in a-YIG leading to an estimate

$$n_+/n_- \approx 1.15 \quad (a\text{-YIG}) \tag{38}$$

in excellent agreement with the prediction of Eq. (2) from the RPHS computer assembly of spheres and quite at odds with the expectation $n_+/n_- \approx 0$ if the local structure of a-YIG were a distorted form of its crystalline equivalent. We have here a direct demonstration of the fact that no vestige of a distorted crystalline environment is retained in the vitreous form of yttrium iron garnet.

Spectral Linewidth Asymmetries

Although the line areas A_i (i = 1,6) in the Zeeman spectrum of Fig. 12 are approximately symmetric in a pattern closely equal to 3:2:1:1:2:3 (a fact which establishes that the frozen spins are themselves randomly oriented in space, as expected for a spin-glass) the individual full RMS linewidths $2W_i$ have an asymmetry as follows:

$$\text{(i)}\ W_6^2 - W_1^2 = 0.0962, \quad \text{(ii)}\ W_5^2 - W_2^2 = 0.0249, \quad \text{(iii)}\ W_4^2 - W_3^2 = 0.0064, \tag{39}$$

all in $(\text{mm/sec})^2$.

From the last two equations, using the relationships (22), we deduce that

$$\mu_N \langle \Delta H(\Delta\delta - \Delta u)\rangle = +0.043\ (\text{mm/sec})^2, \tag{40}$$

$$\langle \Delta\alpha_-(\Delta\delta - \Delta u)\rangle < 0.001\ (\text{mm/sec})^2. \tag{41}$$

The averages $\langle\Delta\alpha_\pm \Delta u\rangle$ are purely quadrupolar and, for random θ, ϕ, can be directly calculated and shown to be zero ($\langle\Delta\alpha_+ \Delta u\rangle$ exactly and $\langle\Delta\alpha_- \Delta u\rangle$ at least to three decimal places). From Eq. (41) it now follows that $\langle\Delta\alpha_- \Delta\delta\rangle$ is also zero to three decimal places. The implication is that all the correlations involving $\Delta\alpha_\pm$ in Eqs. (22) are negligible for a-YIG. Combining the W_6, W_1 linewidth asymmetry result of Eqs. (22) and (39) with that of Eq. (40) now allows us finally to determine the two correlations

$$\mu_N \langle \Delta H\, \Delta\delta\rangle = +0.071\ (\text{mm/sec})^2, \tag{42}$$

and

$$\mu_N \langle \Delta H\, \Delta u\rangle = +0.028\ (\text{mm/sec})^2. \tag{43}$$

The two values (42) and (43) represent a degree of correlation of about +20%

and +5% of their maximum values $\langle \Delta H \Delta \delta \rangle_f$ and $\langle \Delta H \Delta u \rangle_f$ respectively. It can therefore be appreciated that all the various correlations in a-YIG are far from maximal. In particular, heuristic linear relationships of the kind

$$\delta = \delta_0 + aH ,$$

for example, which one often finds in the literature on amorphous magnetic metals, must necessarily be invalid here since, regardless of the magnitude of the correlation parameter a, they necessarily imply full (or 100%) linear correlations.

From this analysis we find that the degree of linear correlation between hyperfine field and isomer shift, namely $\langle \Delta H \Delta \delta \rangle / \langle (\Delta H)^2 \rangle^{\frac{1}{2}} \langle (\Delta \delta)^2 \rangle^{\frac{1}{2}}$, is only about 20%, while that between hyperfine field and the quadrupolar parameter u is even smaller at about 5%. In the crystalline Mössbauer literature it is conventional to use both the magnitude of saturated-spin (i.e. low temperature) hyperfine field and the isomer shift as qualitative measures of covalency, both decreasing as the degree of covalency increases. It is now clear from the work on a-YIG that these two measures can by no means be linearly related, and that other variables must be involved in their association to a very significant degree. From the small but non-zero degree of linear correlation between ΔH and Δu we also learn something quite specific. With $\langle u \rangle = 0$ and $\langle q' \rangle \neq 0$, it follows from Eq. (10) that the angular function in the parameter u averages to zero. Thus, if the hyperfine field fluctuations were isotropic (that is uncorrelated with θ, ϕ) then $\langle \Delta H \Delta u \rangle$ would also be zero. Its non-zero value therefore implies the existence of anisotropic hyperfine field fluctuations, although the small degree of these correlations may imply that such an anisotropy is quite small.

All this information will eventually require more detailed explanation at a fundamental level. Nevertheless, it is clear that rather specific results are now being generated by the use of magnetism as a probe of the amorphous state. If further support for the validity of RPHS computer-based models is forthcoming in this context then the possibility of building a bridge between fundamental chemistry and experimental measurements of this kind becomes a little more realistic. In any event it is rapidly becoming apparent that the densely packed magnetic insulator glasses are a new class of amorphous materials which offer great opportunities for studying details of local structure which were not easily accessible before. If eventually the magnetic Mössbauer studies can be supported by more conventional X-ray or neutron scattering techniques, then further structural model refinement should become possible. In short, the field is still in its infancy, and expectations are high.

Acknowledgement

It is clear from the above discussion that precise analyses of Mössbauer lineshapes and linewidth-inequalities can only be carried out if Mössbauer data of the highest statistical quality are available. It is in this regard that I wish to acknowledge the careful and time-consuming Mössbauer measurements of my colleague M. Eibschutz, without which most of the analyses of sections 4 and 6 would not have been possible.

References

1. Gaskell, P.H., J. Phys., **C12**, 4337 (1979).
2. Finney, J.L. In *The Structure of Non-Crystalline Materials*, (Ed. P.H. Gaskell) Taylor and Francis, London 1977.
3. Vincze, I., Boudreaux, D.S. and Tegzem M., Phys. Rev., **B19**, 4896 (1979).
4. Fujiwara, T. and Ishii, Y., J. Phys., **F10**, 1901 (1980).
5. Gaskell, P.H., J. Non-Cryst. Sol., **32**, 207 (1979).
6. Wagnerm C.N.J., J. Non-Cryst. Sol., **42**, 3 (1980).
7. Waseda, Y., Okazaki, H. and Masumoto, T., J. Mat. Sci., **12**, 1927 (1977).
8. Vertesm A., Korecz, L. and Burger, K., *Mössbauer Spectroscopy*, Elsevier Publ. Co., Oxford (1979).
9. Coey, J.M.D., J. Appl. Phys., **49**, 1646 (1978).
10. Litterst, F.J., J. Phys. Letters, **36**, L197 (1975).
11. Gyorgy, E.M., Nassau, K., Eibschutz, M., Waszczak, J.V., Wang, C.A. and Shelton, J.C., J. Appl. Phys., **50**, 2883 (1979).
12. Matsuzawa, T., Okamuram K., Shishido, T. and Yajima, S., J. de Physique, **40**, C2-151 (1979).
13. See for example: H. Maletta and W. Felch, Phys. Rev., **B20**, 1245 (1979).
14. Simpson, A.W., Phys. Stat. Solidi, **40**, 207 (1970).
15. Ferey, G., LcClere, A.M., de Pape, R., Mariot, J.P. and Varret, F., Sol. St. Commun., **29**, 477 (1979).
16. Ferey, G., Coey, J.M.D., Henry, M., Teillet, J., Varret, F. and Buder, R., J. Mag. Mag. Mat., **15-18**, 1371 (1980).
17. Heger, G., Geller, R. and Babel, D., Sol. St. Commun., **9**, 335 (1971).
18. Dance, J.M., Menil, F., Hanzel, D., Sabatier, R., Tressaud, A., Flem, G. le. and Hagenmuller, P., Physica, **86-88B**, 699 (1977).
19. Jongh, L. de, and Miedema, A.R., Avan. Phys., **23**, 1 (1974).
20. Gupta, G.P., Dickson, D.P.E. and Johnson, C.E., J. de Physique, **40**, C2-304 (1979).
21. Renard, J.P., Maranday, J.P. and Varret, F., Sol. St. Commun., **35**, 41 (1980).
22. Litterst, F.J., Kalvius, G.M. and Boyle, A.J.F., AIP Conf. Proc., **18**, 616 (1973).
23. Boyle, A.J.F., Kalvius, G.M., Gruen, D.M., Clifton, J.R. and McBeth, R.L., J. Phys. Colloq., **32**, C1-224 (1971).
24. Litterst, F.J., Kalvius, G.M. amd Boyle, A.J.F., J. de Physique, **36**, C6-403 (1975).
25. Bernal, J.D., Proc. Roy. Soc. London, **A280**, 299 (1964): Nature, **188**, 910 (1960).
26. Scott, G.D., Nature, **188**, 908 (1960).
27. Finney, J.L., Proc. Roy. Soc. London, **A319**, 479 (1970).
28. Cohen, M.H. and Turnbull, D., Nature, **203**, 964 (1964).
29. See for example: Bennett, C.H., J. Appl. Phys., **34**, 2727 (1972).
30. Barker, J.A., Finney, J.L. and Hoare, M.R., Nature, **257**, 120 (1975).
31. Cochrane, R.W., Harris, R. and Plischke, M., J. Non-Cryst. Sol., **15**, 239 (1974).

32. For a recent review, see Ref. 1.
33. Boudreaux, D.S. and Gregor, J.M., J. Appl. Phys., **48**, 152 and 5057 (1977); Boudreaux, D.S., Phys. Rev. **B18**, 4039 (1978).
34. Jansen, H.J.F., Boudreaux, D.S. and Snijders, H., Phys. Rev., **21**, 2274 (1980).
35. Sadoc, J.F. and Dixmier, J., Mat. Sci. Eng., **23**, 187 (1976).
36. Vincze, I., Kemeny, T. and Arajs, S., Phys. Rev., **B21**, 937 (1980).
37. Mattern, N. and Hermann, H., Phys. Stat. Solidi, **104B**, 546 (1981).
38. Lines, M.E., Phys. Rev., **B21**, 5793 (1980).
39. Lines, M.E., J. Non. Cryst. Sol., **46**, 1 (1981).
40. Lines, M.E., Phys. Rev., **B20**, 3729 (1979).
41. Czjzek, G., Fink, J., Gotz, F., Schmidt, H., Coey, J.M.D., Rebouillat, J.P. and Lienard, A., Phys. Rev., **B23**, 2513 (1981).
42. Eibschutz, M., Lines, M.E., Van Uitert, L.G., Guggenheim, H.J. and Zydzik, G.J., Phys. Rev., **B24**, 2343 (1981).
43. Eibschutz, M. and Lines, M.E., Phys. Rev., **B26**, 2288 (1982).
44. Eibschutz, M., Lines, M.E. and Nassau, K., Phys. Rev., **B21**, 3767 (1980).
45. Sawicki, J. and Hafner, S.S., Phys. Letters, **68A**, 80 (1978).
46. Lines, M.E. and Eibschutz, M., Phys. Rev., **B25**, 6042 (1982).
47. Greenwood, N.N. and Gibb, T.C., *Mössbauer Spectroscopy*, Chapman and Hall Ltd., London (1977).
48. Logan, J. and Sun, E., J. Non. Cryst. Sol., **20**, 285 (1976).
49. Chien, C.L. and Chen, H.S., J. Physique, **C2**, 118 (1979).

Nuclei in Dense Matter

C. J. Pethick
Nordita, Blegdamsvej 17, DK-2100 Copenhagen Ø, Denmark
and
Department of Physics, University of Illinois at Urbana-Champaign, 1110 W. Green Street, Urbana, IL 61801, U.S.A.

ABSTRACT

A review is given of the application of the liquid drop model and extensions of it to calculate properties of nuclei in high density matter encountered in neutron stars and stellar collapse.

The problem I have chosen to address here lies on the boundaries of nuclear physics, condensed matter physics, and astrophysics, and therefore would seem to be a particularly appropriate one in view of Dirk's wide-ranging interests. The observational discovery of neutron stars, as radio pulsars and compact X-ray sources, raised studies of the properties of dense matter from the state of an academic curiosity to one of importance for comparison of theory and observation. In addition, work on stellar collapse and supernovae has led to increased interest in properties of dense matter under conditions different from those in neutron stars.

The nuclei one encounters in astrophysical contexts may be very different from those in the laboratory — much more neutron rich, and larger. In addition they may be immersed in a surrounding nucleon vapour. Consequently it is not possible to obtain good estimates of nuclear energies simply by using mass formulae for laboratory nuclei. On the other hand, microscopic calculations of nuclei using a nucleon-nucleon interaction, can, because of the considerable calculational effort required, be carried out only for a limited number of cases. In this paper I shall describe how many properties of nuclei in dense matter can be understood on the basis of the semi-empirical mass formula for nuclei and its extensions. Such models do not include some effects, such as shell structure, but they give a good quantitative understanding of nuclear properties over a wide range of conditions. The calculations are much less laborious than microscopic ones, and they give valuable insight into the physics determining nuclear properties.

My aim in this paper is to keep the discussion simple, and, therefore, I shall start with the least complicated situation and then introduce additional physical effects piecemeal, rather than go directly to the most general case. First I consider the properties of cold matter at relatively low densities where all nucleons are in nuclei. I then discuss cold matter at higher densities, when one has nuclei immersed in a neutron sea, and finally I consider finite temperatures.

First of all we consider the ground state of matter. The most energetically favourable nucleus at low densities is ^{56}Fe. At the lowest densities some electrons are bound to the nuclei, but when the density is so high that the distance between nuclei is comparable with or less than the radius of the isolated atom, matter becomes pressure ionized. If one takes the Thomas-Fermi radius of an atom, $a_0 Z^{-1/3}$ as an estimate of the atomic size, matter will be completely ionized for $\rho \gtrsim 6\,AZ$ g cm^{-3}. Here A and Z are the mass number and atomic number of the nucleus, and a_0 is the Bohr radius. This density is about 10^4 g cm^{-3} for ^{56}Fe.

At densities above this the electrons are essentially free, since electron-electron interactions are even less important than electron-nucleus interactions. Typical electron-electron interaction energies are $\sim e^2/r_e$, where r_e is a measure of the electron spacing, while the kinetic energy of an electron is $\sim \hbar^2/m_e r_e^2$ non-relativistically, where m_e is the electronic mass. Thus the electron-electron interaction energy is of order $r_e/a_0 \sim e^2/\hbar v_e$ times the kinetic energy, which is small provided $r_e \ll a_0$. Here $v_c \sim \hbar/m_e r_e$ is a typical electron velocity. The quantity $r_s = r_e/a_0$ is the standard dimensionless interaction parameter in studies of the electron gas, and it is small at all densities much above typical terrestrial ones. At densities in excess of $\sim 10^7$ g cm^{-3} the electrons become fully relativistic, and the electron-electron interaction effects are then of order $e^2/\hbar c$ times electronic kinetic energies.

In determining the ground state of matter it is important to specify the subsidiary conditions to be imposed. For the case of neutron stars, the relevant problem is to determine the lowest energy state of a fixed number of nucleons in a given volume. The total energy must include the contribution from electrons (and at higher densities, other particles) that ensure that the matter is electrically neutral. If only nuclei and electrons are present the total energy per unit volume may be written

$$\mathcal{E} = n_N E(A,Z) + \mathcal{E}_e(n_e) \tag{1}$$

where n_N is the number density of nuclei, $E(A,Z)$ is the ground-state energy of the nucleus, and $\mathcal{E}_e$ is the electronic energy density. Charge neutrality ensures that the electron number density is given by

$$n_e = Z n_N \tag{2}$$

and the baryon number density is

$$n = A n_N \ . \tag{3}$$

With increasing density the electron Fermi energy rises, and at a density of about $8 \times 10^6 \text{ g cm}^{-3}$ ^{62}Ni nuclei, which have fewer protons per nucleon and therefore require fewer electrons to ensure electrical neutrality, are more favourable than ^{56}Fe nuclei. The energy of ^{56}Fe matter can be lowered by capturing electrons on protons, and rearranging the nucleons into ^{62}Ni nuclei. This lowers the electronic energy, and raises the nuclear energy by a lesser amount. At higher densities the nuclei become even more neutron rich, and at a density of about $4.3 \times 10^{11} \text{ g cm}^{-3}$ the most energetic neutron in the nucleus becomes unbound. At still higher densities, matter consists of nuclei immersed in a sea of neutrons, as well as a sea of electrons. When the density is about nuclear matter density, the nuclei merge to form a uniform sea of neutrons and protons.

Physically, the condition determining the number of electrons is that it cost no energy to turn a proton and an electron into a neutron, or

$$\mu_p + \mu_e = \mu_n \ , \tag{4}$$

where the μ's are chemical potentials, including rest mass contributions.

In the final stages of collapse of massive stars the conditions determining the electron concentration are somewhat different. Neutrino mean free paths are small compared with the size of the collapsing stellar core, and typical times for a neutrino to diffuse out of the core exceed the collapse time for core densities greater than about $10^{12} \text{ g cm}^{-3}$. As a consequence, neutrinos produced by electron capture on protons are trapped inside the star, and the total lepton number in the star remains fixed for the duration of the collapse. Equilibrium is determined by the condition that it cost no energy to convert an electron and a proton into a neutron and a neutrino or

$$\mu_p + \mu_e = \mu_n + \mu_\nu \tag{5}$$

which differs from (4) only by the presence of the neutrino chemical potential, μ_ν.

At densities below the 'neutron drip' density a good semiquantitative understanding of the dependence of the nuclear size on density may be obtained by using the semiempirical mass formula for the nuclear masses. In the simplest form, one treats nuclei as spherical drops of incompressible matter, and the energy is then given as the sum of bulk, surface and Coulomb energies:

$$E(A,Z) = AW(Z/A) + E_{\mathbf{surf}} + E_{\mathbf{Coul}} \tag{6}$$

where $W(Z/A)$ is the energy per particle of matter with proton concentration Z/A. The nuclear radius is given by

$$r_{\mathrm{N}} = r_0 A^{1/3} \tag{7}$$

where r_0 is related to the saturation density, $n_{\mathbf{s}}$, of matter in the nucleus by

$$n_{\mathbf{s}} \tfrac{4}{3} \pi r_0^3 = 1 . \tag{8}$$

The surface energy is given by

$$E_{\mathbf{surf}} = 4\pi r_{\mathrm{N}}^2 \sigma , \tag{9}$$

where σ is the surface tension.† The Coulomb energy of a uniform sphere is given by

$$E_{\mathbf{Coul}} = \frac{3}{5} \frac{Z^2 e^2}{r_{\mathrm{N}}} . \tag{10}$$

If we imagine the proton concentration to be given, the nuclear size is easily determined by minimizing the energy (1) with respect to A, keeping Z/A and n fixed. In this variation the total number of neutrons, protons and elec-ttrons remains unchanged, and therefore the bulk nuclear energy and the electronic energy are unchanged. The surface energy varies as $n_{\mathrm{N}} A^{2/3}$ or $n A^{-1/3}$, while the Coulomb energy varies as $n_{\mathrm{N}} (Z^2)/A^{1/3}$ or $n\,(Z/A)^2\, A^{2/3}$. Performing the variation, one thus finds the simple result

$$E_{\mathbf{surf}} = 2E_{\mathbf{Coul}} \tag{11}$$

or

$$A = \frac{5\sigma}{2x^2 e^2 n_{\mathbf{s}}} . \tag{12}$$

The equilibrium size is determined by a compromise between the surface energy, which favours large nuclei, and the Coulomb energy, which favours small ones. If this is written in terms of the usual semiempirical mass formula parameters given by

$$E_{\mathbf{surf}} = C_{\mathbf{surf}} A^{2/3} \tag{13}$$

---

† As we shall explain later, what is usually referred to as the surface energy is really the surface thermodynamic potential. (See, e.g., Landau and Lifshitz, 1980, Ch. XV.) Here, however, to avoid confusion we shall retain the conventional terminology.

and

$$E_{\mathrm{Coul}} = C_{\mathrm{Coul}}\, Z^2/A^{1/3} \tag{14}$$

one has

$$A = C_{\mathrm{surf}}/(2x^2 C_{\mathrm{Coul}}) \; . \tag{15}$$

If we take for the coefficients the values $C_{\mathrm{surf}} = 18.9\,\mathrm{MeV}$ and $C_{\mathrm{Coul}} = 0.756\,\mathrm{MeV}$, which correspond to a saturation density $n_s = 0.16 f^{-3}$, and a surface tension $\sigma = 1.149\,\mathrm{MeV}\,f^{-2}$ we find

$$A = 12.5/x^2 \; . \tag{16}$$

If we put x equal to the value for $^{56}Fe_{26}$, we find $A \simeq 58$, which is in satisfactory agreement with the actual value. With increasing density, x decreases and therefore the most favourable value of A increases.

For determining the equilibrium nucleus at higher densities it is necessary to include some other physical effects. One of these is the Coulomb lattice energy. If one calculates the energy in the Wigner-Seitz approximation, in which one treats matter as being made up of spherical cells containing a nucleus plus Z electrons and neglects the interaction between cells, one finds the lattice energy is given for point nuclei by

$$E_{\mathrm{L}} = -\frac{9}{10}\,\frac{Z^2e^2}{r_{\mathrm{c}}} \tag{17}$$

where $r_{\mathrm{c}} = Z^{1/3} r_{\mathrm{e}}$ is the radius of the Wigner-Seitz cell. The corresponding exact results for lattices differ very little from this: for example, for the lattice with the greatest binding, the b.c.c. lattice, one finds

$$E_{\mathrm{L}} \simeq -0.89593\,\frac{Z^2e^2}{r_{\mathrm{c}}} \; . \tag{18}$$

When the finite size of the nucleus is taken into account one finds an additional contribution to the Coulomb energy

$$E_{\mathrm{fin}} = \frac{3}{10}\,\frac{Z^2e^2r_{\mathrm{N}}^2}{r_{\mathrm{c}}^3} \tag{19}$$

in the Wigner-Seitz approximation. The total Coulomb energy in the Wigner-Seitz approximation is given by the sum of (10), (17) and (19), and is

$$E_{\mathrm{Coul}} = \frac{3}{5}\,\frac{Z^2e^2}{r_{\mathrm{N}}}\left(1 - \frac{3}{2}\,\frac{r_{\mathrm{N}}}{r_{\mathrm{c}}} + \frac{1}{2}\left(\frac{r_{\mathrm{N}}}{r_{\mathrm{c}}}\right)^3\right) . \tag{20}$$

Note that the total Coulomb energy includes not only effects from nucleus-nucleus interactions, but also those of interactions of the background electrons with both themselves and with nuclei. When nuclei fill all of space, r_{N} is equal to r_{c} and the total Coulomb energy vanishes, since then the net charge density at every point in space is zero.

The lattice energy plays an important role in determining the optimal size of nuclei in dense matter. In most terrestrial applications lattice energies are small compared with the Coulomb self energy of the nucleus, but when the distance between nuclei is comparable with the nuclear size the two energies are of the same order of magnitude.

Another physical effect that must be taken into account at higher densities, where the nuclei become increasingly neutron rich is the fact that the surface tension depends on the neutron excess. Like the bulk binding energy, the surface tension decreases with increasing neutron excess. When the neutron excess has become so large that there is an appreciable density of neutrons in continuum states outside nuclei, there is little difference between matter inside nuclei and that outside, and the surface tension associated with the interface is even smaller.

A good guide to the size of nuclei in the neutron drip regime may be obtained by a simple generalization of the energy expression (1) by including a term to take account of the energy of the neutrons outside nuclei

$$\mathcal{E} = n_{\mathrm{N}} E(A\ ,\ Z) + \mathcal{E}_{\mathrm{e}}(n_{\mathrm{e}}) + (1-u)\ \mathcal{E}_{\mathrm{n}}(n_{\mathrm{n}})\ , \tag{21}$$

where $\mathcal{E}_{\mathrm{n}}(n_{\mathrm{n}})$ is the energy per unit volume of neutrons of density n_{n}. The quantity $u = (r_{\mathrm{N}}/r_{\mathrm{c}})^3$ is the fraction of space occupied by nuclei, and therefore the fraction of space available to dripped neutrons is $1-u$. The Coulomb energy in the nuclear mass formula is the total Coulomb energy (20), and the surface tension may depend on the neutron excess. The optimal nucleus may be determined by the same procedure as before, except that the total density of neutrons outside nuclei is also held fixed. The total nucleon density is

$$n = A\, n_{\mathrm{N}} + (1-u)\, n_{\mathrm{n}} \tag{22}$$

and therefore the baryon density is unchanged by this variation. The optimal nuclear size is then given by minimizing the surface and Coulomb energies, as before, and one finds

$$A = \frac{5\sigma(x)}{2x^2 e^2 n_{\mathrm{s}}} \frac{1}{\left(1 - \frac{3}{2}\frac{r_{\mathrm{N}}}{r_{\mathrm{c}}} + \frac{1}{2}\left(\frac{r_{\mathrm{N}}}{r_{\mathrm{c}}}\right)^3\right)}\ . \tag{23}$$

The reduction of the total Coulomb energy due to the lattice term leads to an increase in the size of the nucleus. Even at a density one thousandth of nuclear density ($u = 10^{-3}$), the Coulomb energy is reduced by ~ 15% , and therefore A is increased by the same amount. If we neglect the dependence of the nuclear saturation density on x and use empirical values for the coefficients in the mass formula given above, we find

$$A = \frac{12.5}{x^2}\ \frac{\sigma(x)}{\sigma(0.5)}\ \frac{1}{\left(1 - \frac{3}{2}u^{1/3} + \frac{u}{2}\right)}\ . \tag{24}$$

To relate the density of exterior neutrons to other nuclear properties one demands that the energy be stationary when a neutron is transferred from the exterior to the interior, with the number of protons in the nucleus held fixed. This is the condition that the chemical potentials of neutrons inside and outside nuclei be the same:

$$\mu_n^{(i)} = \mu_n^{(0)} \tag{25}$$

where $\mu_n^{(i)}$ and $\mu_n^{(0)}$ are the neutron chemical potentials inside and outside nuclei, respectively. For the energy expression (21) it is

$$\frac{\partial E(A/Z)}{\partial A}\bigg|_Z = \frac{\partial \mathcal{E}_n}{\partial n_n} \; . \tag{26}$$

The bulk properties of nuclear matter have little effect on the nuclear size, but they are important for calculating properties such as the equilibrium proton concentration and the pressure. The nuclei of interest can have proton concentrations as low as ~ 0.1, which means that it is inadequate to use expressions for the bulk energy derived from empirical properties of laboratory nuclei, which have roughly equal numbers of protons and neutrons. Another physical effect important when there are neutrons outside the nuclei is the pressure exerted on the nuclei by the neutrons. To allow for both of these effects Baym, Bethe and Pethick (1971) devised an expression to describe the bulk energy of matter as a function of proton concentration and density. The expression was designed to agree with empirical results for the binding energy, saturation density, compressibility and symmetry energy of symmetrical nuclear matter, and at low density it went over to the result for a free Fermi gas. To make the interpolation formula applicable to very neutron rich matter, for which no experimental data are available, the expression was designed to give at low proton concentrations results calculated from microscopic theory. Similar energy expressions have also been obtained by suitably modifying Skyrme interactions which give good results for ordinary nuclei (Vautherin and Brink, (1970)) so that they also yield results which agree with microscopic calculations for pure neutrons (Lattimer and Ravenhall, (1978)). The bulk energy of both the exterior neutrons and the matter in nuclei are given by the same algebraic function, and this ensures that the two sorts of nuclear matter, which can be very similar when the nuclei are very neutron rich, are treated consistently.

By allowing the density of matter in nuclei to be variable, we are able to take into account the effect of the external nuclear pressure on the nuclear properties. The equilibrium nuclear density is determined by the condition that the energy be stationary with respect to the nuclear size, which is equivalent to equality of the pressure of the nucleus and the pressure of the external neutrons.

Even when the finite compressibility is allowed for, the nuclear size is still determined by (23), but with n_s replaced by the density of matter n_i inside the nucleus:

$$A = \frac{5}{2}\,\frac{\sigma(x)}{x^2 e^2 n_i}\,\frac{1}{1 - \frac{3}{2}\frac{r_N}{r_c} + \frac{1}{2}\left(\frac{r_N}{r_c}\right)^2} \quad . \tag{27}$$

To make numerical estimates of A, it is important to have reliable values of the surface tension coefficient $\sigma(x)$. Empirical information about this is available only for nuclei with roughly equal numbers of neutrons and protons, but that is insufficient for our purposes, since in neutron stars A may be as much as ten times Z. Qualitatively speaking, one would expect the surface tension to decrease as the proton fraction decreases, just as the bulk binding energy does. When the proton fraction becomes so low that dripped neutrons are present, matter inside nuclei and that outside become increasingly similar and the surface tension decreases even more rapidly. Simple estimates of the surface tension were made by Baym, Bethe and Pethick (1971), but these underestimated the reduction of the surface tension for very neutron rich nuclei, and therefore led to overestimates of the nuclear size. Subsequently, microscopic calculations of plane interfaces were carried out by Ravenhall, Bennett and Pethick (1972), using a Skyrme Hamiltonian in the Hartree-Fock and Thomas-Fermi approximations. One of the important points to come out of these calculations is that it is essential to use the correct definition of the surface tension. Consider first of all a plane surface separating two phases of a one-component system. One constructs the surface tension by subtracting a bulk energy from the total energy. The bulk energy is usually defined by dividing the system into two volumes containing matter at the densities of the two bulk phases far from the interface, the volumes being chosen such that the total volume and the total number of particles are the same as in the actual system. This is equivalent to 'squaring off' the density profile, as illustrated in Fig. 1. When one deals with systems with more than one component, a complication enters because there is no unique way of dividing the system into two bulk phases. In the case of neutron-rich nuclei, the neutron distribution extends further than the proton one and therefore the way in which the total volume is divided up into two volumes, as described above, depends on whether one squares off the distribution of protons, or the distribution of neutrons (or that of the total density of nucleons). It turns out that it is easiest to work with the thermodynamic potential

$$\Omega = E - \mu_n N_n - \mu_p N_p \ , \tag{28}$$

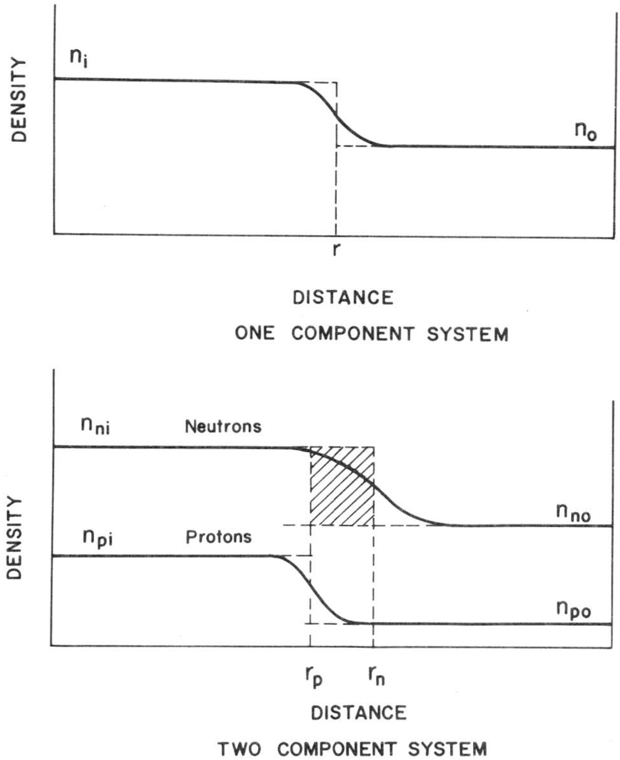

Fig. 1 Density profiles for one and two component systems. In the case of the one component system a unique position for the surface may be defined by squaring off the density distribution. For the two component system, the position of the surface obtained depends on which of the density profiles for the two components one squares off. The number of excess neutrons per unit area of the surface is given by $\nu_s = (n_{n_i} - n_{n_0})(r_n - r_p)$

where E is the total energy, and N_n and N_p are the total numbers of neutrons and protons. The thermodynamic potential in the absence of surface tension is just the sum of bulk contributions

$$\Omega_0 = V_1\omega_1 + V_2\omega_2 \tag{29}$$

where ω_1 and ω_2 are the thermodynamic potentials per unit volume of the two bulk phases, and V_1 and V_2 are their volumes. In the bulk phases, ω_1 and ω_2 are just equal to $-P$, when P is the bulk pressure, which is the same on both sides of the interface. Thus the surface tension, the surface thermodynamic potential per unit area, is given by

$$\sigma = \frac{\Omega - \Omega_0}{\mathcal{A}} = \frac{\Omega + PV}{\mathcal{A}} \ , \tag{30}$$

where $\mathcal{A}$ is the area of the surface, and $V = V_1 + V_2$ is the total volume of the system. Note that because $\omega_1 = \omega_2 = - P$, Ω_0 is independent of how one formally assigns the total volume to the two bulk phases.

For symmetric nuclear matter, the neutron and proton density profiles are the same (to the extent that one neglects the neutron-proton mass difference and the Coulomb interaction) and the surface tension is then given by squaring off any of the nucleon distributions. If, for asymmetric matter, one uses the surface energy so defined, with the volumes of the two phases fixed by squaring off the total nucleon density, one finds the surface energy initially *increases* as the nuclei become more neutron-rich, starting from the symmetrical case. However, as we mentioned before, the surface tension actually *decreases* with increasing neutron richness.

The surface tension gives information about the number of neutrons in the neutron 'skin' of nuclei. The surface tension, being a thermodynamic potential, is most naturally regarded as a function of the neutron and proton chemical potentials. However, only one of them is independent, since the conditions for phase equilibrium give an extra restriction. The number of excess surface neutrons per unit area is given by

$$\nu_s = - \frac{\partial \sigma}{\partial \mu_n} \, . \tag{31}$$

A more detailed discussion of nuclear surface properties may be found in Ravenhall, Pethick and Lattimer (1983) and Lamb *et al.* (1984).

We now return to the question of the nuclear size, determined by (27). At a given density, r_N/r_c is essentially fixed if there are no neutrons outside the nuclei. As the proton fraction decreases, A first of all increases, due to the x^2 term in the denominator. The x dependence of the surface tension is not important for nearly symmetrical nuclei since the leading corrections to σ are of order $(x - 0.5)^2$, but for smaller x, σ varies more rapidly. The surface tension has been calculated for a Skyrme interaction in the Hartree-Fock and Thomas-Fermi approximations by Ravenhall, Bennett and Pethick (1972), and over the range $0.1 \leqslant x \leqslant 0.3$ one finds $\sigma \sim x^3$. Thus $A \sim x$ and the size decreases with decreasing x. The effect of dripped nucleons is to reduce r_N/r_c somewhat as x decreases, which leads to A depending less strongly than linearly on x at fixed total density.

Under conditions of beta equilibrium with no trapped neutrinos, the case relevant for neutron stars, x decreases with increasing density, as we argued earlier, and competition between the surface and Coulomb energies leads to Z being about 40 over most of the region where one has nuclei immersed in a sea of neutrons (Ravenhall, Bennett and Pethick, (1972)).

Up to now we have considered mainly the problem of cold matter. However, in stellar collapse, temperatures reach some tens of MeV, which is significant on the scale of nuclear excitation energies. The first question we wish to address is, when does one expect nuclei to be present in hot dense matter. To do this, we generalize our earlier discussion to finite temperatures. Instead of working with the energy as a function of the baryon density, for a system at a fixed temperature T one must work with the Helmholtz free energy. The free energy may then be expressed in terms of bulk, surface, Coulomb and other contributions, just as the energy was at zero temperature. The simplest approximation is to neglect all contributions to the free energy except bulk ones. The problem of determining when nuclei are present then reduces to finding the conditions under which hot nuclear matter can be in equilibrium with a nucleon vapour. At a sufficiently low (but finite) temperature, matter forms a vapour at low densities. As the density is increased, a point will be reached at which drops of nuclear matter begin to condense. With further increase in density, the fraction of matter in the vapour decreases, and eventually nuclear matter fills all of space. Condensation will not occur if the temperature is higher than some temperature which is usually close to the critical temperature. The conditions for coexistence of two bulk phases is that the neutron chemical potentials in the two phases be equal, that the proton chemical potentials be equal, and that the pressures be equal. A review of such calculations is given by Lattimer (1981), and in Fig. 2, we show, as the line marked 'bulk equilibrium', the boundary of the two-phase region calculated for the Skyrme I' interaction by Lattimer and Ravenhall (1978). Notice that at densities approaching nuclear density, the two-phase region extends to temperatures of order 20 MeV. In nuclear physics language, the persistence of nuclei to such high temperatures is due to the importance of nuclear excited states, which increase the statistical weight of the droplets of nuclear matter (see, e.g., Mazurek, Brown and Lattimer, (1978); Bethe, Brown, Applegate and Lattimer, (1979)). In the bulk equilibrium calculation, such effects are automatically included through the finite-temperature contributions to the nucleon chemical potentials of the denser phase.

The bulk equilibrium calculations are a good guide to the conditions under which nuclei are likely to be present, but do not give any indication of the nuclear size, which, as in the zero temperature case, is determined by finite nucleus effects. There are quite a number of these, as is discussed in detail by Lamb, Lattimer, Pethick and Ravenhall (1978, 1984), but at low

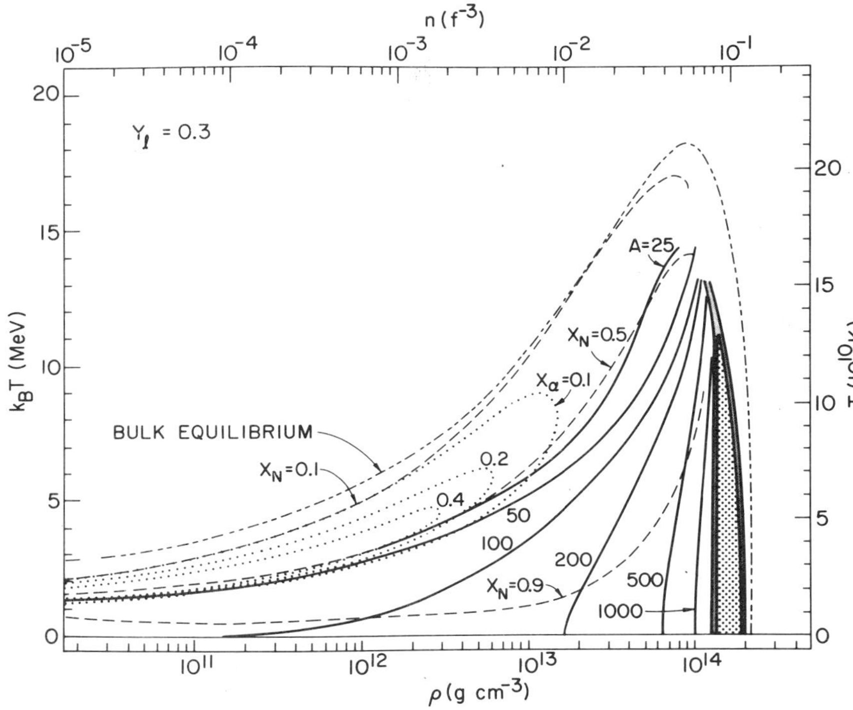

Fig. 2 Properties of hot dense matter at subnuclear densities for a lepton fraction $Y_\ell = 0.3$. The line labelled 'bulk equilibrium' denotes the boundary of the two-phase region when surface and Coulomb effects are ignored. Finite nucleus effects were calculated using the methods of Lamb, Lattimer, Pethick and Ravenhall (1978, 1984). Contours of constant nuclear mass number A are shown, as well as the mass fractions of nucleons in nuclei. X_α is the fraction of baryons in alpha particles, and X_N the fraction in heavy nuclei (including alpha particles). Bubbles are present in the dotted region, and mixtures (nuclei-bubbles, bubbles-uniform matter, or nuclei-uniform matter) in the surrounding grey region.

temperatures, the most important finite nuclear size contributions to the free energy are those from the Coulomb interaction and the nuclear surface. As long as the nuclei are highly correlated, that is the Coulomb lattice energy $\sim Z^2e^2/r_c$ is large compared with the temperature T, or equivalently that the dimensionless coupling constant $\Gamma = Z^2e^2/(r_cT)$ is large compared with unity, one may neglect finite temperature contributions to the Coulomb free energy, which is therefore given by (20). The surface thermodynamic potential is given by the analogue of (9) :

$$\Omega_{\mathbf{surf}} = 4\pi r_N^2 \sigma \ , \tag{32}$$

where the surface tension now depends on temperature. With increasing temperature, the nuclear surface becomes increasingly diffuse, which leads to a decrease in the surface tension. Calculations of the dependence of the surface tension on proton concentration and temperature have been carried out in the Thomas-Fermi approximation by Ravenhall, Pethick and Lattimer (1983), using the Skyrme I' nucleon-nucleon interaction.

The optimal nuclear size is given by the condition

$$\Omega_{\mathbf{surf}} = 2E_{\mathbf{Coul}} \tag{33}$$

which is a generalization of Eq. (11), or with the values of the coefficients given before

$$A = \frac{12.5}{x^2} \ \frac{\sigma(x,T)}{\sigma(0.5\,,T=0)} \ \frac{n_{\mathbf{i}}(x=0.5)}{n_{\mathbf{i}}(x)} \ \frac{1}{1-\frac{3}{2}u^{1/3}+\frac{u}{2}} \tag{34}$$

which is a generalization of Eq. (24). This result enables us to understand the variation of A with density and temperature obtained in the detailed calculations of Lamb, Lattimer, Pethick and Ravenhall (1978, 1984), which are displayed in Fig. 2. With increasing density, u increases, which leads to a decrease in the Coulomb energy and an increase in A. As the temperature increases, the surface tension decreases, and therefore A decreases.

So far we have assumed nuclei to be spherical, but that ceases to be the case at densities approaching nuclear saturation density, since the Coulomb potential due to neighbouring ions has the symmetry of the crystal lattice and is therefore aspherical. This leads to a number of interesting consequences for nuclear shapes. For simplicity, we consider the case of zero temperature, and take into account only bulk, surface and Coulomb contributions to the energy. We also consider proton concentrations so high that there are no neutrons outside the dense nuclear matter phase. When nuclei fill half of space $(u=\frac{1}{2})$, the Coulomb and surface energies are both unchanged if nuclei are turned inside out, so that matter consists of a lattice of spherical voids or bubbles in otherwise uniform nuclear matter (Baym, Bethe and Pethick, (1971); Lamb, Lattimer, Pethick and Ravenhall,(1978, 1983)). When $u<\frac{1}{2}$, the state with nuclei has the lower energy, while when $u>\frac{1}{2}$, the bubble phase is energetically more favourable. The transition between the nuclei and bubble states occurs when $u \simeq \frac{1}{2}$, or $n \simeq n_s/2$. In Fig. 2, the region where the bubble state is favoured is also shown. Detailed calculations of the nuclei-bubble transition have been carried out by Lamb, Lattimer, Pethick and Ravenhall (1983).

At this point it is perhaps useful to investigate how the results of the generalized liquid-drop model approaches compare with those of other calculations. Calculations of nuclei in dense matter have been carried out in the Thomas-Fermi approximation; this takes into account the curvature of the nuclear surface and therefore implicitly takes into account terms in the nuclear mass formula of higher order than those usually considered. Such calculations have been carried out by Arponen (1972) and by Barkat, Buchler and Ingber (1972) for neutron star matter, and by Buchler and Epstein (1980) and by Ogaswara and Sato (1982) for neutron star matter, and the results are in good agreement with those of the compressible liquid drop model. Hartree-Fock calculations have also been carried out: by Negele and Vautherin (1973) for neutron star matter, and by Bonche and Vautherin (1981, 1982) for hot dense matter. These avoid the statistical assumption of the Thomas-Fermi approach, and therefore allow for nuclear shell effects to be included. Because the nuclei of interest are rather large, shell effects are not a dominant effect, and the results of the compressible liquid drop model calculations agree well with the Hartree-Fock ones. For neutron star matter, for example, Negele and Vautherin (1973) find that the optimal Z is 40 up to $\rho \sim 4 \times 10^{12}$ g cm^{-3}, and then increases to 50 up to $\rho \sim 4 \times 10^{13}$ g cm^{-3}, and subsequently decreases slightly at higher densities. On the other hand, Ravenhall, Bennett and Pethick (1972) found Z to be ~ 37 at all densities above neutron drip. The fact that Negele and Vautherin (1973) found Z to be 40 and 50 reflects the fact that these are magic numbers even for neutron-rich nuclei.

Recently it has been found that a number of other shapes are energetically favoured at certain densities. The first possibilities considered (Ravenhall, Pethick and Wilson, (1983)) were rod-like and plate-like nuclei, which for obvious reasons have been dubbed spaghetti and lasagna, respectively, together with the bubble analogue of spaghetti. (Since the bubble analogue of lasagna is still lasagna, it does not represent a new possibility.) With increasing density, the sequence of most favourable nuclear states is spherical nuclei, spaghetti, lasagna, the bubble analogue of spaghetti, spherical bubbles, and, finally, uniform nuclear matter. The densities at which the various states are energetically favoured are shown by the dotted curve in Fig. 3, where N stands for nuclei and B for bubbles. 3 denotes spheres, 2 denotes rods, and 1 denotes plates. U stands for uniform nuclear matter. The curve is symmetrical about $n/n_s = \frac{1}{2}$ if the nuclear compressibility is neglected.

One of the surprising conclusions of these calculations is that very aspherical nuclei are favourable at densities as low as $\sim 0.1\ n_s$. The basic

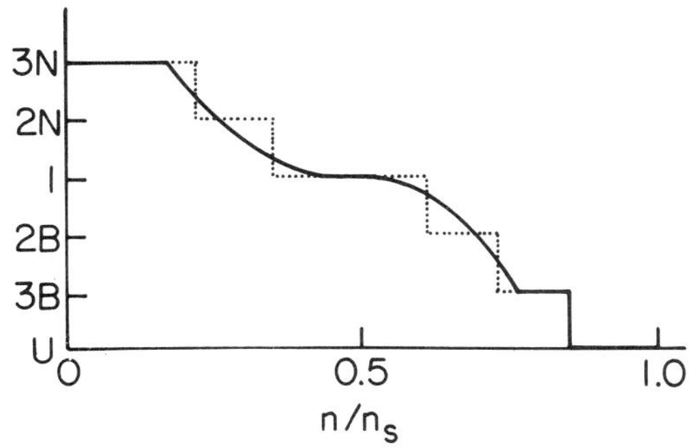

Fig. 3 Optimal dimensionality of nuclei as a function of density. The continuous curve is what is obtained if d is treated as a continuous variable. If d is allowed to take on only the discrete values 1 (plates), 2 (rods) and 3 (spheres) one obtains the dotted line. U denotes uniform matter.

physical reason for this is that with increasing density, the nuclear size increases, due to the reduction of the nuclear Coulomb energy coefficient. Eventually, the nuclear size becomes so large that the nuclei are unstable to spontaneous elongation, as was considered in the classic fission calculation of Bohr and Wheeler (1939). The essential point is that the lattice energy has a large effect on the equilibrium nuclear size, but has much less influence on the fission condition (Pethick and Ravenhall, (1983)).

The simple shapes considered above are clearly over-simplifications of the actual nuclear shapes. For example, the cross section of 'spaghetti' will not be circular, due to the Coulomb interaction with other strands. To attempt to model these more complicated situations, we regard matter as being made up of objects whose shape is specified by a parameter, d, which gives the number of spatial dimensions in which the nuclear matter density varies. The states with nuclei, spaghetti and lasagna therefore correspond to $d = 1, 2$ and 3, respectively. To allow for intermediate shapes, we treated d as a continuous variable, whose value is to be determined by minimizing the energy with respect to d. The optimal value of d is shown in the continuous curve in Fig. 3.

Much work remains to be done on these more exotic phases. As yet, the zero temperature problem of determining the equilibrium shape of incompressible nuclear matter under the influence of Coulomb interactions and surface tension has not been solved completely, even though our simplified calculations have given some clues as to the sort of answer to expect. Once the ground state has been elucidated, it will be possible to embark upon studies of finite temperature properties. We remark in passing that the exotic states bear a number of similarities to some liquid crystal states. For example, the spaghetti state resembles a discotic liquid crystal. This consists of long molecules which are ordered on a lattice in the plane perpendicular to their length, and which are relatively free to move along their length.

To conclude I want to consider the nuclear contribution to the equation of state in stellar collapse. At first one may think that this is unimportant because most of the pressure is provided by degenerate electrons. To understand why the nuclear contribution matters we note that if the equation of state is $P \propto \rho^{\Gamma}$, stable stars can exist in Newtonian gravity only if Γ exceeds 4/3. Similarly, as one would expect, the value $\Gamma = 4/3$ is a critical one in stellar collapse. Now degenerate electrons, which are ultrarelativistic at the densities of interest here, have a pressure which varies as $\rho^{4/3}$, since they behave as an ideal Fermi gas, and therefore have an equation of state which yields stars on the border of instability. (See, e.g., Landau and Lifshitz, (1980), Ch. XI). Consequently, contributions to the equation of state from other sources, such as nucleons, have a decisive influence on the outcome of stellar collapse.

To illustrate the effects of the nuclear pressure on the equation of state we show, in Fig. 4, results calculated using the compressible liquid drop model (Lamb, Lattimer, Pethick and Ravenhall, (1983)). Finite temperature effects are included in the calculations, and the plot is for an entropy per nucleon of 1.0 k_B, which is typical of what is expected in stellar collapse. At nuclear densities this corresponds to a temperature of about 6 MeV. The results are for a fixed *electron* fraction $Y_e = 0.3$, with no neutrinos. Roughly speaking, this corresponds over the densities of the phase changes to a lepton fraction $Y_\ell \sim 0.35$. The only possible configurations of the nuclear matter taken into account are spherical nuclei and spherical bubbles. Over the density range shown, the electrons provide most of the pressure. In the nuclei phase, however, the nuclear pressure is negative at the higher densities, due mainly to the Coulomb lattice effect; the total pressure is reduced by as much as 6%. When there are few nucleons between nuclei, as is the case for the conditions considered here, the nuclear pressure at low temperatures is dominated by the lattice energy. If the atomic number Z were independent of density, the lattice contribution to the nuclear pressure (which is negative) would behave as $\rho^{4/3}$, and therefore the total pressure would behave as $\rho^{4/3}$. However, as we have seen, Z increases as the density increases. Therefore, the magnitude of the lattice pressure term increases more rapidly than $\rho^{4/3}$,

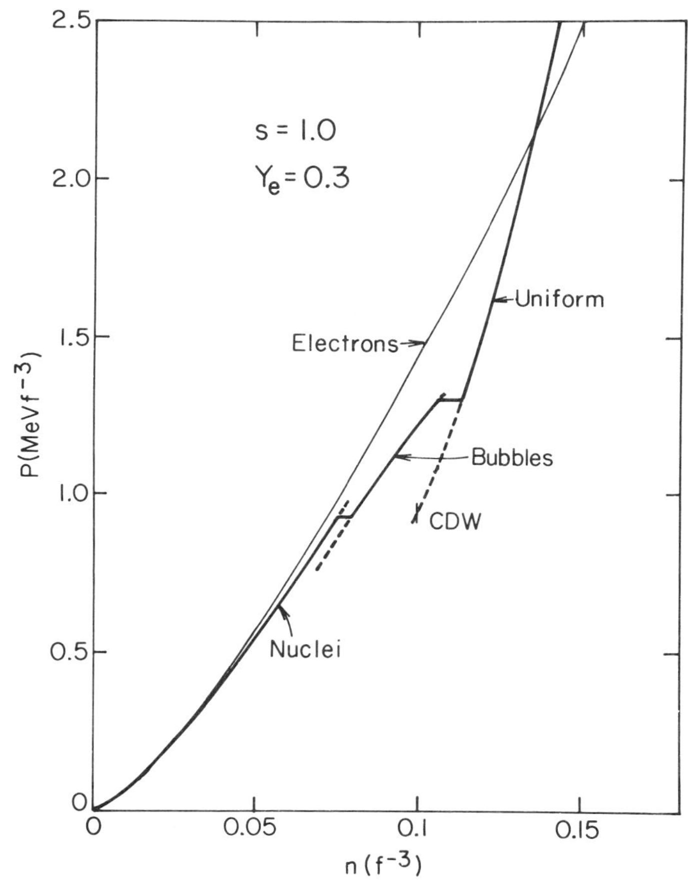

Fig. 4 Pressure versus density for hot dense matter on the adiabat corresponding to an entropy of 1.0 k_B per nucleon and for an electron fraction Y_e = 0.3. (Neutrinos are neglected in this calculation). The dashed portions of the curves correspond to metastable regions, and the vertical line marked 'CDW' denotes the point at which the uniform phase becomes unstable to small distortions leading to a charge density wave. (Pethick, Ravenhall and Lattimer, 1983).

and the total pressure increases less rapidly than $\rho^{4/3}$. In the bubbles phase the negative nuclear pressure is even larger, and can amount to almost 17% of the electronic pressure. This is due to the fact that both the Coulomb energy, which is lowered by making the charge distribution as uniform as possible, and the surface tension tend to favour shrinking the bubbles, and therefore lead to a negative pressure for the nuclear matter. The fact that bubbles can exist only in nuclear matter under tension also accounts for the fact that bubbles disappear at densities somewhat less than the nuclear saturation density. As the density increases further, the pressure rises much more rapidly, due to the large bulk modulus of the uniform nuclear matter. When the more exotic nuclear shapes are allowed for, the single nuclei-bubbles transition will be replaced by a series of rather smooth transitions through the new phases.

So far we have concentrated on astrophysical problems. However, a number of the ideas described above are useful in describing aspects of physics in the laboratory. Astrophysical studies not only benefit from laboratory physics studies, but also provide a stimulus to laboratory physics by expanding the range of conditions under which properties must be understood. As one example of this influence we may cite the renaissance in the study of homogeneous dense matter following the efforts to understand the properties of neutron matter in the context of neutron stars. (For a review, see Pandharipande and Wiringa, (1979)). In the future a number of aspects of physics first investigated in connection with the equation of state in stellar collapse may well be useful in studies of heavy-ion collisions, where the conditions can be rather similar to those in stellar collapse. One of the properties of interest in heavy ion collisions is the fragment size distribution, which is determined by, among other things, the energy of nuclei immersed in a nucleon vapour. In making such calculations the properties of hot uniform matter, and the energy of interfaces between nuclear matter and a vapour are important ingredients. Also the exotic equilibrium shapes of nuclei in hot dense matter may well have consequences for heavy-ion fragmentation.

In summary, the liquid drop model and extensions of it are extremely valuable for describing properties of nuclei under a wide range of conditions. They are sufficiently flexible to take into account a wide range of physical effects, but are at the same time relatively simple computationally. In addition, since contributions to the nuclear energy are directly associated with physical features of the nucleus, the models are valuable for obtaining insight into the physical effects that determine nuclear properties.

Acknowledgements

I am grateful to Jim Lattimer and Geoff Ravenhall for helpful comments in the preparation of this article. I also wish to thank Hanne Bergen, Lise Madsen, and Carol Monsrud for typing the manscript.

The work described in this article was supported in part by the U.S. National Science Foundation Grant NSF PHY 80-25605.

References

Arponen, J., Nucl. Phys., **A 191**, 257 (1972).
Barkat, Z.K., Buchler, J.R. and Ingber, L., Ap. J. **176**, 723 (1972).
Baym, G., Bethe, H.A. and Pethick, C.J., Nucl. Phys., **A 175**, 225 (1971).
Baym, G., and Pethick, C.J., Ann. Rev. Nucl. Sci., **25**, 27 (1975).
Bethe, H.A., Brown, G.E., Applegate, J. and Lattimer, J.M., Nucl. Phys., **A324**, 487 (1979).
Bohr, N., and Wheeler, J.A., Phys. Rev., **56**, 426 (1939).
Bonche, P. and Vautherin, D., Nucl. Phys., **A372**, 496 (1981).
Bonche, P. and Vautherin, D., Astron. and Ap., **112**, 168 (1982).
Buchler, J.R. and Epstein, R.I., Ap. J. Letters, **235**, L91 (1980).
Lamb, D.Q., Lattimer, J.M., Pethick, C.J. and Ravenhall, D.G., Phys. Rev. Letters, **41**, 1623 (1978).
Lamb, D.Q., Lattimer, J.M., Pethick, C.J. and Ravenhall, D.G., Nucl. Phys., **A411**, 449 (1983).
Lamb, D.Q., Lattimer, J.M., Pethick, C.J. and Ravenhall, D.G., (1984). (in preparation).
Landau, L.D. and Lifshitz, E.M., *Statistical Physics*, Part I, Pergamon, Oxford, (1980).
Lattimer, J.M., Ann. Rev. Nucl. Part. Sci., **31**, 337 (1981).
Lattimer, J.M. and Ravenhall, D.G., Ap. J., **223**, 314 (1978).
Mazurek, T.J., Lattimer, J.M. and Brown, G.E., Ap. J., **229**, 713 (1979).
Negele, J.W. and Vautherin, D., Nucl. Phys., **A207**, 298 (1973).
Ogasawara, R., and Sato, K., Prog. Th. Phys., **68**, 222 (1982).
Pandharipande, V.R. and Wiringa, R.B., Rev. Mod. Phys., **51**, 821 (1979).
Pethick, C.J. and Ravenhall, D.G., In *Numerical Astrophysics*, (eds. J. Centrella, J. LeBlanc, M. LeBlanc and R.L. Bowers) (1983).
Pethick, C.J., Ravenhall, D.G. and Lattimer, J.M., Nucl. Phys. A, (1983). (in press).
Ravenhall, D.G., Bennett, C.D. and Pethick, C.J., Phys. Rev. Lett., **28**, 978 (1972).
Ravenhall, D.G., Pethick, C.J. and Lattimer, J.M., Nucl. Phys., **A407**, 571 (1983).
Ravenhall, D.G., Pethick, C.J. and Wilson, J.R., Phys. Rev. Lett., **50**, 2066, (1983).
Vautherin, D. and Brink, D.M., Phys. Lett., **B32**, 149 (1970).

Solar Neutrinos and Comets

A. J. R. Prentice
Department of Mathematics, Monash University, Clayton, Victoria, 3168, Australia

1. INTRODUCTION

It is more than a decade since Prentice and ter Haar (1971) showed how, on the basis of the solid H_2 grain-cloud theory of Reddish & Wickramasinghe (1968, 1969), that the gravitational sedimentation of grains in interstellar clouds may play an important role in the processes of star formation. Prentice and ter Haar found that a cloud of settling H_2 grains is able to shed nearly 99% of its angular momentum through resistive interaction with the uncondensed gases of the cloud. This overcame the so-called angular momentum problem which had plagued earlier purely gaseous theories (Mestel and Spitzer, 1956). In 1974 Duley, however, pointed out that the temperature inside a typical interstellar cloud of hydrogen atom density $n_H \sim 10^4$ atom per cm^3 would probably never fall low enough ($\leqslant 5$ K) for the direct condensation of solid H_2. Instead, one would expect temperatures low enough (10 — 20 K) only for the condensation of various ices of C, N, and O, admixed with rocky silicates.

By the mid 1970's it thus became clear that although the angular momentum problem of star formation may still remain unresolved, one would still expect a chemical fractionation to take place within the dense pockets of interstellar clouds, with the selective gravitational settling of grains rich in CNO towards local centres of gravitational attraction, and the consequent enhancement of elements heavier than H and He at the centres of these pockets. Using this idea, along with some ideas of planetary formation put forward by ter Haar (1972) and Prentice (1974a), I proposed that the primordial sun was born with a small metal-rich core of mass 2—3% $M_\odot$

($M_{\odot}$ = solar mass = 1.989×10^{30} kg) surrounded by a homogeneous envelope of normal solar composition (Prentice 1974b, 1976).

The Solar Neutrino Problem

The suggestion that the sun may possess a small metal-rich core represented a substantial departure from the conventional, initially chemically homogeneous solar model. It was therefore, decided to explore the structural and evolutionary consequences of such a non-standard model. These investigations were carried out in Oxford in 1975 under the generous hospitality and support of Dr ter Haar. Whilst pursuing these investigations, I was particularly mindful of the discrepancy which existed between the solar neutrino flux which was predicted by the standard initially chemically homogeneous solar models and the value which had been measured experimentally by R. Davis Jr. Davis *et al.* (1971), using a Cl^{37} detector located some 1500 m underground in an abandoned lead mine in South Dakota, obtained a solar neutrino count rate of 1.5 ± 1 SNU (1 SNU = 10^{-36} captures/Cl^{37} atom/s). This fell very much short of the theoretically expected value of 9.0 SNU obtained for an initially chemically homogeneous model (Bahcall and Ulrich, 1971). Considerable excitement was thus aroused by the possibility that the new class of chemically inhomogeneous models might be able to resolve the solar neutrino problem.

Oxford Calculations

The outcome of the 1975 Oxford calculations (Prentice, 1976) was as follows. Because of high metal mass fraction $Z_{\mathbf{core}}$ of the central core, it was assumed that the core would be highly convectively unstable and that a large turbulent stress $\langle \rho_t v_t^2 \rangle$ was created by the motions of supersonic rising convective elements. Here ρ_t denotes the smeared-out mass density of the moving material and v_t the turbulent velocity (cf. Prentice, 1973a). The effect of the turbulent stress is to cause the central regions of the core to become much hotter than those of the standard Z model, whilst at the same time causing the layers outside the core to become correspondingly cooler. As a result, especially because of the enhanced mass fraction of CNO, the convective core consumes its store of **H** quite rapidly compared to the regions outside the core. At solar age, a configuration similar to that proposed by Prentice (1973b) is attained consisting of a burnt-out core of mass 2—3% $M_{\odot}$ surrounded by a barely consumed H-rich envelope and having a greatly reduced neutrino flux of order 1.5 SNU. In 1973 I had pointed out that as most of the sun's neutrinos came from the inner 3% of solar mass, then if this region were burnt out the flux might be substantially reduced.

Recent Developments

Since 1976 two independent developments have occurred in the solar neutrino puzzle. First, and more importantly, the discrepancy between standard theory and observation has not disappeared. At last count, Cleveland *et al.* (1981) obtained a value of 2.1 ± 0.3 SNU (1 σ error) whilst the predicted capture rate for the best standard model is 7.6 ± 1.1 SNU (1 σ error) (Bahcall *et al.* 1982). Second, although numerous suggestions have been put forward to explain this discrepancy, so far none of these has been found to be physically satisfactory (Rood, 1978; Bahcall, 1979). In particular, although Rood assessed Prentice (1976) as an exciting approach to the solar neutrino puzzle, there were 3 major aspects of the model required further study before it could be regarded as a viable solution. These questionable aspects were;

(i) The existence of a turbulent stress equal to an order of magnitude or larger than the gas pressure.

(ii) The assumption that a region of high Z is necessarily convective.

(iii) The use of simplified polytropes in calculating the solar structure.

In this paper, written in honour of Dirk ter Haar, I present a brief account of some calculations which have been performed over the past nine months. These calculations address themselves to criticisms (i) and (ii) raised by Rood (1978) above and form the basis of a new solution to the solar neutrino puzzle. They incorporate new data on the solar composition that has been garnered over the past seven years, both directly from spectroscopic studies of the solar surface and, indirectly, from the NASA Pioneer and Voyager space missions to Jupiter and Saturn. The calculations also employ improved data on nuclear reaction rates.

Quite unexpectedly, the new calculations lead to the realization that there is a direct connection between the solar neutrino puzzle and the formation of comets. It is a fitting recognition of the broad range of ter Haar's scientific interests, that the ideas of star formation which he and I put forward nearly thirteen years ago (Prentice and ter Haar, 1971) should provide the cornerstone to a possible solution to both these cosmogonic problems.

2. Assumptions of the New Solar Models

Consider now in detail the assumptions underlying the construction of the new class of chemically inhomogeneous solar models.

(i) The initial solar model consists of a small metal-rich core of fractional mass q_{core} and metal fraction Z_{core} surrounded by an envelope having

H mass fraction $X_{env} = 0.7835$, He mass fraction $Y_{env} = 0.1963$, and metal fraction $Z_{env} = 0.0202$. These values follow from the solar abundance compilation of Ross and Aller (1976), augmented with more recent measurements of Lambert (1978) and Lambert and Luck (1978). It is worth noting that Y_{env} here is appreciably smaller than the value 0.25 ± 0.01 that is derived by theoretical means from conventional standard solar models (cf. Bahcall *et al.* 1982). Nevertheless, it is precisely in line with the value obtained by the Voyager IRIS experiment at Jupiter, as well as with other direct solar measurements (Gautier *et al.* 1981).

(ii) The core is assumed to have formed from solid grains consisting of rocky silicates and CNO ices. All elements, with the exception of the uncondensing H_2, He and Ne, are assumed to be present in the same relative proportions as in the envelope. Both He and Ne are absent, whilst H is present only through its molecular incorporation in CNO ices. Three different classes of CNO ices, with correspondingly different H fractions X_{core} and metal fractions Z_{core}, are considered, as detailed in Table 1.

Table 1

Possible Initial Chemical Compositions of the Solar Core

Mix	Chemical Constituents	H fraction X_{core}	Metal fraction Z_{core}
	Anhydrous rock (MgO, SiO_2, FeO, ...,)		
1	H_2O, CH_4, NH_3, Ar	0.123	0.877
2	Rock; H_2O, C, N_2, Ar	0.051	0.949
3	Rock; H_2O, CO, N_2, Ar	0.0165	0.9835

(iii) The core is assumed *a priori* to be convectively unstable, with density ρ and temperature T related polytropically by the equation $\rho = K_{core} T^{n_{core}}$, where K_{core} is a constant and $n_{core} < 1.5$ is the core polytropic index. The envelope is assumed to be radiatively stable, with ρ and T related polytropically by $\rho = K_{env} T^{n_{env}}$. Again K_{env} is a constant and $n_{env} > 1.5$.

(iv) Next, it is assumed that if the core is strongly convectively unstable, that a powerful turbulent stress is created by the motions of overshooting convective elements. This stress is given by

$$\langle \rho_t v_t^2 \rangle = \beta \rho G M(r)/r \ , \qquad (1)$$

where $M(r)$ is the mass interior to radius r and β is the so-called turbulence parameter (Prentice, 1973a; see also ter Haar, 1950; Prentice, 1978a, 1978b; Prentice and ter Haar, 1979a, 1979b).

Prentice and ter Haar (1979a, b) pointed out that the physical and chemical structure of the Galilean satellite system of Jupiter can be understood if one takes into account the phenomenon of supersonic convective turbulence within the proto-planetary cloud which contracted gravitationally to form this planet. The detailed chemistry of the Saturn system can also be understood in terms of this phenomenon (Prentice 1981, 1983). In fact, Prentice (1980) predicted, prior to the Voyager encounters at Saturn, that the regular satellites of this planet would show the same general compositional trends as those seen in the Galilean moon and planetary systems. In particular, one would expect the regular satellites to become steadily rockier moving towards the planet. This viewpoint was strengthened by the Voyager 2 encounter in August 1981 (Tyler *et al.* 1982; Stone and Miner, 1982) although it had earlier appeared, on the basis of Voyager 1 and ground-based data alone, that the opposite trend was true (cf. Smith *et al.* 1981).

Another theoretical prediction which is based on ideas of turbulent stress and which has been confirmed very recently, concerns the planet Neptune. According to Prentice and ter Haar (1979a, b) Neptune, unlike Jupiter, Saturn and Uranus, should not possess a ring system. On 15th June, 1983 a rare central occultation occurred between Neptune and a distant red star of nearly the same magnitude as that planet. The event took place under ideal observing conditions in Australia and was recorded by many astronomers. They included a large group from the University of Arizona, led by W.B. Hubbard, as well as a group at the Anglo-Australian Telescope at Coonabarabran, N.S.W., led by D.A. Allen, who were working at both optical and infrared wavelengths. None of these observers found any evidence for rings around Neptune (see Marsden, 1983).

In conclusion, therefore, we feel that there is strong observational support for admitting the possibility that turbulent stress may play an important role in cosmogony.

Returning now to the problem in hand, we suppose that once the rising metal-heavy convective elements have left the core they proceed to decelerate to rest through the negative buoyancy force created by the metal-light environment of the radiatively stable envelope. It is assumed that $\langle \rho_t v_t^2 \rangle$ declines almost linearly with increasing fractional mass $q(r) = M(r)/M$ in the interval $q_{\mathbf{core}} < q < q_{\mathbf{osh}}$ in the manner described by Stokes (1977), where $q_{\mathbf{osh}}$ is the point where all motion ceases. Stokes considered the case of a two-layered star, of uniform densities, with the convective element having a uniform density intermediate to that of the two layers.

(v) The remaining equations governing the model are the same as in

Prentice (1976). Nuclear reaction rates are taken from Clayton (1968) and revised, where possible, with Bahcall *et al.* (1982).

3. Calculations and Results

The Standard Solar Model

In order to determine the envelope polytropic index n_{env}, an initially chemically homogeneous model having the same chemical composition as that of Bahcall *et al.* (1982), namely $X = 0.7322$, $Y = 0.2499$, $Z = 0.0179$, was evolved. To achieve solar luminosity $\mathcal{L}_{\odot} = 3.86 \times 10^{33}$ erg/s at solar age $t_{\odot} = 4.6 \times 10^{9}$ yr we require $n_{env} = 3.064$.

Chemically Inhomogeneous Models

Having fixed n_{env}, a series of chemically inhomogeneous models was evolved for the three compositional cases given in Table 1, and for different choices of the core size q_{core} and overshoot point q_{osh}. For each model the parameters β and n_{core} are fixed from the conditions of correct final luminosity and that the core burns itself out at time $t_U = 4.03 \times 10^{9}$ yr, determined by the base of the Cambrian Period in the Earth's geological record (see Prentice, 1976). Typically we find $\beta \simeq 0.2$ and $n_{core} \simeq 0.4$.

Figure 1 gives a plot of the total neutrino flux at solar age $F_{\nu\odot}$ for a Cl^{37} detector versus size of the core for the case $Z_{core} = 0.95$. We see that $F_{\nu\odot}$ falls very sharply with increasing core size q_{core}. In order to account for the 2.1 ± 0.3 SNU measurement of Cleveland *et al.* (1981) we require $q_{core} \simeq 0.01$. This result is very insensitive to the choice of q_{osh}, provided $q_{osh} \geqslant 0.1$.

Validity of the Assumptions

For the models with $q_{core} = 0.01$ and $q_{osh} = 0.25$, the turbulent stress acquires a maximum value of about 0.2 times the gas pressure at the edge of the convective core. This overcomes Rood's (1978) first criticism, mentioned above. Moreover, the fact that the turbulent stress is everywhere much less than the thermal gas pressure implies that the convective motions are probably not supersonic. Independent calculations of P. Cleary of Monash University support this conclusion.

As the numerical computations show that $n_{core} \lessdot 1.5$, the core is clearly strongly superadiabatic. Even so, the absolute test for convective instability rests with the Schwarzschild criterion (Cox and Guili, 1968). Figure 2 gives a plot of the ratio S of the radiative temperature gradient to the adiabatic gradient versus the cube root of the fractional mass q in the star for the three cases of Z_{core}. We see that in all cases $S_{core} \gg 1$, indicating that the core is indeed very convectively unstable, as was assumed in the model.

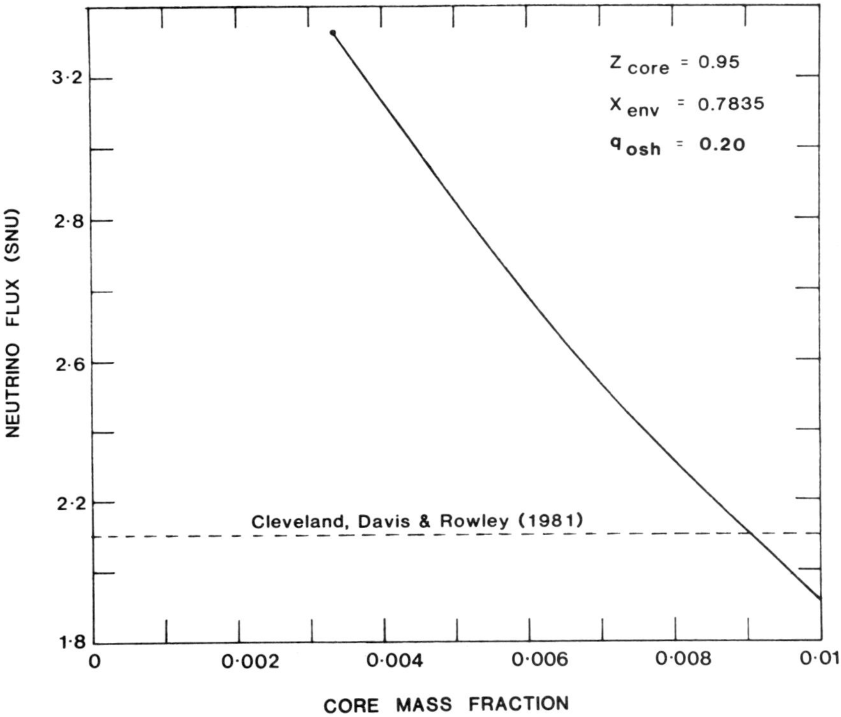

Fig. 1 The expected neutrino capture for the chemically inhomogeneous solar models, plotted against the mass fraction q_{core} covered by the central burnt-out 'cometary' core. The core metal fraction is $Z_{core} = 0.95$ and the overshoot mass point is 0.20. No models can be constructed if $q_{core} \leqslant 0.0035$. All neutrino fluxes have been standardized against those of an evolved chemically homogeneous polytropic solar model which has the same initial elemental abundances, initial luminosity $\mathcal{L}_0 = 0.6943\ \mathcal{L}_\odot$, and final neutrino flux $F_{\nu\odot}$ = 7.64 SNU as that of the detailed standard solar model of Bahcall *et al.* (1982). The broken line indicates the neutrino capture rate measured in the Davis experiment, for a Cl^{37} detector,

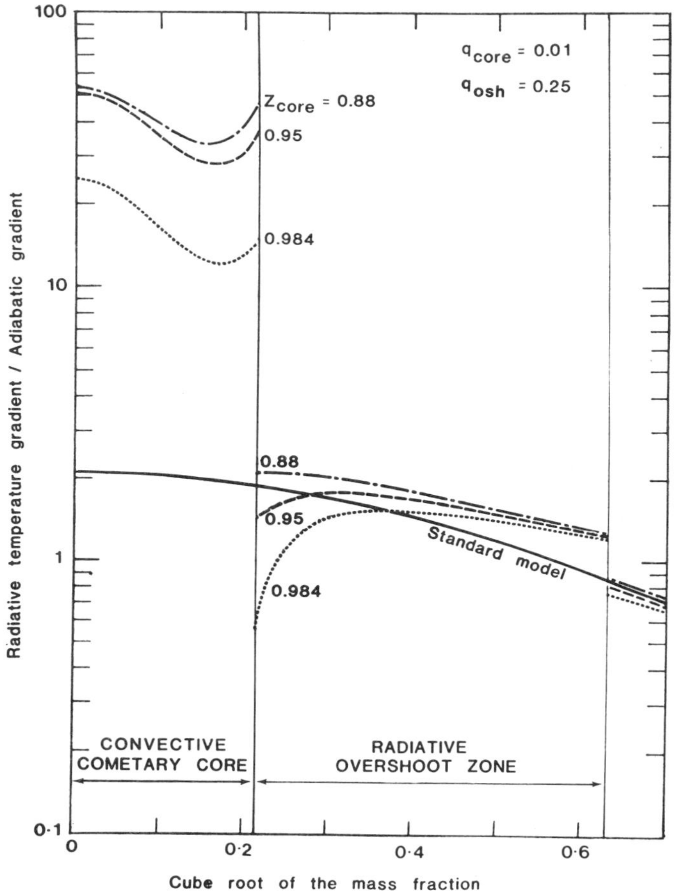

Fig. 2 Test of the validity of the assumptions of convective instability and radiative stability of the core and overshoot zone of the initial (zero-age) chemically inhomogeneous solar models. The core mass fraction is set at $q_{core} = 0.01$ and the overshoot mass point is $q_{osh} = 0.25$. The broken lines give the ratio S of the radiative temperature gradient to the adiabatic temperature gradient for the three different cases of Z_{core} defined in Table 1. The solid line is the value of S for the initial chemically homogeneous polytropic solar model described in the text and Fig. 1. S has been calculated using the stellar opacity approximation formulae assembled by Cox and Guili (1968, Ch. 16).

The run of S with $q^{1/3}$ for the initial (t = 0) standard homogeneous model is shown by the heavy line in Fig. 2. Comparing this with the runs for the inhomogeneous models we see that the models with $Z = 0.95$ and 0.984 are more stable just outside the core edge than the standard model, whilst the 0.88 model is less stable. Bearing in mind that the main sequence standard model is known from more detailed studies (J. Lattanzio, private communication) to be convectively stable, the present calculations indicate that the 0.95 model, and certainly the 0.984 model, satisfy the necessary assumption of radiative stability outside the core edge. The core thus remains chemically intact. For the 0.88 model at the point $q = q_{core}$, however, $S_{env}(q_{core}, t = 0) = 2.1$ and $S_{env}(q_{core}, t)$ increases with time t to a maximum value of 2.8 at 2×10^9 yr. For the other models S_{env} decreases with time. The 0.88 model thus does not satisfy the stability criterion and so must be rejected. In physical terms, the envelope and core of this model would mix together at their common boundary, thus dissolving the proposed initial chemical inhomogeneity.

4. Conclusions and Discussion

We have seen how a new class of chemically inhomogeneous solar models can be constructed which leads to a present-day sun possessing a small burnt-out core consisting of at least 95% of elements heavier than He. Such models with a core mass fraction of 1% have a neutrino flux of order 2 SNU, in line with the observed value.

We have shown how these models overcome some of the difficulties that surrounded the earlier inhomogeneous models of Prentice (1976), as voiced by Rood (1978). Nonetheless, we feel that the continued use of polytropic relations between density and temperature in the stellar structure equations, as a means for facilitating the numerical modelling, is still justified in view of the pioneering nature of this work and the need for rapidly evaluating the role played by the various parameters which enter this problem.

The hypothesis that the sun may contain a small central core containing at least 95% metals has very important implications regarding the very early stages of the sun's formation. We have shown that in the initial main-sequence sun, the core has a chemical composition equivalent to that of a rock-ice mix in which C and N are present in very primitive molecular forms, as CO or N_2, rather than as the hydrogenated structures CH_4 and NH_3. This suggests an origin for the core material in the cool and tenuous environment of an interstellar cloud, rather than in the warm and dense regions of the proto-solar cloud (Prentice, 1983). The existence of such a chemically inhomogeneous solar model is consistent with theories of star formation which are based on the gravitational sedimentation of solid grains rather than simple collapse of gaseous fragments (Prentice & ter Haar, 1971; Krautschneider 1977; see also McCrea and Williams, 1965).

The Role and Origin of Comets

How then do comets fit into this picture? Comets have an optical spectrum indicating a composition very similar to that we have found for the solar core (cf. Greenberg, 1982). It is known that they occupy a spherical region of space which is very similar in dimension to that of interstellar clouds (Oort, 1950). This leads us to the conclusion that the comets may have formed through gravitational fragmentation of the original cloud of dirty CNO grains which we have suggested sedimented to form the sun's core. Such gravitational fragmentation of compressed grain clouds is known to occur in the related study of planetesimal formation (Hourigan and Prentice, 1979). After fragmentation has occurred, the majority of 'cometesimals' would continue to co-settle towards the centre of the cloud. Presumably, however, through processes of collisional and gravitational interactions, a small handful might acquire sufficient random orbital angular momentum to avoid accumulation with the main cometary mass. These 'strays' would form the population of comets which occupy the Oort cloud.

If these ideas should prove to be tenable, it is amazing to consider that the stray comets which visit our solar system today are the long-lost brothers of those early home-seeking comets. Even so, those early comets are not forgotten as their ashes today make their presence felt in a greatly reduced neutrino flux.

Acknowledgements

I thank P. Cleary for assistance with some of the numerical computations and Monash University for the award of a Special Research Grant. I also thank K. Hourigan, M.J. Krautschneider, J. Lattanzio and K.C. Westfold for useful discussions.

References

Bahcall, J.N., Space Sci. Rev., **24**, 227 (1979).

Bahcall, J.N. and Ulrich, R.K., Astrophys, J., **170**, 593 (1971).

Bahcall, J.N., Huebner, W.F., Lubow, S.H., Parker, P.D. and Ulrich, R.K., Rev. Mod. Phys., **54**, 767 (1982).

Clayton, D.D., *Principles of Stellar Evolution and Nucleosynthesis*, McGraw-Hill (New York, 1968).

Ckeveland, B.T., Davis, R., Jr. and Rowley, J.K., *Weak Interactions as Probes of Unification*, In AIP Conference Proceedings No.72, (eds. G.B. Collins, L.N. Chang and J.R. Fience), p.322 (1981).

Cox, J.P. and Guili, R.T., *Principles of Stellar Structures*, Vol.1, Gordon and Breach (New York, 1968).

Davis, R., Jr., Rogers, L.C. and Radeka, A., Bull. Amer. Phys. Soc., **16**, 631 (1971).
Duley, W.W., Astrophys. Space Sci., **26**, 199 (1974).
Gautier, D., Conrath, B., Flaser, M., Hanel, R., Kunde, V., Chedin, A., and Scott, N., J. Geophys. Res., **86**, 8713 (1981).
Greenberg, J.M., In *Comets*, (ed. L.L. Wilkening) University of Arizona Press (Tucson), p.131 (1982).
Haar, D. ter, Astrophys. J., **111**, 179 (1950).
Haar, D. ter, In *Symposium on the origin of the Solar System*, (ed. H. Reeves) Centre National de la Recherche Scientifique (Paris), p.71 (1972).
Hourigan, K., and Prentice, A.J.R., Proc. Astron. Soc. Aust., **3**, 389 (1979).
Krautschneider, M.J., Astron. Astrophys., **57**, 291 (1977).
Lambert, D.L., Mon. Not. R. Astron. Soc., **182**, 249 (1978).
Lambert, D.L., and Luck, R.E., Mon. Not. R. Astron. Soc., **183**, 79 (1978).
McCrea, W.H. and Williams, I.P., Proc. Roy. Soc., **A287**, 143 (1965).
Marsden, B.G., I.A.U. Circular No. 3831 (1983).
Mestel, L. and Spitzer, L., Mon. Not. R. Astron. Soc., **116**, 503 (1956).
Oort, J.H., Bull.Astron.Inst. Neth., **408**, 91 (1950).
Prentice, A.J.R., Astron. Astrophys., **27**, 237 (1973a).
Prentice, A.J.R., Mon. Not. R. Astron. Soc., **163**, 331 (1973b).
Prentice, A.J.R., In *In the Beginning ..... the Origin of Planets and Life*, (ed. J.P. Wild), Australian Acad. Sci. (Canberra), p.15 (1974a).
Prentice, A.J.R., Proc. Astron. Soc. Aust., **2**, 263 (1974b).
Prentice, A.J.R., Astron. Astrophys., **50**, 59 (1976).
Prentice, A.J.R., In *The Origin of the Solar System*, (ed. S.F. Dermott), Wiley (London), p.111 (1978a).
Prentice, A.J.R., The Moon and the Planets, **19**, 341 (1978b).
Prentice, A.J.R., Phys. Letts., **80A**, 205 (1980).
Prentice, A.J.R., JPL Publication 81-79, Jet Propulsion Laboratory (Pasadena) (1981).
Prentice, A.J.R., Aust. Phys., **20**, 37 (1983).
Prentice, A.J.R., and Haar, D. ter, Mon. Not. R. Astron. Soc., **151**, 177 (1971).
Prentice, A.J.R., and Haar, D. ter, Nature, **280**, 300 (1979a).
Prentice, A.J.R., and Haar, D. ter, The Moon and the Planets, **21**, 43 (1979b).
Reddish, V.C., and Wickramasinghe, N.C., Nature, **218**, 661 (1968).
Reddish, V.C., and Wichramasinghe, N.C., Mon. Not. R. Astron. Soc., **143**, 189 (1969).
Rood, R.T., In *Proceedings of the Brookhaven Solar Neutrino Conference*, (ed. G. Friedlander, Vol. 1, p.175 (1978).
Ross, J.E., and Aller, L.H., Science, **191**, 1223 (1976).
Smith, B.A., and Voyager Imager Team, Science, **218**, 169 (1981).
Stone, E.C., and Miner, E.D., Science, **215**, 499 (1982).
Tyler, G.L., Eshleman, V.R., Anderson, J.D., Levy, G.S., Lindal, G.F., Wood, G.E., and Croft, T.A., Science, **215**, 553 (1982).

Origin and Propagation of Cosmic Rays

S. Ramadurai
Tata Institute of Fundamental Research, Homi Bhabha Road, Bombay 400 005, India

1. Introduction

Dirk ter Haar has taken active interest in the study of cosmic radiation, whose extraterrestial nature was established by a series of balloon flights by Victor Hess and coworkers during the period 1910 – 1920, just prior to Dirk's birth. He was the first person to speculate on the possibility of molecule formation around novae (ter Haar, 1946). These molecules form into grains which are thought to be accelerated and which may serve as a source of cosmic rays (Epstein, 1980; Tarafdar and Apparao, 1981). During the late forties and early fifties, the origin of cosmic rays was investigated by him. Recently the problem of propagation of these particles has been engaging his interest (Carvalho and ter Haar, 1979). The acceleration of these particles is most likely due to some form of plasma turbulence, a subject which is very close to his present interest. Hence it is a great pleasure for me to discuss this subject in the article for his Festchrift.

Until the large accelerators were built in the early fifties, cosmic rays remained the only source of high energy particles. Hence emphasis in the study of cosmic radiation until that time was on the nature and study of the elementary particles and their interactions. However, even at that time, acceleration and origin of these particles were being discussed in the appropriate astrophysical context. At present except for the highest energy regions ($> 10^{12}$ eV or so) accelerator study has taken over the role of cosmic rays in high energy physics and cosmic ray studies have taken their rightful place as an integral part of high energy astrophysics. Current studies concentrate on delineating the composition, energy spectrum, energy requirements and acceleration and propagation processes of this radiation. These in turn yield information about the nature of the sources of these particles. The astrophysical investigations on the nucleosynthetic processes as well as the various radiative and dynamical processes taking place in stars, interstellar medium, galaxies and beyond, help us to identify possible sources of cosmic rays. In this article a brief overview of these interesting investigations will be given.

2. Experimental Information

After nearly seven decades of investigation, it is now firmly established that cosmic rays consist of charged particles — electrons, protons, nuclei up to uranium and beyond. Similarly it is now clear that cosmic rays have almost a continuous energy spectrum right up to 10^{19} eV and probably higher energies. Naturally, in order to obtain this information, a variety of experimental techniques were developed, like photographic nuclear emulsions, cloud chambers, Cerenkov-counters, dE/dX counters, plastics etc. Since our main purpose is to understand the origin and propagation of these particles, we will restrict ourselves to the discussion of only the results using these techniques. Since the nature of the energy spectrum is the first one to be determined historically, we will consider that first and then go on to the elemental and isotopic composition.

2.1 *Energy Spectrum*

The energy spectrum yields two important pieces of information. First, the shape of the spectrum gives us a clue to the propagation of these particles from their source to the earth. This is especially true for the energy region below 1 GeV where the presence of solar wind determines the propagation through interplanetary space. Second, the nature of acceleration and escape of the particles from the confinement region is directly related to the energy spectrum. Thus if there is a break in the energy spectrum, it might mean a change in the composition or contributions from different sources etc. Since the flux is larger at lower energies, it is possible to collect enough particles in a balloon/space flight spanning a reasonable period. Hence fairly detailed investigations are accomplished in this region. At energies above 10^{14} eV or so the flux is obtained by using extensive air-shower arrays (Sreekantan, 1972). Consequently what is known reliably at present beyond this energy is the flux of total number of nuclei and not the individual nuclei as at lower energies. In Fig. 1, we have plotted the differential energy spectrum of p, He, C, O and Fe up to about 10 GeV/nucleon. Above 10^{12} eV, the integral energy spectrum is known and it is given in Fig. 2. The energy spectra clearly reveal the following features. There is a maximum

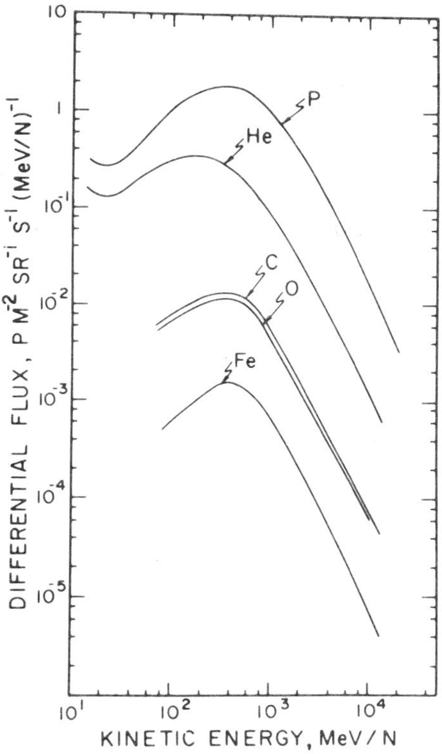

Fig. 1 Differential energy spectra of proton, helium, *C*, *O*, *Fe*, nuclei near solar minimum. (Biswas & Durgaprasad, 1980)

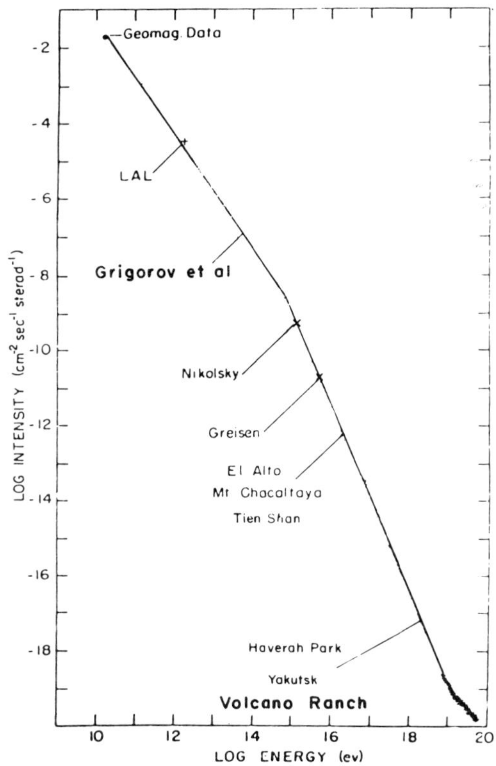

Fig. 2 Integral energy spectrum of cosmic ray nuclei.

in the differential energy spectrum around 400 MeV/n. This is due to the solar modulation, whose broad features are fairly understood. Below about 50 MeV/n there is a break in the spectrum; the spectral slope changes from positive to negative index. This is taken as an indication of the presence of 'anomalous' component (Biswas and Durgaprasad, 1980).

There is a clear break in the integral energy spectrum at about 10^{15} eV as seen in Fig. 2. This break may be due to the change in composition, or due to change in the interaction characteristics which are used in arriving at the flux from the EAS data. At energies beyond 10^{17} eV there are great differences in data. While most of the data seem to indicate a flattening of the spectrum, the Sydney group data indicate that there is no flattening and the spectrum continues to have the same slope (Linsley, 1981). The break will imply the presence of a new component, most probably extragalactic and hence a great interest in this part of the spectrum. The energy beyond 10^{19} eV is subject to a lot of uncertainties. There are claims that the energy spectrum extends up to 10^{20} eV and there is no indication of the presence of a steepening of the energy spectra expected on the basis of photonuclear reactions with the microwave photons (Greisen, 1966; Zatsepin and Kuzmin, 1966). But there are contrary claims to the effect that the spectrum beyond 10^{19} eV shows a steepening. In summary one can say that the energy spectrum can be represented by

$$N(>E) \propto E^{-\alpha} , \qquad (1)$$

where $\alpha \sim 1.6$ up to energies of $\sim 10^{15}$ eV, $\alpha \sim 2.2$ from $10^{15} - 10^{19}$ eV and $\alpha \sim 1.5$ or less beyond 10^{19} eV. Below a few GeV, the spectral shape is determined by solar modulation resulting in a maximum in the energy spectrum around 400 MeV/n. Below 50 MeV/n, there is another feature referred to as anomalous component with a positive spectral slope (Biswas and Durgaprasad, 1980).

2.2 *Elemental Composition*

The clues to the understanding of the origin and propagation of cosmic rays came mainly from the study of the elemental composition of the radiation. It was only in 1949 that nuclei heavier than helium in cosmic rays were first identified. Since then, using detectors borne aboard balloons and spacecrafts, nuclei up to uranium have been detected. As a result, detailed comparison with the elemental abundances determined in various astronomical bodies is now possible. In Tables 1 and 2 the cosmic ray abundances determined at the surface of the earth are given along with the universal abundance (Cameron, 1980). It is immediately clear from a comparison of the two abundances that the light elements like *Li*, *Be*, *B* are enriched by several orders of magnitude compared to the universal abundance. This gives us important information

Table 1

Chemical Abundance of Cosmic Rays and Solar System abundances (taken from Longair, 1981)

Element	z	Cosmic Rays	Solar System
H	1	26000	270000
He	2	3600	18728
Li	3	18.0 ± 2.0	4.2×10^{-4}
Be	4	10.5 ± 1.0	6.9×10^{-6}
B	5	28.0 ± 1.0	3.0×10^{-3}
C	6	100	100
N	7	25.0 ± 2.0	31.7
O	8	91.0 ± 4.0	182
F	9	1.7 ± 0.4	2.1×10^{-2}
Ne	10	16.0 ± 2.0	29.2
Na	11	2.7 ± 0.4	0.51
Mg	12	19.0 ± 1.0	8.99
Al	13	2.8 ± 1.0	0.72
Si	14	14.0 ± 2.0	8.47
P	15	0.6 ± 0.2	8.1×10^{-2}
S	16	3.0 ± 0.4	4.24
Cl	17	0.5 ± 0.2	4.83×10^{-2}
A	18	1.5 ± 0.3	0.99
K	19	0.8 ± 0.2	3.6×10^{-2}
Ca	20	2.2 ± 0.5	0.611
Sc	21	0.4 ± 0.2	3.0×10^{-4}
Ti	22	1.7 ± 0.3	2.35×10^{-2}
V	23	0.7 ± 0.3	2.22×10^{-3}
Cr	24	1.5 ± 0.4	0.108
Mn	25	0.9 ± 0.2	7.88×10^{-2}
Fe	26	10.8 ± 1.4	7.03
Co	27	< 0.2	1.87×10^{-2}
Ni	28	0.4 ± 0.1	0.407

Table 2

Chemical Abundances of Cosmic Rays and Solar System Abundances

Charge	Cosmic Rays near Earth*	Cosmic Rays Source*	Solar System Cameron (1980)
26	6.3×10^6	10^6	10^6
34	433	53 ± 15	83
36	199	19 ± 12	54
38	270	42 ± 9	31
40	110	17 ± 13	17
42	71	6 ± 4	4.9
44	63	5.7 ± 3.8	2.3
46	36	2.1 ± 3.1	1.9
48	51	6.0 ± 3.5	2.1
50	37	-0.2 ± 3.1	4.4
52	41	5.3 ± 2.3	8.1
54	28	-0.7 ± 2.8	7.4
56	52	7.5 ± 3.0	5.8
58	120	1.5 ± 1.9	1.6
60	14.0	1.1 ± 1.7	0.98
62	11.1	-0.6 ± 1.4	0.32
64	19.4	2.0 ± 1.7	0.56
66	18.5	3.5 ± 1.8	0.50
68	0.8	-1.6 ± 1.1	0.33
70	9.3	1.1 ± 1.3	0.26
72	8.1	0.4 ± 1.3	0.22
74	11.1	2.4 ± 1.5	0.37
76	5.6	0.1 ± 1.3	1.20
78	19.3	5.2 ± 2.0	2.08
80	6.0	1.2 ± 1.6	0.46
82	10.7	3.4 ± 1.7	3.07
84	2.4	0.7 ± 1.2	0.08
88	1.0	0.3 ± 0.3	—
90	2.0	0.7 ± 0.5	0.08

* **Cosmic ray abundances are averages of Fowler *et al.* (1981) and Binns *et al.* (1981, 1983)**

about the changes taking place during the propagation of cosmic rays from source to the earth. The propagation has a direct bearing on the abundances of higher charge species, as can be seen from the source abundances shown in Table 2 compared to the observed abundances. This is because the interaction mean free path of the ultra heavy nuclei are comparable to or less than the amount of matter traversed by cosmic rays from the source to the earth. Another aspect which is obvious from Table 2 is the enrichment in cosmic rays of $74 \leqslant Z \leqslant 84$ compared to the solar system abundance. All the data shown above in Table 1 refer to energies below 500 MeV/n. At higher energies the ratio of secondaries like *Li*, *Be*, *B*, Sub-*Fe* group etc. to the primaries decrease. The energy dependence of the secondary to primary ratio is shown in Fig. 3. There is a clear indication for the decrease of pathlength with increasing energy. This has a bearing on the confinement of cosmic rays in the galaxy. These propagation features will be discussed later in this article.

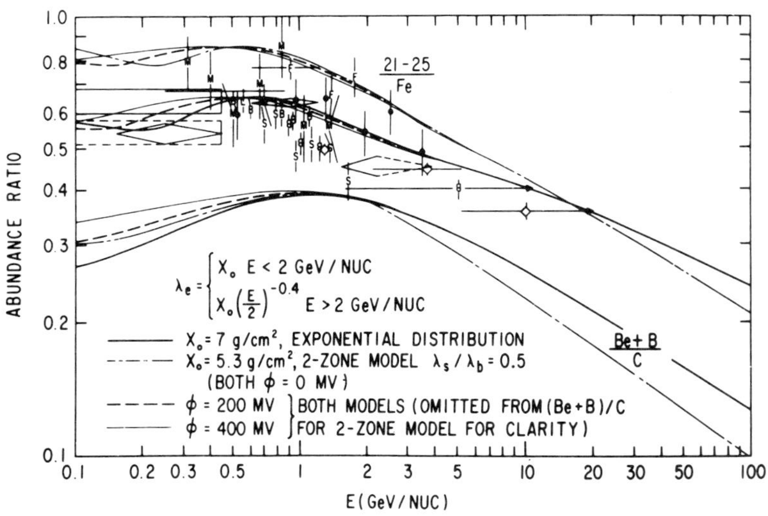

Fig. 3 **The variation with energy of the ratios $(Be + B)/C$ and $(21 \leqslant Z \leqslant 25)/Fe$ for the standard model and for the nested leaky box model with $\lambda_s/\lambda_c = 0.33$. (Protheroe *et al.* 1981)**

While detailed composition is available at low energies, at energies greater than 10^{14} eV one is not certain about the composition. This is unfortunate because the composition would be able to resolve controversies regarding the confinement volume. The steepening in the energy spectrum, observed around this region, can possibly be attributed to the magnetic confinement of cosmic rays which will lead to the escape of high rigidity particles from the Galaxy. Hence there will be a progressive enrichment of *Fe* as energy increases until eventually all the *Fe* too escape (Peters, 1961). However, if photonuclear processes are operative, then heavier nuclei are preferentially destroyed and hence the composition will be enriched in lighter species. So far inferences from the experiments are contradictory (Hillas, 1981). While the atmospheric Cerenkov and hadron measurements yield an enrichment of heavies, the muon experiments in general lead to a near normal composition continuing up to 10^{15} eV and enrichment of lighter species thereafter (Acharya, 1983).

2.3 *Isotopic Composition*

Ever since the discovery of ^3He isotopes in Cosmic Rays (Apparao, 1961), much effort has been expended to determine the isotopic abundances. This is due to a variety of reasons. The isotopes of hydrogen and helium, together with the determination of the shapes of the energy spectra at low energies (< 200 MeV/N), were first used to determine the source spectral shapes (Meyer, 1970; Biswas *et al*, 1967). The isotopic analysis will constrain the acceleraion and propagation models better than the elemental abundances (Reeves, 1981; Meyer, 1981). Further the dominance or otherwise of the neutron rich isotopes will give us clues about the nature of the nucleosynthetic sites of origin of these particles. The results of several experiments in this regard are given in Table 3.

Table 3

Isotopic Ratios in Cosmic Rays

Ratio	Near Earth*	'Source'*	Solar System*
$^3He/He$	~ 0.10	~ 0	$\sim 10^{-5}$
$^{13}C/^{12}C$	0.07	0.035	0.01
$^{15}N/^{14}N$	1.35	0	0.005
$^{17}O/^{16}O$	0.018	0	—
$^{22}Ne/^{20}Ne$	0.55	0.43	0.1
$^{25}Mg/^{24}Mg$	0.24	0.16	0.15
$^{26}Mg/^{24}Mg$	0.27	0.22	0.18
$^{29}Si/^{28}Si$	0.11	0.08	0.08
$^{30}Si/^{28}Si$	0.08	0.05	0.05
$^{54}Fe/^{56}Fe$	~ 0.1	—	0.08
$^{57}Fe/^{56}Fe$	< 0.09	—	0.02
$^{58}Fe/^{56}Fe$	< 0.08	—	0.005

* Experimental ratios are an average of Berkeley, Caltech and New Hampshire measurements. Solar values are taken from Reeves (1981).

From a look at Table 3, it is immediately clear that there are important differences between the cosmic ray abundances and the Solar System abundances, especially in the case of *Ne*. Actually this anomaly together with the near normal ratios for the other isotopes indicates the possible sites for its origin. Even though the over-abundance of ^{22}Ne is not so high as in the case of meteorites, the high ratio implies a common origin for both (Reeves, 1981; Olive and Schramm, 1981; Ramadurai, 1982).

2.4 *Evidence for Antimatter*

Search for antinuclei in cosmic rays using magnetic spectrographs began several years ago. So far the only positive evidence is for the presence of antiprotons (Bogomolov *et al.* 1979; Golden *et al.* 1979; Buffington *et al.* 1981) The experimental results are summarised in Fig. 4. The antiprotons are believed to arise from the collisions of high energy protons with the interstellar medium. Since the threshold for their production is very high, the flux of antiprotons at low energies would be expected to be very small. On the contrary the experimental results show a very high value. This gives us a clue to the nature of the sources. Is there really a source of antimatter? It is possible that the secondary antiprotons produced in the sources undergo adiabatic deceleration in the source (Eichler, 1982). This brings us once again to the acceleration, deceleration and propagation characteristics of the cosmic ray sources.

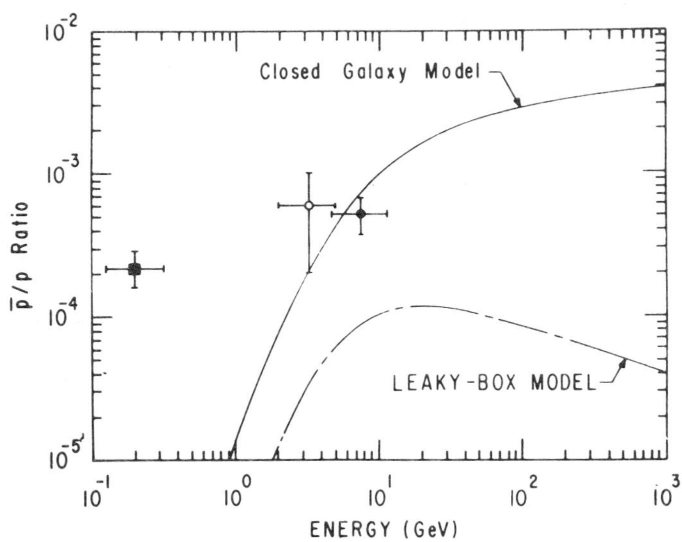

Fig. 4 Experimental observations of $\bar{p}/p$ ratio compared with the propagation models.

2.5 *Isotropy*

One sure way to pinpoint the source of cosmic rays is to determine the direction of arrival of the particles. This essentially means to look for anisotropy. However at energies below 10^{14} eV, the solar wind scatters the cosmic ray particles, thus destroying any anisotropy originally present. At higher energies there are some measured upper limits shown in Table 4. It is clear from a study of this table that the arrival directions are isotropic even at very high energies. This data provides some information on the confinement volume as well as the effects of cosmic ray streaming through the galaxy.

Table 4

Anisotropy Measured. (Longair, 1981)

Energy	Anisotropy δ *
3×10^{14} eV	$<0.1\%$
$1\times 10^{15}-2\times 10^{15}$	$<0.2\%$
10^{16} eV	$<1.5\%$
10^{17} eV	$<2.0\%$
2×10^{18} eV	$13 \pm 6\%$

* **Anisotropy δ is defined as**

$$\delta = \frac{I_{max} - I_{min}}{I_{max} + I_{min}}$$

Recently the anisotropy measurements have been reviewed by Lloyd-Evans and Watson (1982). From the harmonic analysis of the distribution of showers in the right ascension (RA) at different energies, it is found that the amplitude of the first harmonic is constant (about 0.05%) up to an energy of 10^{14} eV. Beyond this energy it increases with energy, approximately as $\sqrt{E}$ up to the highest energies. The phase of the maximum is found to change from 212 ± 17 RA to 30 ± 19 RA between 10^{17} eV and 10^{18} eV. There seems to be an evidence for an enhancement of cosmic rays in the energy range 5×10^{17} eV $- 10^{19}$ eV from southern galactic latitude, mainly from the direction of spiral out and near the South galactic pole. Also there is a tendency for cosmic rays of energy $>10^{19}$ eV to arrive from Northern galactic latitudes, from the direction of Virgo cluster for example.

2.6 *Electron Spectrum*

So far we have been discussing the nucleonic components. The electron spectrum provides a very valuable data to determine the propagation characteristics. Further, since the background radio emission in our galaxy and elsewhere is due to synchrotron radiation, it is tempting to compare the experimentally determined cosmic ray electron spectrum and the electron spectrum deduced from radio observations. Several experiments have indicated that the electron spectrum exhibits the spread due to solar modulation as in the case of nucleonic components, as seen in Fig. 5. Further the spectra at high energies show some evidence for steepening beyond 100 GeV (Stephens, 1981). The steepening is expected on the basis that the synchrotron and inverse Compton losses suffered by electrons should lead to a steepening corresponding to the confinement volume, and lifetime of these particles in the confinement volume. Thus once again we are able to get information about the propagation of cosmic rays from this data.

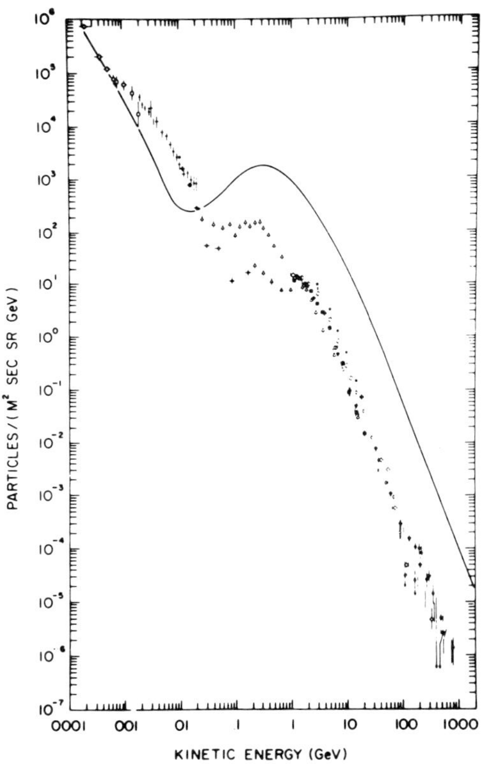

Fig. 5 Energy spectrum of electrons along with the spectrum of protons, (continuous curve). (Meyer, 1975).

Since positrons are expected to be secondaries from the interaction of nucleons with the medium, a study of e^+/e^- ratio as a function of energy will give valuable information about the propagation characteristics. However, if it is found that the secondaries cannot account for all the positrons observed, then the sources themselves may have positrons. Since positrons are also subject to radiative losses like the electrons, the interpretation of the ratio becomes complicated. So far experimental uncertainties are extremely large in determining this quantity and consequently the conclusions are only suggestive (Stephens, 1981).

2.7 *Gamma Ray Astronomy*

Cosmic ray nuclei during their traversal collide with the gas in the medium. These collisions lead to the production of pions. The neutral pion decays into gamma rays. In general gamma rays above a 100 MeV or so are produced by this process, (Wolfendale, 1981). Hence a mapping of our galaxy in gamma rays is a map of the product of cosmic ray intensity and gas density. Thus it is possible to extract information on both the quantities using the distribution of galactic gamma rays. This has been done by several authors. (Stecker, 1977; Kniffen *et al.* 1977; Paul *et al.* 1976; Carvalho and ter Haar, 1977). The gamma ray intensity is shown to have a peak around 4 – 6 kpc from the galactic centre, (Fichtel *et al.* 1975; Puget *et al.* 1976; Mayer Hasselwander *et al.* 1982). With the currently available data about the interstellar hydrogen density distribution, this leads to the conclusion (Tan and Ng, 1983) that cosmic ray intensity distribution is identical to that of supernovae remnants in our galaxy, (Kodaira, 1974).

Information about the sources of cosmic ray particles can also be obtained by measuring very high energy gamma ray intensity. Since gamma rays are not subject to the scattering by interstellar magnetic fields, it is possible to identify the sources of these high energy gamma rays. Since these high energy gamma rays result from interaction of high energy nucleons with the matter present, or from the inverse Compton scattering of high energy electrons, identification of a gamma ray source is an identification of a possible cosmic ray source too. There are experimental indications for the detection of these ultrahigh energy gamma rays (10^{12} eV) from the direction of Crab and Vela Pulsars, (Ramana Murthy, 1981).

3. Models of Propagation

So far we have assembled all the experimental data. In order to arrive at the nature of the sources, we have to unravel the story of propagation of these particles through our galaxy and solar system. Since the basic characteristics of propagation of cosmic rays through interplanetary space are well understood (Ramadurai, 1981) on the solar wind model of Parker (1963) with diffusion-convection with adiabatic deceleration, we will not be concerned with this aspect here. The propagation through interstellar space will form the main basis for discussion here (Daniel and Stephens, 1976). Since the details of calculations of the extrapolation from abundances measured at earth to the sources are involved and lengthy, in this article only the salient features of the various models are given. For illustration, the slab model is discussed in some detail. Even here the detailed equations and their solutions are not given. Further ionization losses are completely ignored in this discussion. Since it is dealt with in great detail in almost any reference on the electron spectrum, the propagation of electrons is not discussed at all. The complete transport equation is written in the last section, without fully explaining all the terms, leaving the reader to consult the appropriate references.

3.1 *Slab Model*

The first picture one can easily visualise is the one in which all cosmic rays arising from a source travel through a certain amount of matter and this is known as the 'Slab model'. Obviously this picture is an oversimplification, but in order to understand the physical processes taking place during propagating, this will be useful. Ignoring energy loss one can write the transfer equation as (Longair, 1981)

$$\frac{dJ_s(x)}{dx} = -\frac{J_s(x)}{l_s} + \sum_{p>s} \frac{P_{ps}}{l_p} J_p(x) \tag{2}$$

where $J_s(x)$ is the flux of species s after traversing x g.cm^{-2} of matter. The right-hand side represents the destruction by interaction and addition due to spallation from heavier species. l_s and l_p represent the interaction mean free paths of the species s and the heavier species p. P_{ps} represents the probability of the interaction of p leading to the production of s. In the case of heavier nuclei the destruction term dominates and one can write the solution as

$$J_p(x) = J_p(0) \exp(-x/l_p) \ . \tag{3}$$

Similarly one can write for the secondaries

$$\frac{J_s(x)}{J_p(x)} = \frac{P_{ps}\, l_s}{(l_s - l_p)} \left[\exp(x/l_s - x/l_p) - 1\right] \tag{4}$$

In the case of Li, Be, and B, l_s is about 10 g.cm^{-2} and the main contribution comes from the breakup of C, N and O nuclei. Hence l_p can be taken as about 6.7 g.cm^{-2}. The experimentally measured partial cross sections in this case yield a value of about 0.35 for P_{ps}. From Table 1 the ratio of

$(Li + Be + B)/(C + N + O)$ is seen as 0.25. Using equation (4) one gets a value of

$$x = 4.3 \text{ g.cm}^{-2}$$

for the pathlength of cosmic rays. This value of pathlength is in good agreement with the pathlength derived using the isotopic ratios $^3He/(^3He + ^4He)$ as well as the individual nuclei of *Li*, *Be* and *B* (Ramadurai and Biswas, 1974; Biswas *et al.* 1966; 1967). However as one goes to higher charge species, for example *Fe* with its l_p of about 2.6 g.cm^{-2}, discrepancies arise. If all the cosmic rays pass through the same 4 – 5 g.cm^{-2} of matter, then *Fe* will be depleted very much and the nuclei beyond *Fe* will not be present at all. Hence one infers that the *Fe* secondaries and ultraheavy secondaries will dominate over the parent primaries. From a look at Table 1 and 2, the *Fe* secondaries as well as the ultraheavy secondaries do not dominate. Hence the slab model has to be abandoned in favour of a physically more meaningful model. Essentially what is required is a distribution of pathlengths instead of a single value for all the species. The pathlength distribution is a natural consequence of the physics of propagation of particles from the source to the earth by some form of diffusion and confinement. Depending on the set of assumptions regarding these one will be able to obtain different distributions of pathlengths. We will see the various models so far proposed to account for the observations summarised earlier.

3.2 *Diffusion Model*

Let us assume that there is a source situated at a mean pathlength of $\bar{x}$ g.cm^{-2} from the observer. Particles from this source reach the observer by pure diffusion. This situation will lead to a more or less Gaussian distribution of pathlength. The probability of a particle traversing the amount of matter between $x + dx$ is given by the normalised Gaussian distribution (Balasubrahmanyan *et al.* 1964; Ramadurai and Biswas, 1974)

$$P(x)\,dx = \left\{(5/\pi)^{\frac{1}{2}}/1.886\,\bar{x}\right\} \exp\left\{(-5\,(x-\bar{x})^2/4\,\bar{x}^2)\right\} dx\,. \tag{5}$$

If the flux of particles at any distance x from the source be represented by $N(x)$, then the flux after a mean amount of matter traversal of $\bar{x}$ g.cm^{-2} is given as

$$N(\bar{x})\,dE = dE \int_0^\infty N(x)\,P(x)\,dx\,. \tag{6}$$

However even this picture is an oversimplification in the sense that it does not take into account the confinement volume of cosmic ray particles. It is likely that cosmic ray particles will be confined within a boundary by the galactic magnetic fields and hence it is more meaningful to take this fact into account. This is done in the leaky box model which is described next. But it should be borne in mind that Gaussian distribution gives a better agreement with experimental results on nuclei up to *Fe*. But this fails to account for the abundancies of post *Fe* nuclei in cosmic rays.

3.3 *Leaky Box Model*

Let us assume that cosmic rays are confined to a volume V and then escape from the confinement region after a time T. In the steady state

$$\frac{\partial N}{\partial t} + \frac{N}{T} = 0 \ , \tag{7}$$

which immediately leads to the distribution of lifetime/pathlength given by

$$P(x)\ dx = \exp\left(-x/\bar{x}\right)\ dx \ . \tag{8}$$

This distribution has been extensively used since it was first used by Cowsik *et al.* (1967) to determine the nuclear abundances of cosmic rays during propagation. The same leaky box model is used to calculate the electron energy spectrum too. The detailed investigations in this regard have been done by Shapiro and Silberberg (1974) for the elements up to *Fe*. The value for $\bar{x}$ they deduce is

$$\bar{x} = 4\ \text{g.cm}^{-2}\ .$$

There are two aspects of the observations which are hard to explain in terms of the simple leaky box model outlined above. Firstly, the secondary/primary ratio like $(Be+B)/C$ and Sub*Fe*/*Fe* shown in Fig. 3 imply that the escape mean free path has an energy dependence of the form shown in Fig. 3. This feature is very hard to interpret on a pure leaky box model. Secondly the data seem to agree with the predictions of a two zone model where the pathlength distribution has an exponential character with truncation at small pathlengths (Shipiro and Silberberg, 1974). The data on ultraheavy nuclei (post *Fe*) also demand this truncation. This second point finds a ready explanation in the nested leaky box model discussed next.

3.4 *Nested Leaky Box Model*

In this model (Cowsik and Wilson, 1975) cosmic rays take a finite time to leak out of their sources thereby travelling a finite amount of matter in the source regions themselves. Once they are injected into the interstellar medium the leaky box model is adopted. Essentially in this model there are two parameters — the amount of matter traversed within the source, λ_s, and the mean amount of matter traversed, λ_b. In Fig. 3 is plotted the results of a recent calculation by Protheroe *et al* (1981). One can see from this that the energy dependence of escape mean free path fits well with the observations. This quantity can be better determined with the ratio of $(Be+B)/C$. Hence a measurement of this ratio at high energies will be able to settle this question unambiguously.

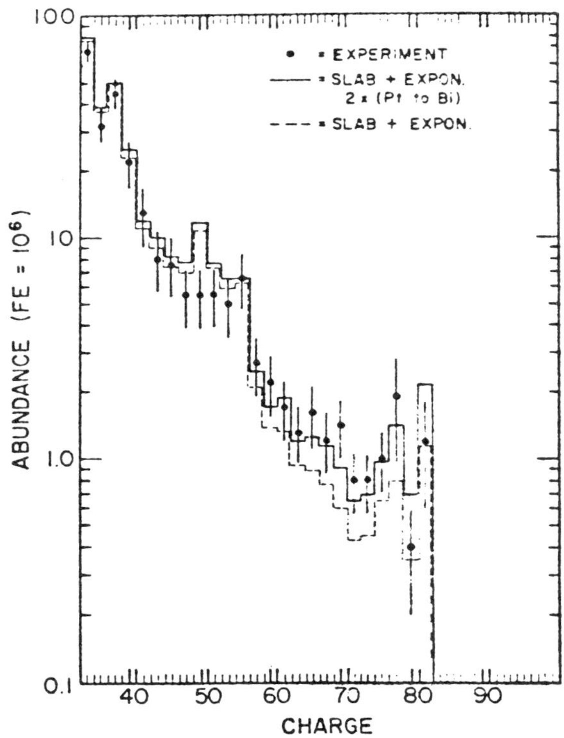

Fig. 6 Relative abundances with a 1 g/cm^2 slab at the source followed by a 5 g/cm^2 exponential distribution. The solid line illustrates doubling of nuclei with $76 \leqslant Z \leqslant 83$. (Tsao *et al*. 1983).

The importance of ultraheavy nuclei in determining the propagation characteristics can hardly be overemphasized. Actually a recent calculation by Tsao *et al* (1983) concentrating on the nuclei with $Z > 34$ have come to remarkable conclusions. A slab of 1 g.cm^{-2} of matter at the source followed by an exponential distribution with $\bar{x}$ = 5 g.cm^{-2} seem to fit the observations reasonably well up to $Z = 60$ or so (Fig. 6). Beyond that it requires an enhancement of solar system abundances of (*Pt* to *Bi*) by a factor of two. This is remarkable because it throws light on the nature of cosmic ray sources and the nucleosynthetic processes responsible for cosmic ray particles. This fact has been brought forth with the analysis done using the HEAO 3 data mentioned earlier (Binns *et al*, 1981; 1983; Brewster *et al*, 1983).

Having surveyed the various models for explaining the nucleonic components, we have not discussed two observations mentioned earlier — namely, the electron spectrum and the antiproton observations. The electron spectrum is explained by the nested leaky box model well. However the antiprotons pose severe problems. This may be explained by the Closed Galaxy model mentioned next.

3.5 *Closed Galaxy Model*

This model arose out of the desire to explain the high energy observations of near normal composition up to the highest energies for which the composition measurements can be relied upon. This implies as mentioned earlier that there is no leakage from the galaxy in certain models. Hence Rasmussen and Peters (1975) tried also to understand the low energy observation using the same assumptions — namely cosmic rays are effectively confined within our galaxy. This, combined with a lifetime of the order of 10^7 years, imply a pathlength traversal of more than 100 $g.cm^{-2}$. Hence all the heavy nuclei are effectively destroyed and some of the protons also get destroyed. In order to explain the observations of all nuclei in cosmic rays, a local source/sources nearby, travelling very little amount of matter is postulated. Later on Peters and Westergaard (1977) have modified this model in order to get good agreement with the detailed charge spectrum mentioned in Table 1. Essentially they come to the conclusion that roughly 12% of the components are old. Further one more point in favour of this model is the predicted low abundances of hydrogen and helium compared to the cosmic abundances as seen in Table 1. However when antiproton measurements were made the proportion of the old components was revised to be about 50% for protons and progressively smaller for the higher charges (Stephens, 1981a). In the revised closed galaxy model, the old component exists in addition to the nested leaky box model. Thus while the leaky box or nested leaky box models do not fit with the observations, the revised closed galaxy model fits with the high energy observations of antiprotons. However even this model cannot account for the low energy observations as seen in Fig. 4. Perhaps the explanation lies in the adiabatic deceleration suffered in the sources themselves (Eichler, 1982).

3.6 *Energy Changes During Propagation*

In considering the various models of propagation so far the energy changes taking place have not been considered at all. The heavy nuclei suffer ionization losses in the medium during traversal. Further there may be adiabatic deceleration if the cosmic ray particles partake in the sudden expansion of the supernova envelopes. In addition, the electrons suffer radiative losses due to the interaction with the interstellar magnetic fields, and the photon field due to microwave background radiation. Hence a proper formulation of the propagation problem could be to write the transport equation as

$$\frac{\partial N}{\partial t} - \nabla\cdot(D\nabla N) + \frac{\partial}{\partial E}\left(\frac{\partial E}{\partial t}\,N\right) + \frac{N}{T} = Q(\boldsymbol{r},t,E)\,. \qquad (9)$$

Here the particle density $N(\boldsymbol{r},t,E)$ at any point $\boldsymbol{r}$ at time t will be governed by the combined effects of diffusion determined by the diffusion coefficient D (which may depend on energy and position), and the escape from the confinement region as well as the energy changes taking place during propagation. The source term too is taken as a function of position, time and energy as several sources are likely to contribute to cosmic ray particles. The diffusion coefficient can be determined from measuring the spectrum of magnetic irregularities present in the medium of propagation. The solution of this equation for application to cosmic ray propagation has been given in detail by Ginzburg and Syrovatsky (1974). The pathlength distributions arising out of this have already been discussed. The energy changes for electrons determined by the synchrotron losses and more importantly due to interaction with microwave background photons with the assumed lifetime for the cosmic rays determine the break in the electron spectrum. Alternatively, the break in the electron spectrum will imply a lifetime which will be treated next.

3.7 *Lifetime of Cosmic Rays*

The first indication about the lifetime comes from the determination of the mean amount of matter traversed, $\bar{x}$, mentioned earlier. Assuming an average density of the medium of propagation $\bar{\rho}$ the mean lifetime of cosmic rays is

$$T = \bar{x}/\bar{\rho}c = 3\times10^{6}\ \text{years}$$

if $\bar{\rho}$ is taken as given by 1 hydrogen atom/cm^3 and $\bar{x} = 5\,g.\,cm^{-2}$. Obviously this method is not very satisfactory, as it depends on the ill-defined parameter of density of medium. For example instead of taking the value of 1 hydrogen atom/cm^3 as in the galactic disc, if one takes a value of $10^{-2}/cm^3$ as in the galactic halo, the lifetime is 3×10^{8} years. Hence one has to look at alternative ways of determining this quantity.

One method will be to use the radioactive clocks like ^{10}Be. Since all the isotopes are produced during the traversal from the sources to earth, it is easy to infer the lifetime from the measurement of the radioactive to stable isotopes of Be. Essentially what is required is to add one more term, J_s/τ_s, to equation (2) due to the radioactive decay. In the leaky box model it can be shown that the ratio of a radioactive species abundance N_i to the stable species abundance N_j can be given by (Longair, 1981)

$$\frac{N_i}{N_j} = \frac{(1/T) + Nc\sigma_j}{(1/T) + Nc\sigma_i + (1/\tau_i)} \cdot \frac{C_i}{C_j} \tag{10}$$

where $NcT = \bar{x}$; σ_j and σ_i are the production cross-sections; C_i and C_j are the production terms given by the second term on the right in equation (2). ^{10}Be has a halflife (τ_s) of 3×10^6 years. Thus by determining the abundance of this isotope in cosmic rays the lifetime can be unambiguously determined. Of course experimentally to separate the isotopes of *Be* is a monumental task. This has been done by the Chicago group (Garcia Munoz *et al.* 1977), who find a value of 0.028 for the ratio $^{10}Be/Be$, corresponding to a lifetime of 1.7×10^7 years. If this lifetime is used then the mean density turns out to be about $0.2/\text{cm}^3$. This in turn means that particles might have spent most of the time outside the disc, perhaps forming a cosmic ray halo. Of course it should be clear that this lifetime refers to the traversal from the source to the earth. Thus it is not possible to rule out a model of cosmic rays filling the entire halo leading to lifetime of about 10^8 years derived using the path-length and halo matter density. This is because that lifetime may refer to the time taken for cosmic rays to diffuse out of the halo, which is different from the time taken for cosmic rays to travel from source to earth. One important inference one can make from the lifetime measurements and the path-lengths is that cosmic rays have necessarily to diffuse through the medium: propagation is not along a straight line as in slab model. This is clear from the fact that the straight line propagation time from the source to the earth will be of the order of 10^{4-5} years. During that time the matter traversed will be less than 0.05 g.cm^{-2} contradicting the secondary/primary ratio discussed earlier.

As mentioned earlier the electron spectrum can be used to obtain the lifetime. Unfortunately this method suffers from the uncertainty with regard to the location of the break. If the break occurs close to the region affected by solar modulation, ~ 3 GeV, then the lifetime T is given by the formula (Longair, 1981)

$$T = \frac{2.8 \times 10^{12}}{\{(B/B_0)^2 + 1.1\,(U/U_0)\}\,\gamma} \text{ years ,} \tag{11}$$

where B and U represent the magnetic field and photon field strengths respectively and B_0 and U_0 correspond to 3 μG and 0.25 eV cm^{-3} corresponding to the microwave background energy density. γ is the energy at which the break occurs, expressed in terms of the rest mass of the electron. Hence for $\gamma = 6 \times 10^3$ the lifetime turns out to be 2.2×10^8 years, assuming $B = B_0$ and $U = U_0$. This is clearly in contradiction to the lifetime derived on the basis of the radioactive clock ^{10}Be. However, as mentioned already, this lifetime should not be taken as the lifetime for all cosmic rays. It is possible that cosmic ray electrons fill up the halo while the primary species producing ^{10}Be might have traversed for only 17 million years. If, however, the break is seen to occur about 100 GeV the lifetime will be of the order of 10^7 years, almost in good agreement with the *Be* clock. So in summary one can say that the cosmic ray has a lifetime of about 10^7 years. This is used in deriving some information about the sources, which is treated next.

3.8 *Energy Requirements of Cosmic Ray Sources*

In order to arrive at any conclusions regarding the energy requirements of cosmic ray sources, it is necessary to know the mean lifetime of cosmic rays. Since the mean cosmic ray intensity has remained constant over the past million years, it is possible to draw conclusions regarding the energy requirements of cosmic ray sources. This then can be compared with the astronomical energy machines like stars, novae, supernovae, pulsars, galactic nuclei, and extragalactic radio sources including quasars. Thus some speculations regarding the origin of cosmic rays can then be made.

The energy density of cosmic rays derived from observation corresponds to $\varepsilon_{CR} = 1\,eV.cm^{-3}$. This is comparable to the energy density of the interstellar magnetic field and the turbulent energy of the interstellar medium (Fermi, 1949). So the energy requirement E_{CR} of the probable sources of cosmic rays with a mean density of $\rho_{sources}$ should be such that

$$\varepsilon_{CR} = E_{CR} \cdot \rho_{sources} . \qquad (12)$$

This brings us to the nature of the sources as well as their distribution. From the astronomical observations ranging from radio to gamma ray regions, it is possible to estimate the energy output in cosmic ray particles from various sources. This will be outlined in the next sections. But from equation (12) it is clear that the source density too is an important quantity. This in turn depends on the confinement volume of cosmic rays. For example if cosmic rays are confined to the volume of the galactic disc, with a thickness of 700 pc and radius of 10 kpc, then the total cosmic ray energy required is clearly $\varepsilon_{CR} \cdot 6 \times 10^{66}$ ergs. From the last section we saw that the mean lifetime of cosmic rays is 10^7 years. Hence the energy output in cosmic rays E_S is

$$E_S = \text{total energy/lifetime} = 10^{55}/10^{15} = 10^{40} \text{ erg/sec}.$$

Compared to this we know the energy output in cosmic rays from the sun is about 10^{24} erg/sec. So even if all the stars in our galaxy numbering 10^{11} contribute cosmic rays it will be a negligible fraction of the cosmic ray energy required. So this energy output per second in cosmic rays is an important clue to the probable sources. This will be dealt with after considering the acceleration of particles to these high energies.

4. Acceleration of Cosmic Rays

From the experimental observations the twin requirements of acceleration processes are known; (i) the energy spectrum emerging out of the accelerating regions is a power law given by (1) with a spectral idex close to 1.5; (ii) cosmic ray particles have energies up to 10^{20} eV. From laboratory experience we know the problems associated with acceleration of charged particles. The particles have to be confined to the accelerating volume for sufficient time. Further the processes are mainly electromagnetic. On the other hand the cosmic ray beam is essentially low density compared to the laboratory plasmas. As a result there are additional processes, dynamical and hydrodynamical, which may be responsible for accelerating cosmic ray particles. We will give a bare outline of all these processes below. But it is necessary to remember that present day observations extend to *in situ* measurements of particles being accelerated in the terrestrial and Jovian magnetospheres and in the interplanetary medium. The shock acceleration in interplanetary regions were directly observed by instruments aboard helios, Pioneers 10 and 11, (Barnes and Simpson, 1976). These have observed in the corotating interaction regions in the solar wind the double peaked structure associated with the forward-backward shock pairs. In the earth's magnetic tail the magnetic field reconnection is important. Acceleration due to the presence of electric fields parallel to the magnetic field occurs in the Auroral zones. Magnetospheric rotation is seen to accelerate particles in Jupiter. The stochastic acceleration due to interaction with waves occur in the magnetosphere. All these processes play a role in accelerating cosmic rays.

4.1 *Fermi Mechanism*

The first person to give a quantitative step in the acceleration process was Fermi (1949), who proposed the acceleration of cosmic ray particles as a result of collisions with moving magnetic clouds in interstellar space. In a head-on collision with the magnetic cloud, energy is gained, and in an overtaking collision, energy is lost. It can be shown that the probability of head-on collision is more than the overtaking collisions and hence there is a net gain of energy. The rate of energy gain is

$$\frac{dE}{dt} = \alpha E \ , \tag{13}$$

where α is dependent on the cloud velocity and the number of collisions per second. Suppose the particles are confined in this region for time τ; then in a situation where there are no sources and there is no diffusion, Eqn (9) reduces in a steady state to

$$\frac{\partial N}{\partial E} = -\left(1 + \frac{1}{\alpha\tau}\right)\frac{N}{E}, \tag{14}$$

which yields a spectrum of the form

$$N(E) = \text{constant} \times E^{-(1 + 1/\alpha\tau)} \tag{15}$$

which is very close to the required form (1).

However this mechanism faces the difficulties due to the low velocities prevalent in the general interstellar clouds. Further the losses due to ionization restrict the acceleration to those above the threshold, since the energy gain is low. One way in which the energy gain can be increased is by ensuring that the collisions are mostly only head-on. This is achieved by having the acceleration taking place in plasma/magnetohydrodynamic turbulence which can be generated by large scale mass motions. However, effectiveness of this process depends on the intensities of these modes of turbulence about which little information is available. Further since the large scale mass motions themselves can directly accelerate particles, effectiveness of this mechanism is unclear.

4.2 *Hydrodynamical Processes*

This has been suggested by Colgate and Johnson (1969); outer layers of a supernova are raised to relativistic energies due to the passage of shock generated in the core. The basic idea is that during a supernova collapse the core deposits so much energy in the shock waves that as the shock moves outward, it gets strengthened as the density of the medium is progressively decreasing. It is possible to get an energy of about 10^{17} eV in regions where the densities are as low as 10^{-12} g.cm^{-3}. But the basic difficulty in this process is that the adiabatic losses due to expansion will make this process very ineffective. The dissipation of the shock will lead to only a modest gain of energy in this process.

4.3 *Shock Acceleration*

The main distinguishing feature of this process-class is that the cosmic rays are scattered effectively by the Alfven waves produced by the strong shocks. There is a minimum energy needed so that the particles can move upstream. Roughly the picture is as follows; whenever cosmic ray particles cross a shock front there is a net gain of energy. Since the shock itself moves in the low density medium, the shock is very strong and the scattering by cosmic rays with the shock front produces always a net energy gain. If the medium is highly ionized and of low density, damping is minimal. It is shown by Axford (1981) and Blandford and Ostriker (1980) that a power law spectrum is a natural result. The details of the process taking into account the energy dependence of the diffusion coefficient, the anisotropy of the diffusion coefficient, the energy losses etc. is an active area of investigation at present (Volk *et al.* 1981; 1982; Webb, 1983). The varuous types of shocks as well as the regions where they may exist are also important aspects of this problem pursued actively by several groups (Cowsik and Lee, 1982; Casse *et al.* 1979; Volk and Forman, 1982). We will briefly mention the probable acceleration sites in the next section.

4.4 *Acceleration in Neutral Sheets*

This process is important in the Earth's geomagnetic tail and in solar flares. Whenever magnetic field lines run in opposite directions the region in between, called a neutral sheet, experiences an induced electric field. This induced electric field can accelerate particles to high energies. Essentially the dissipation of magnetic energy in the form of Ohmic losses may be responsible for solar flares. By extension one can speculate the process occurring in many magnetic stars as well as galaxies. But very little work is done in this field.

4.5 *Acceleration in Pulsar Magnetospheres*

The pulsars are known to be rotating neutron stars. It is known that the avaialble rotational energy in pulsars can be converted into particle energy with an efficiency of about 10%. The slowing down of the pulsars fits in with the assumption of large magnetic dipole radiation from the pulsar surface. The propagation of this radiation is very peculiar as the frequencies involved are below the plasma frequency of the medium. This strange situation has been exploited to accelerate particles to relativistic energies (Arons, 1981). From the early works of Gunn and Ostriker several attempts have been made to understand the details of this acceleration process. In principle there is no difficulty in accelerating particles to even energies of about 10^{20} eV. But the situation is so complex that it is not possible to assert the effectiveness of the mechanism.

4.6 *The Plasma Turbulent Reactor*

This process, advocated by Tsytovitch (1972) school, relies on the completely different nature of turbulence in a magnetoactive plasma. The plasma waveforms serve as eddies in the medium. Since there are several types of plasma wavemodes, the interaction among them can lead to an equilibrium spectrum of turbulence known as Langmuir turbulence. The relativistic particles interacting with this equilibrium spectrum of Langmuir turbulence is shown to yield a power spectrum of the form (1). As emphasized by Longair (1981), in order that the equilibrium is established the medium should be optically thick to synchrotron radiation. Hence the source should be extremely compact. So when the particles are ejected the adiabatic losses would be severe, as in the supernova hydrodynamic models mentioned earlier.

Before concluding this sketchy description of the various acceleration processes, it should be mentioned that the isotopic measurements give an estimate of the time which has elapsed between the acceleration and production in the sources. This is because if the nuclei decay by *K* capture, like Mn^{54} or ^{56}Ni, then if they are accelerated immediately after production, they will lose the electron shell and hence *K* capture is not possible. Thus a determination of over-abundance of these isotopes will be a direct proof that acceleration took place immediately before the decay of the isotopes. So far all indications are that there are no significant anomalies and hence the acceleration has taken place after a lapse of 10^5 years.

5. Sources of Cosmic Rays

We have already seen that the basic energy requirement of cosmic rays is that the source should be able to generate cosmic rays with an output rate of about 10^{40} erg.sec^{-1}. In this respect normal stars like the sun will not be able to sustain cosmic rays. Even if one takes magnetic variables with thousand-fold increase in the magnetic field strength, it is seen that the source strength falls short by two orders of magnitude (Ginzburg and Syrovatsky, 1964). Other suggestions in this regard are the novae (Durgaprasad, 1977), stellar winds from *O* and *B* stars (Casse *et al.* 1979; Montmerle, 1980). *OB* associations (Montmerle, 1980; Volk and Forman, 1982; Olive and Schram, 1980), supernovae (Ginzburg and Syrovatsky, 1964). In all cases it is seen that the energetic arguments are satisfied. However, the models which are most favoured are those where the supernovae-induced shocks accelerate material thrown out of the explosion in the hot interstellar medium or in the *OB* associations (Axford, 1981; Olive and Schram, 1982; Montmerle, 1980).

In the discussion of sources above, only galactic sources were considered. There are compelling arguments in favour of galactic models for major sources of cosmic rays (Ginzburg, 1979). However the observed near complete isotropy of the radiation is a strong argument in favour of a universal hypothesis. However the scanty evidence in favour of a small anisotropy in the direction of the Virgo Supercluster as well as the indication of a break in the spectrum at energies greater than 10^{19} eV suggest the presence of an extragalactic component in cosmic rays near earth. However, until the anisotropy is more firmly established than now, it is premature to conclude in favour of this model.

6. Future Outlook

In spite of the enormous progress in the last seven decades since their discovery, the origin of cosmic rays is still not completely solved. The following questions naturally arise:

(1) What are the acceleration processes which can accelerate particles to energies as great as 10^{20} eV ?

(2) Is there direct evidence for the presence of recently synthesised nuclei in cosmic rays ?

(3) What is the confinement volume of cosmic rays ? Is it galactic or confined to local supercluster or universal?

(4) What is the relationship between the supernovae and the cosmic ray sources ? Are they the main sources of cosmic rays ?

(5) What is the exact nature of propagation ? In this respect it is interesting to determine the diffusion coefficients from the dynamical considerations of the interstellar medium. Work in this regard has to be intensified.

(6) Can we learn much more about cosmic ray sources from gamma ray work ?

The above are some of the general questions of interest to everybody in the field. Let me conclude this with a statement from Dirk ter Haar written nearly four decades ago (1949): "Indeed, it may well be that, as in so many instances, a more acceptable theory will only be formed when ideas from all the different papers will have been melded together into one large compromise."

7. Acknowledgements

The author wishes to thank Professors S. Biswas, R.R. Daniel and P.V. Ramana Murthy for critical reviews of the manuscript.

References

Acharya, B.S., Ph.D. Thesis, University of Bombay, (1983).

Apparao, M.V.K., Phys. Rev., **123**, 295 (1961).

Arons, J., IAU Symposium 94, *Origin of Cosmic Rays*, D. Reidel Publishing Co., Dodrecht, p.175 (1981).

Axford, W.I., IAU Symposium 94, *Origin of Cosmic Rays*, D. Reidel Publishing Co., Dodrecht, p.339 (1981).

Balasubrahmanyan, V.K., Boldt, E.A. and Palmei. ra, R.A.R., Phys. Rev., **140**, 1157 (1964).

Barnes, G.W. and Simpson, J.A., Astrophys. J., **210**, L91 (1976).

Binns, W.R., Fickle, RK.K., Garrard, T.L., Israel, M.H., Klarmann, J., Stone, E.C. and Waddington, C.J., Astrophys. J. (Letters), **247**, L115 (1981).

Binns, W.R., Fickle, RK., Garrard, T.L., Israel, M.H., Klarmann, J., Krombel, K.E. Stone, E.C. and Waddington, C.J., Astrophys. J., **267**, L93 (1983).

Biswas, S. and Durgaprasad, N., Sp. Sci. Rev., **25**, 285 (1980).
Biswas, S., Ramadurai, S. and Sreenivasan, N., Phys. Rev., **149**, 1037 (1966).
Biswas, S., Ramadurai, S. and Sreenivasan, N., Phys. Rev., **159**, 1063 (1967).
Blandford, R.D. and Ostriker, J.P., Astrophys. J., **237**, 793 (1980).
Bogomolov, E.A., Lubyanaya, N.D., Romanov, V.A., Stepanov, S.V. & Shulakova, M.S., Proceedings of International Conference on Cosmic Rays, Kyoto, **1**, 330 (1979).
Brewster, N.R., Freier, P.S. and Waddington, C.J., Astrophys. J., **264**, 324 (1983).
Buffington, A., Schindler, S.M. and Pennypacker, C.R., Astrophys. J., **248**, 1179 (1981).
Cameron, A.G.W., *Essays in Nuclear Astrophysics*, (ed. C. Barnes, D.D. Clayton and D.N. Schramm), Cambridge University Press, Cambridge (1980).
Carvalho, C. and ter Haar, D., Mon. Not. Roy. Astr. Soc., **180**, 45 (1977).
Carvalho, J.C. and ter Haar, D., Mon. Not. Roy. Astr. Soc., **187**, 23 (1979).
Casse, M. and Paul, J., Astrophys. J., **237**, 236 (1979).
Cesarsky, C.J., Ann. Rev. Astron. Ap., **18**, 289 (1980).
Colgate, S.A. and Johnson, M.H., Phys. Rev. Letters, **5**, 235 (1960).
Cowsik, R. and Wilson, L.W., Proc. Int. Conf. on Cosmic Rays, Munich, **2**, 475 (1975).
Cowsik, R., Pal, Y., Tandon, S.N. and Verma, R.P., Phys. Rev., **158**, 1238 (1967).
Cowsik, R. and Lee, M.A., Proc. Roy. Soc. (London), **A383**, 409 (1982).
Daniel, R.R. and Stephens, S.A., Sp. Sci. Rev., **17**, 45 (1976).
Durgaprasad, N., Astrophys. Sp. Sci., **47**, 435 (1977).
Eichler, D., Nature, **295**, 391 (1982).
Epstein, R.I., Mon. Not. Roy. Astr. Soc., **193**, 723 (1980).
Epstein, R.I., IAU Symposium 94, *Origin of Cosmic Rays*, D. Reidel Publishing Co.m Dodrecht, p.109 (1981).
Fermi, E., Phys. Rev., **75**, 1169 (1949).
Fowler, P.H., Walker, R.N., Masheder, M.R., Moses, R.T. and Worley, A., Nature, **291**, 45 (1981).
Fichtel, C.E., Hartman, R.C., Kniffen, D.A., Thompson, D.J., Bignami, G.F., Ogelman, H.B., Ozel, M. and Turner, T., Astrophys. J., **198**, 163 (1975).
Garica Munoz, M., Mason, G.M. and Simpson, J.A., Astrophys. J., **217**, 859 (1977).
Ginzburg, V.L., Soviet Phys. Uspekhi, **21**, 155 (1979).
Ginzburg, V.L. and Syrovatskii, S.I., *Origin of Cosmic Rays*, Pergamon Press, Oxford (1964).
Golden, R.L., Horan, S., Mauger, B.G., Badhwar, G.D., Lacy, J.L., Stephens, S.A., Daniel, R.R. and Zipse, J.E., Phys. Rev. Letters, **43**, 1196 (1979).
Greisen, K., Phys. Rev. Letters, **16**, 748 (1966).
ter Haar, D., Mon. Not. Roy. Astr. Soc., **106**, 283 (1946).
ter Haar, D., Science, **110**, 285 (1949).
Hillas, A.M., Proc. Int. Conference on Cosmic Rays, Paris, **13**, 69 (1981).
Kniffen, D.A., Fichtel, C.E. and Thompson, D.J., Astrophys, J., **215**, 765 (1977).
Kodaira, K., Pub. Astr. Soc. Japan, **26**, 255 (1974).
Longair, M.S., *High Energy Astrophysics*, Cambridge University Press, Cambridge (1981).
Linsley, J., IAU Symposium 94, *Origin of Cosmic Rays*, (Eds. G. Setti, G. Spadha and A.W. Wolfendale) Reidel Publishing Co., Dordrecht, p.53 (1981).
Lloyd-Evans, J. and Watson, A.A., Review paper presented at VIII European Cosmic Ray Symposium, Rome, 1982.
Montmerle, T., Astrophys. J., **231**, 95 (1980).
Mayer Hasselander, H.A. *et al.*, Astron. Ap., **105**, 164 (1982).
Meyer, P., IAU Symposium 94, *Origin of Cosmic Rays*, D. Reidel Publishing Co., Dodrecht, p.339 (1981).

Meyer, J.P., Astrophys. Lett., **7**, 61 (1970).
Morrison, P., Handbuch der Physik, XLVI, 1 (1961).
Olive, K.A. Schramm, D.N., Astrophys. J., **257**, 276 (1982).
Parker, E.N., *Interplanetary Dynamical Processes*, Interscience Publishers, New York (1963).
Paul, J., Casse, M. and Cesarsky, C.J., Astrophys. J., **207**, 62 (1976).
Peters, B., Nuovo Cimento, **22**, 800 (1961).
Peters, B. and Westergaard, N.J., Astrophys. Sp. Sci., **48**, 21 (1977).
Protheroe, R.J., Ormes, J.F. and Comstock, G.M., Astrophys. J., **247**, 362 (1981).
Puget, J.L., Ryter, C., Serra, G. and Bignami, G., Astron. Ap., **50**, 247 (1976).
Ramadurai, S., Proc. Ind. Nat. Sci. Acad., **47A**, 92 (1981).
Ramadurai, S., Mon.Not.Roy.Astr.Soc., **200**, 123 (1982).
Ramadurai, S. and Biswas, S., Astrophys. Sp. Sci., **30**, 187 (1974).
Ramana Murthy, P.V., *Nonsolar gamma rays*, (Eds. R. Cowsik and R.D. Wills) Pergamon Press, Oxford, p.70 (1980).
Rasmussen, I.L. and Peters, B., Nature, **258**, 412 (1975).
Reeves, H., IAU Symposium 94, *Origin of Cosmic Rays*, (Eds. G. Setti, G. Spada and A.W. Wolfendale) D. Reidel Publishing Co. Dodrecht, p.23 (1981).
Shapiro, M.M. and Silberberg, R., Phil. Trans. Roy. Soc., **227**, 319 (1974).
Sreekantan, B.V., Sp. Sci. Rev., **14**, 103 (1972).
Stecker, F.W., Astrophys. J., **212**, 60 (1977).
Stephens, S.A., Proc. Int. Conf. on Cosmic Rays, Paris, **13**, 89 (1981).
Stephens, S.A., Nature, **289**, 267 (1981a).
Tan, L.C. and Ng, L.K., Astrophys. J., **269**, 751 (1983).
Tarafdar, S.P. and Apparao, M.V.K., Astrophys. Sp. Sci., **77**, 521 (1981).
Tsao, C.H., Silberberg, R., Adams, J.H. Jr. and Letaw, J.R., Proc. Int. Conf. on Cosmic Rays, Bangalore, **2**, 225 (1983).
Tsytovich, V.N., *An Introduction to the Theory of Plasma Turbulence*, Pergamon Press, Oxford (1972).
Volk, H.J. and Forman, M.A., Astrophys. J., **253**, 188 (1982).
Volk, H.J., Morfill, G.E. and Forman, M.A., Astrophys. J., **249**, 161 (1981).
Webb, G.M., Astrophys. J., **270**, 319 (1983).
Wolfendale, A.W., IAU Symposium 94, *Origin of Cosmic Rays*, D. Reidel Publishing Co., Dodrecht, p.309 (1981).
Zatsepin, G.T. and Kuzmin, V.A., Sov. Phys. JETP., **4**, 114 (1966).

Stochastic Parameters in Quantum Mechanical Systems

Stig Stenholm
Research Institute for Theoretical Physics, University of Helsinki, Helsinki, Finland

1. Personal Recollections

Almost twenty years ago, during the autumn of 1964 I was starting my first year as a student at Oxford University. Dr ter Haar was my supervisor and he was guiding my first steps into the mysteries of many-body physics and the theory of superfluidity. To complete my studies at my home university, the Helsinki University of Technology, I needed to submit a Licentiate Thesis. This is part of the Nordic way of initiation into the brotherhood of researchers and was earlier a necessary prerequisite when one wanted to submit a doctoral thesis. As things turned out, I went through with the prerequisite, but submitted my doctoral thesis in Oxford. The work itself, however, became significant for me in a different way.

When asking Dr ter Haar for a problem suitable for my thesis, I was suggested the derivation of Bloch's equation by a coarse graining in time. This was in accordance with his ability to keep up an interest in many different topics simultaneously. From him I learned much about the charm in following various fields of physics, their parallel developments, separations by methods and conflicts of goals. The problem itself was based on ter Haar's earlier lectures [1]. A spin precessing in an external magnetic field is also affected by a randomly fluctuating effective field. The work was related to the pioneer research by Wangsness and Bloch [2] and Redfield [3] where the Bloch equation was derived from a very general point of view. They, however, used ensemble averages over the degrees of freedom of the bath, a point discussed by Argyres and Kelley [4] in some detail. The suggestion, put forward by ter Haar, was that a time average would give the same result assuming that the averaging period exceeds the correlation time of the random fields.

The work carried out was published in [5] and also in [6]. I did, however, continue my interest in derivations of phenomenological equations describing irreversible time evolution. I got acquainted with the projection method introduced by Zwanzig [7] and developed further by Fano [8]. This was applied to a variety of problems by Argyres [9]. It has been developed into a general technique for the treatment of non-equilibrium phenomena, see Jancel [10]; for a more recent review of applications, see Grabert [11].

Attending the Brandeis Summer School in 1966, I learned about a different line of development. The lectures by Mel Lax [12] introduced me to the then rather new field of fluctuations in laser systems. Especially the quantum noise work by Lax [13], and the work by H. Haken and his collaborators in Stuttgart [14] were generalizations of the random field approaches I knew from nuclear magnetic resonance theory. Most of the work was, however, already well developed and I had little to add to it.

After my work at Oxford I came into laser research myself and got the fortunate start of one year's collaboration with Willis Lamb Jr. The reason behind this was that ter Haar recommended me strongly to Lamb and subsequently recommended strongly that I accept Lamb's offer to work on laser physics. The choice of field turned out to be a very propitious one.

My laser work was concerned with strong saturation features and nonlinear laser spectroscopy. In these problems usually the fluctuations play very little role; in experiments they are avoided as far as possible.

When laser spectroscopy turned to the use of dye lasers, even theoreticians had to admit that laser light contains fluctuations. Then the theory of fluctuating external fields came back into vogue. Probably the earliest revival of this theory in its new clothes was the work by Eberly [15], but the stage was set much earlier in the work by Burshtein [16]. Later a large body of literature has developed about these problems; *vide infra*.

For a long time I thought that the theory of field fluctuations was well taken care of by others. A couple of years ago I found that some questions concerning the relative contributions from coherent and incoherent processes were not answered by the existing papers. Some partial answers were given by Salomaa [17]. I looked into the published theories and recognized much of the approach from earlier work on random processes. In the limit of fast fluctuations it turned out that my problems could be discussed in a rather straightforward way [18]; see also the discussion in my forthcoming book [19], Part 5. The theory of laser fluctuations in nonlinear spectroscopy has not yet been exploited completely.

When I was asked to write this contribution to the "Festschrift" for my former teacher and long time friend, D. ter Haar, I found it an instance of synchronicity that my interests had just closed the circle back to our first research project together. The topic of the present essay was hence chosen for me and I only had to accept the challenge.

I will try to summarize the problems and methods related to the occurrence of a fluctuating parameter in quantum mechanical systems. The topic is much larger than my treatment suggests; neither the purely mathematical aspects nor all relevant applications are presented. The work reviewed mirrors basically my own knowledge and interests. In a brief review like this I cannot present the detailed consequences of the theory, but it must suffice to include the basic models and outlines of the methods developed to treat them.

Part 2 of the essay discusses formal aspects, presents the philosophy of approach and its relation to observables. Part 3 discusses special cases of physical interest and Part 4 presents some mathematical details used in the treatment of noise in quantum systems. Finally, Part 5 discusses the situation and presents some concluding comments.

2. Formal Developments

2.1 *Probability and Quantum Mechanics*

The most general description of physical observables equates them with the elements of an algebra. The specific features of quantum mechanics are due to the fact that these elements do not commute, which is the mathematical counterpart of the complementarity concept introduced by Heisenberg and Bohr.

von Neumann[20] introduced the most general expectation value as a linear functional of the observables. This defines the state space, where the observables act as operators and the functional is represented by a statistical or density operator ρ by the relation

$$\langle A \rangle = Tr \rho A \quad . \tag{1}$$

Because the observables do not commute, the operator ρ is to be represented by a density matrix ρ_{ij} with essential information residing also in the off-diagonal elements. These elements can be interpreted as the expectation values of an operator with the elements $\delta_{ij}^{k\ell}$. These can be made into observables in a trivial way by taking the Hermitean and anti-Hermitean parts. This is a brief summary of the mathematical approach to the structure of quantum mechanics as we know it. It has been argued that the additivity assumption for observables transcends our heuristic interpretation of complementarity; see Chapter 7 of Jammer[21]. This is the clue to most hidden variable theories and a starting point for much of the present work on quantum mechanics as we do not yet know it.

Here we accept fully the quantum mechanical features of the description, but add another aspect which usually has received attention only in special cases. The fact that quantum mechanical probability is properly accounted for does not relieve us of the burden forced upon us by old-fashioned stochasticity. The physical problem may still contain classical parameters that change in time in such a fashion that they are best described by a stochastic process. Then the average over the quantum states in (1) must be supplemented by an average over the random variables. In the general case then, the density operator must also assume the role of the classical distribution function. If ζ denotes the set of stochastic parameters we generalize (1) to

$$\overline{\langle A\rangle} = \int d\zeta \; Tr\, \rho(\zeta)\, A(\zeta), \tag{2}$$

where a bar is used to denote the stochastic average. This stochastic aspect differs from the ensemble view based on a collection of equally likely initial states. Here stochasticity is fed into the system continuously through ζ.

The fact that ζ is a stochastic parameter is here taken in the classical sense that its time dependence cannot be ascribed to any Hamiltonian system. It can only be described in a stochastic way. The pure mathematical description is in terms of measure on the space of all functions $\zeta(t)$. Under quite general conditions this description is equivalent with one using transition probabilities; if these are stationary and Markovian, the theory simplifies considerably. We shall, for simplicity, make all these assumptions in this essay.

The classical view is that stochasticity emerges as the effect of many unobserved degrees of freedom. Recent work on nonlinear systems, especially the strange attractors, has shown that highly nonpredictable behaviour can result even in systems with but a few degrees of freedom. At present it is unclear if a basis for stochasticity can be built in this way, but our understanding of stochastic phenomena is firm enough to be unshaken by discussions about the conditions for its validity. The ordinary physical view is that when enough information is lost to unobserved degrees of freedom the behaviour becomes stochastic. At some point we also want to invoke the universality of the Gaussian distribution. This philosophy is embodied in the physicists' use of Langevin equations for random parameters. As an example we set

$$\dot{\zeta} = F(\zeta) + L(t) , \tag{3}$$

where the parameter ζ is driven by the deterministic force $F(\zeta)$ and perturbed by the random fluctuation $L(t)$, the Langevin force. As this represents the influence of a multifarious environment, it is assumed to have a

Gaussian probability distribution. The environment continuously sucks out information which prevents predictability. Whether this is lost by dilution into a bath of many degrees of freedom or by sheer dissipation by a few is immaterial.

What characterizes the Markovian stochastic processes is that they can be described by the use of master equations; see, e.g., van Kampen[22]. Here we have to combine this feature with the essential quantum mechanical off-diagonality. We start with the density matrix element $\rho_{\mathbf{ij}}(\zeta,t)$, which we first treat as a simple probability. The stochastic influence is an instantaneous perturbation which occurs with the probability $\gamma_{\mathbf{ij}}(\zeta)$. For a stationary process no time dependence is assumed. When a perturbation has occurred, the stochastic parameter ζ has acquired a redistribution of its probability. But, in the general case, the perturbation may also change the quantum mechanical state of the system. Hence there is a probability for the state $\rho_{\mathbf{ij}}$ to be redistributed over the states $\rho_{\mathbf{k}\ell}$. For an infinitesimal time dt we hence generalize the ordinary result for stochastic processes to

$$\rho_{\mathbf{ij}}(\zeta\,,t+dt) = (1-\gamma_{\mathbf{ij}}(\zeta)\,dt)\;\rho_{\mathbf{ij}}(\zeta,t) + \sum_{\mathbf{k}\ell\mu} dt\,\gamma_{\mathbf{ij}}(\zeta)\,\Pi^{\mathbf{k}\ell}_{\mathbf{ij}}(\zeta|\mu)\,\rho_{\mathbf{k}\ell}(\mu,t)\,. \tag{4}$$

We have treated all variables by discrete sums; when needed integrals can be substituted. This relation is equivalent with a master equation of the form

$$\dot{\rho}_{\mathbf{ij}}(\zeta\,,t) = \gamma_{\mathbf{ij}}(\zeta)\left\{\sum_{\mathbf{k}\ell\mu}\left[\Pi^{\mathbf{k}\ell}_{\mathbf{ij}}(\zeta|\mu) - \delta^{\mathbf{k}\ell}_{\mathbf{ij}}\,\delta_{\zeta\mu}\right]\rho_{\mathbf{k}\ell}(\mu,t)\right\}\,. \tag{5}$$

The operator on the right is a linear operator on the density matrix and hence

$$\dot{\rho} = \boldsymbol{W}\rho\,. \tag{6}$$

The normalization of ρ under the averaging procedure (2) is guaranteed if we require

$$\sum_{\zeta\mathbf{i}} \gamma_{\mathbf{ii}}(\zeta)\,\Pi^{\mathbf{k}\ell}_{\mathbf{ii}}(\zeta|\mu) = \delta_{\mathbf{k}\ell}\,\gamma_{\mathbf{kk}}(\mu)\,. \tag{7}$$

Physically this relation states that what disappears from the state $\rho_{\mathbf{k}\ell}(\zeta)$ must appear somewhere else. The formal solution of (6) is given by

$$\rho(t) = e^{\boldsymbol{W}t}\,\rho(0)\,. \tag{8}$$

For the cases where an asymptotically stable state is reached, the eigenvalues of the operator $\boldsymbol{W}$ must all be non-positive. I do not try to find the most general case when this can be proven.

If we introduce the eigenoperators $\Phi_{\mathbf{n}}$ and their eigenvalues $\lambda_{\mathbf{n}}$ we write

$$\boldsymbol{W}\Phi_n = -\lambda_n \Phi_n \, . \tag{9}$$

Then all values λ_n are non-negative. The formal solution (8) can then be written

$$\rho(t) = \rho_0 \Phi_0 + \sum_{\lambda_n \neq 0} e^{-\lambda_n t} \rho_n \Phi_n \, . \tag{10}$$

The component ρ_0 corresponding to the eigenvalue zero is the steady state solution approached asymptotically in time. General conditions for the existence and completeness of these eigenfunctions are not sought here. We will discuss some specific examples later.

2.2 *Steady State Calculations with Noise*

The density matrix normally evolves under the influence of a Hamiltonian $\mathcal{H}$ and it may be affected by some relaxation mechanisms not connected with the stochastic process under investigation. These can be represented by a linear operator R on ρ. In addition steady state operation often implies the inclusion of an inhomogeneous pumping Λ_{ij} into some states ρ_{ij}. The quantum mechanical part of the equation of motion is then given by

$$\begin{aligned} i\hbar\dot{\rho} &= \boldsymbol{\Lambda} + [\mathcal{H}, \rho] + R(\rho) \\ &= \boldsymbol{\Lambda} + \hbar\boldsymbol{\Omega}\rho \, , \end{aligned} \tag{11}$$

where the linear operator $\boldsymbol{\Omega}$ is taken to contain the relaxation terms too.

When both quantum mechanics and stochasticity are present we write the equation of motion as a combination of (11) and (6) in the form

$$i\hbar\dot{\rho} = \boldsymbol{\Lambda} + \hbar\boldsymbol{\Omega}\rho + i\hbar\boldsymbol{W}\rho \, . \tag{12}$$

Introducing the Laplace transform $\tilde{\rho}$ of ρ we find

$$i\hbar s\tilde{\rho} = i\hbar\rho(t=0) + \frac{\boldsymbol{\Lambda}}{s} + \hbar[\boldsymbol{\Omega} + i\boldsymbol{W}]\tilde{\rho} \, . \tag{13}$$

The steady state solution can be obtained by setting $\rho(t=\infty) = \lim_{s\to 0} s\rho(s)$ and the result is

$$\rho(t=\infty) = -\frac{i}{i\boldsymbol{\Omega} - \boldsymbol{W}} \frac{\boldsymbol{\Lambda}}{\hbar} \, . \tag{14}$$

This follows immediately from (12) in steady state. For transients we assume no pumping, $\Lambda = 0$ and obtain

$$\tilde{\rho}(s) = \frac{1}{s + i\boldsymbol{\Omega} - \boldsymbol{W}} \rho(t=0) \, . \tag{15}$$

Thus the properties of the system are determined by the spectrum of the linear operator, $i\boldsymbol{\Omega} - \boldsymbol{W}$. The static properties follow directly from (14). The populations on the various levels are given by

$$\overline{\rho_{jj}(t=\infty)} = -\frac{i}{\hbar} \sum_{\zeta} \left\langle j \left| \frac{1}{i\boldsymbol{\Omega}(\zeta) - \boldsymbol{W}(\zeta)} \boldsymbol{\Lambda}(\zeta) \right| j \right\rangle , \tag{16}$$

where the dependence on ζ is indicated formally.

In spectroscopy the exchange of energy between the random field and the atomic matter is proportional to the average of the field times the induced dipole moment, namely

$$\begin{aligned}\overline{\zeta\mu} &= \overline{e\zeta Tr\rho r} \\ &= -\frac{ie}{\hbar}\sum_{\zeta}\sum_{jk} r_{jk}\left\langle k \left| \frac{\zeta}{i\boldsymbol{\Omega}(\zeta) - \boldsymbol{W}(\zeta)} \boldsymbol{\Lambda}(\zeta) \right| j \right\rangle . \end{aligned} \tag{17}$$

For the results (16) and (17) the steady state solution (14) suffices and no dynamics is required.

In order to obtain the spectrum of the system we have to evaluate the quantity

$$S_{ij}(\omega) = \frac{1}{\pi} \operatorname{Re} \int_0^{\infty} e^{i\omega\tau} \overline{\langle \mu_i(0)\, \mu_j(\tau) \rangle}\, d\tau \, . \tag{18}$$

This generalizes the Wiener-Khintchine theorem to the present case. The operators $\{\mu_j\}$ are those observables used to probe the system. Here we mostly think of the induced dipole μ of (17) and omit the indices.

It is often claimed that the spectrum (18) must be calculated in the Heisenberg picture. This is clearly wrong because, as Fano [8] points out, we can write

$$\begin{aligned}\langle \mu_i\, \mu_j(\tau) \rangle &= Tr[\rho\, \mu_i\, e^{iH\tau/\hbar}\, \mu_j\, e^{-iH\tau/\hbar}] \\ & Tr[\mu_j\, e^{-iH\tau/\hbar}\, \rho\, \mu_i\, e^{iH\tau/\hbar}] \\ & Tr[\mu_j\, \rho_{\mu_i}(\tau)] \, . \end{aligned} \tag{19}$$

This is the time dependent expectation value of μ_j with respect to a Schrödinger picture density matrix with the initial condition

$$\rho_{\mu_i}(t=0) = \rho\mu_i \, , \tag{20}$$

and the customary equation of motion. If we take over this result to include the stochastic time evolution we can directly obtain the spectrum in terms of the Laplace transform (15). There are some problems of justification connected with this; the Hamiltonian part of the time evolution can easily be transferred between the observables and the density matrix, but the stochastic part relates to the latter in an essential way. To employ (18) and (19) under these circumstances requires some physical analysis of the observations, but here we adopt the procedure without pinpointing its range of validity. With the results (18), (19) and (15) we now find the relation

$$S_{ij}(\omega) = \frac{1}{\pi} \operatorname{Re} \overline{\left\langle \mu_j \frac{1}{-i\omega + i\boldsymbol{\Omega} - \boldsymbol{W}} \rho(t=0)\, \mu_i \right\rangle} \, , \tag{21}$$

which expresses the fluctuation spectrum in terms of the spectrum of the linear operator $i\mathbf{\Omega}-\mathbf{W}$.

To evaluate the dynamic properties one has to solve the quantum mechanical evolution with noise. For time dependent problems Keldysh [23] has developed a diagrammatic perturbation theory, but as far as I know it has not yet been combined with stochastic time evolution. The superoperator formalism was introduced by Klein [24] and later anew by Kubo [25]. It has been developed extensively by Zwanzig [7], and it is directly suggested by our notation. For spectroscopic applications the quantum mechanical Hilbert space is often truncated to a few states only, and then the operator $\mathbf{\Omega}$ becomes rather easy to handle; we can add the noise complications directly to known results in nonlinear spectroscopy.

The formal solution (15) is analogous in form to the result of the memory function formalism, see, e.g., Forster [26] Chapter 5. There the dynamic operator $\mathbf{\Omega}$ contains the Markovian part of the time evolution and the rest depends on the Laplace variables to indicate memory effects. These are due to the evolution within an unobserved set of dynamic variables. Here our operator $\mathbf{W}$ has a different origin; it is present because of a set of variables postulated to have stochastic dynamics. When exactly the former description can be replaced by the latter one is a far-reaching problem which is outside the scope of the present treatise.

In many cases the operator $\mathbf{W}$ in Eq. (6) becomes independent of the quantum states, i.e. we have

$$\gamma_{ij} \equiv \gamma ; \quad \Pi_{ij}^{k\ell} = \Pi\, \delta_{ik}\, \delta_{j\ell} . \tag{22}$$

The stochastic process evolves independently of the quantum evolution, and the whole situation becomes much simpler. The stochastic operator $\mathbf{W}$ commutes with the dynamic operator $\mathbf{\Omega}$ and the stochastic problem can be treated independently of the quantum mechanical one. In this case all techniques for ordinary stochastic processes can be taken over, and in particular the eigenfunction expansion (10) provides a very powerful technique.

3. Special Cases

3.1 *Scattering Theory*

When a radiating atom encounters a perturber atom, the scattering is a random event normally lasting a short time only, and it can be regarded as an instantaneous perturbation. This is the impact approximation. The probability of a scattering process, the scattering cross-section, depends on the state. During the collision usually both the internal atomic state and the atomic velocity change. Hence the most general version of our theory applies.

The actual collision time τ is taken to be so short that the change of the states during a collision is given by the S-matrix through

$$\psi(t+\tfrac{1}{2}\tau) = S\,\psi(t-\tfrac{1}{2}\tau) \tag{23}$$

as noted by Alekseev, Andreeva and Sobel'man [27]. If we introduce the T-matrix by setting

$$T = S - 1 \tag{24}$$

we find the change in the density matrix during one collision to be given by

$$\begin{aligned}\delta\rho_{\mu\mu'}(t+\tfrac{1}{2}\tau) &= \left[\sum_{\nu\nu'} S_{\mu\nu}\,S^{\dagger}_{\mu'\nu'}\,\rho_{\nu'\nu}(t-\tfrac{1}{2}\tau) - \rho_{\mu\mu'}(t-\tfrac{1}{2}\tau)\right] \\ &= \sum_{\nu'} T^{\dagger}_{\mu'\nu'}\,\rho_{\mu\nu'} + \sum_{\nu} T_{\mu\nu}\,\rho_{\nu\mu'} \\ &\qquad + \sum_{\nu\nu'} T_{\mu\nu}\,T^{\dagger}_{\mu'\nu'}\,\rho_{\nu\nu'}\,. \end{aligned} \tag{25}$$

Because the collisions are supposed to be independent, we can average each one over all its parameters characterizing the event and (25) becomes expressed in terms of the averages $\overline{T}$, $\overline{T^{\dagger}}$ and $\overline{TT^{\dagger}}$. Berman [28] has analyzed these expressions applied to the diagonal elements in the translational velocity representation $v = \hbar k/m$ and found that all these averages become proportional to the time variable and can be expressed in terms of the quantum mechanical scattering amplitudes $f_{ab}(v_i \to v_f)$ from the initial state a with velocity v_i into the state b with velocity v_f. Then the collision interval τ may approach zero and (25) gives the collisional contribution to the equation of motion

$$\begin{aligned}\frac{\partial}{\partial t}\rho_{ab}(v,t) &= \sum_{cd}\left[t_{ac}(v)\,\delta_{bd} - t^{*}_{bd}(v)\,\delta_{ac}\right]\rho_{cd}(v,t) \\ &\quad + \sum_{cd}\int d^3v'\,W^{cd}_{ab}(v'\to v)\,\rho_{cd}(v',t)\,. \end{aligned} \tag{26}$$

Here $t_{ab}(v)$ is the element of the $\overline{T}$-matrix with the velocity dependence taken out and W is proportional to the collisional average of $f_{ac}(v'\to v)\,f^{*}_{bd}(v'\to v)$. The details of the derivation are given by Berman [28].

We now define the line-width parameters,

$$\gamma^{cd}_{ab}(v) = \int d^3v'\,W^{cd}_{ab}(v\to v') \tag{27}$$

$$\Gamma^{cd}_{ab}(v) = -\,t_{ac}(v)\,\delta_{bd} + t^{*}_{bd}(v)\,\delta_{ac} - \gamma^{cd}_{ab}(v)\,. \tag{28}$$

In terms of these expressions, the equation (26) becomes

$$\frac{\partial}{\partial t}\rho_{ab}(v,t) = -\sum_{cd}\Gamma^{cd}_{ab}(v)\,\rho_{cd}(v,t) + \sum_{cd}\int d^3v'\Big[W^{cd}_{ab}(v'\to v)\,\rho_{cd}(v',t) - W^{cd}_{ab}(v\to v')\,\rho_{cd}(v,t)\Big]. \tag{29}$$

The first term comes from the forward scattering amplitude $f_{ab}(v \to v)$ and gives a generalized non-diagonal relaxation term in the equation of motion. The second term is a master equation term of a structure clearly related to the form we introduced directly in Eq. (5). The random parameter is, in this case, the velocity v, which becomes redistributed in each collision. In fact the form (28) is slightly more general than (5) because it also contains the possibility that $\gamma_{ij}(\zeta)$ in (5) is a matrix, γ^{ab}_{ij}, in the atomic state indices. The form (28) is, however, a consequence of scattering theory and the assumption of binary collisions of short duration. The stochastic aspect enters through the uncontrolled initial and final states of the independent perturber atoms.

The physical implications of the form (28) have been investigated extensively by Berman. For a simple version see Ref. 29 and for more sophisticated considerations, Ref. 30.

3.2 *The State Independent Perturbation Probability*

If we assume that the perturbations occur totally at random but with an average time T between them we obtain a generalization of the Poisson process. The probability of no collision in a time t is then $e^{-t/T}$ and between the collisions the time evolution of the density matrix becomes

$$\rho(t) = e^{-(t-t_0)/T}\,\boldsymbol{U}(t-t_0)\,\rho(t_0)\,, \tag{30}$$

where the operator $\boldsymbol{U}(t)$ is the solution of the quantum mechanical part of the time evolution in Eq. (11) namely

$$i\frac{\partial}{\partial t}\boldsymbol{U}(t) = \boldsymbol{\Omega}\,\boldsymbol{U}(t)\,, \tag{31}$$

with the initial condition

$$\boldsymbol{U}(t=0) = \mathbf{1}\,. \tag{32}$$

When the perturbation acts during a short interval only we can neglect the influence of $\boldsymbol{\Omega}$ during this interval and assume that the effect of the perturbation is

$$\rho(t+) = \boldsymbol{S}\,\rho(t-)\,. \tag{33}$$

We now consider a time interval (t_0, t) and form the probabilities of no collisions, one collision etc. in this interval.

For no collisions, we use directly the result (30). For one collision in the interval occurring at t_1 we find

$$\rho^{(1)}(t) = \int_{t_0}^{t} e^{-(t-t_1)/T} \boldsymbol{U}(t-t_1)\, \boldsymbol{S}\, e^{-(t_1-t_0)/T} \boldsymbol{U}(t_1-t_0)\, \rho(t_0) \frac{dt_1}{T}, \quad (34)$$

because the probability to have one collision in dt_1 is (dt_1/T). We do not care when the collision occurred, $t_1 \in (t_0, t)$. We average over the parameters in the collision and introduce the averaged collision operator $\bar{\boldsymbol{S}}$; we do not assume that $\boldsymbol{U}$ and $\bar{\boldsymbol{S}}$ commute. This means that the parameters changed in the collision will effect the propagation during the subsequent interval (t_1, t).

For two collisions we can perform the averages independently of each other as the perturbations are assumed to be well separated compared with their duration; that is, they are statistically independent. We find

$$\overline{\rho^{(2)}} = \int_{t_0}^{t} \frac{dt_2}{T} \int_{t_0}^{t_2} \frac{dt_1}{T} e^{-(t-t_2)/T} \boldsymbol{U}(t-t_2)\, \bar{\boldsymbol{S}}\, e^{-(t_2-t_1)/T}$$
$$\times\, \boldsymbol{U}(t_2-t_1)\, \bar{\boldsymbol{S}}\, e^{-(t_1-t_0)/T} \boldsymbol{U}(t_1-t_0)\, \rho(t_0). \quad (35)$$

This way we can easily continue and add up the probabilities for all possible collisional histories. We obtain the result

$$\rho(t) = \sum_{n=0}^{\infty} \overline{\rho^{(n)}}(t) = e^{-(t-t_0)/T} \sum_{n=0}^{\infty} T^{-n} \int_{t_0}^{t} dt_n \int_{t_0}^{t_n} dt_{n-1} \ldots \int_{t_0}^{t_2} dt_1$$
$$\times\, \boldsymbol{U}(t-t_n)\, \bar{\boldsymbol{S}}\, \boldsymbol{U}(t_n-t_{n-1})\, \bar{\boldsymbol{S}} \ldots\ldots \bar{\boldsymbol{S}}\, \boldsymbol{U}(t_1-t_0)\, \rho(t_0). \quad (36)$$

An instructive way to look at this expression is to write it as

$$\rho(t) = \sum_{n=0}^{\infty} P_n(t) \int_{t_0}^{t} dt_n \ldots \int_{t_0}^{t_2} dt_1\, W_n(t_n, \ldots, t_1, t)$$
$$\times\, \boldsymbol{U}(t-t_n)\, \bar{\boldsymbol{S}} \ldots \boldsymbol{U}(t_1-t_0)\, \rho(t_0),$$

where

$$P_n(t) = \frac{e^{-(t-t_0)/T}}{n!} \left(\frac{t-t_0}{T}\right)^n \quad (37)$$

is the Poisson probability for n collisions in the interval (t_0, t) when the average collision separation is T. The remaining integration measure

$$W_n(t_n, \ldots, t_1, t)\, dt_n \ldots dt_1 = n! \frac{dt_1}{(t-t_0)} \cdots \frac{dt_n}{(t-t_0)} \quad (38)$$

is the probability that given n collisions in (t_0, t) these occur in the intervals $dt_1 \ldots dt_n$. As the occurrence is fully random the result (38) follows; the factor $n!$ comes from the fact that the occurrences in (37) are unlabelled events but the times $t_1 \ldots t_n$ are assumed to be ordered. From this consideration we see exactly in what sense the present case is a generalization of the simple Poisson process. If we take the derivative of Eq. (36) with respect to t we find three types of terms:

$$\frac{d\rho(t)}{dt} = e^{-(t-t_0)/T} \sum_{n=0}^{\infty} T^{-n} \int_{t_0}^{t} dt_n \cdots \int_{t_0}^{t_2} dt_1,$$
$$\times \frac{d}{dt} \boldsymbol{U}(t-t_n)\,\bar{\boldsymbol{S}} \ldots \boldsymbol{U}(t_1-t_0)\,\rho(t_0)$$
$$+ e^{-(t-t_0)/T} \sum_{n=1}^{\infty} T^{-n} \int_{t_0}^{t} dt_{n-1} \cdots \int_{t_0}^{t_2} dt_1\,,$$
$$\times\ \bar{\boldsymbol{S}}\,\boldsymbol{U}(t-t_{n-1})\,\bar{\boldsymbol{S}} \ldots \boldsymbol{U}(t_1-t_0)\,\rho(t_0) - \frac{1}{T}\rho(t)\,,$$
$$= -i\boldsymbol{\Omega}\rho + \frac{1}{T}(\bar{\boldsymbol{S}}\rho - \rho)\,, \tag{39}$$

where the equation (31) has been used. If in this equation we identify T^{-1} with the state dependent γ in Eq. (5) and $\bar{\boldsymbol{S}}$ with the operator Π we have derived that equation. Setting again

$$\boldsymbol{W} = \frac{1}{T}(\bar{\boldsymbol{S}} - 1) \tag{40}$$

the equation (39) agrees exactly with Eq. (12).

Another derivation of the same result is also instructive. We write the evolution operator formally

$$\boldsymbol{U}(t) = e^{-i\boldsymbol{\Omega}t} \tag{41}$$

in agreement with (31). Then the result (36) can be written

$$\rho(t) = \sum_{n=0}^{\infty} \int_{t_0}^{t} dt_n \int_{t_0}^{t_n} dt_{n-1} \cdots \int_{t_0}^{t_2} dt_1\, e^{-(i\boldsymbol{\Omega}+\frac{1}{T})(t-t_n)}$$
$$\times \frac{\bar{\boldsymbol{S}}}{T} e^{-(i\boldsymbol{\Omega}+\frac{1}{T})(t_n-t_{n-1})} \frac{\bar{\boldsymbol{S}}}{T} \cdots \frac{\bar{\boldsymbol{S}}}{T} e^{-(i\boldsymbol{\Omega}+\frac{1}{T})(t_1-t_0)}\rho(t_0)\,. \tag{42}$$

This is a convolution and after a Laplace transform it becomes a product. We have

$$\int_0^{\infty} e^{-st}\, e^{-(i\boldsymbol{\Omega}+\frac{1}{T})t}\, dt = \frac{1}{i\boldsymbol{\Omega}+\frac{1}{T}+s}\,. \tag{43}$$

The transform of (36) becomes

$$\tilde{\rho}(s) = \sum_{n=0}^{\infty} \frac{1}{s+i\boldsymbol{\Omega}+\frac{1}{T}} \frac{\bar{\boldsymbol{S}}}{T} \frac{1}{s+i\boldsymbol{\Omega}+\frac{1}{T}} \cdots \frac{\bar{\boldsymbol{S}}}{T} \frac{1}{s+i\boldsymbol{\Omega}+\frac{1}{T}}\rho(t_0)$$
$$= \frac{1}{s+i\boldsymbol{\Omega}+\frac{1}{T}}\left[1 - \frac{\bar{\boldsymbol{S}}}{T}\,\frac{1}{s+i\boldsymbol{\Omega}+\frac{1}{T}}\right]^{-1}\rho(t_0)$$

$$= \frac{1}{s + i\boldsymbol{\Omega} - \frac{1}{T}(\bar{\boldsymbol{S}} - 1)} \rho(t_0)$$

$$= \frac{1}{s + i\boldsymbol{\Omega} - \boldsymbol{W}} \rho(t_0) \; . \tag{44}$$

This result is identical with the one in (15) and, of course, with the equation of motion (39). For the calculations of the spectrum in (21) we directly need the Laplace transform and hence the form (44) turns out useful in many cases.

The derivations in the present section are based on two articles: The one, Stenholm [31], used this formalism for the collision problem in atomic spectroscopy. The collisions take place at random times and the parameter which is redistributed is the velocity. The time evolution operator $\boldsymbol{U}(t)$ was taken to include the propagation of the atom from point to point in space, and hence a change in velocity affected the propagator. The drawback of this model is that the real scattering probability is not independent of the state of the atom. Hence the present model is too crude in representing it by one single parameter T. The more exact treatment worked out by Berman shows how the state dependence of scattering is to be included; see the previous section.

Clauser and Blume [32] assume a perturbation potential V_i which acts at random times $\{t_i\}$. The perturbing part of the Hamiltonian is then

$$\mathcal{H}_{int} = \sum_i V_i \, \delta(t - t_i) \; . \tag{45}$$

This is the quantum theoretic generalization of the shot noise process. The treatment above goes through with

$$\bar{\boldsymbol{S}} \rho \propto \overline{e^{iV_i} \rho e^{-iV_i}} \; . \tag{46}$$

Clauser and Blume [32] derive the equation (44), Stenholm [31] derives (39). We have seen that they are fully equivalent.

Clauser and Blume were interested in magnetic resonance calculations. The applications of this formalism to relaxation in Mössbauer spectra had earlier been considered by Blume [33] and Blume and Tjon [34]. They assume the spectrum to be perturbed by a Hamiltonian of the form

$$\mathcal{H}_{int} = \sum_n V_k f_k(t) \; , \tag{47}$$

where the operators V_k are noncommuting spin operators, and the functions $f_k(t)$ jump at random between the values zero and one. Then the spectrum (21) can be evaluated exactly by the method introduced originally by Anderson [35] and which consists of taking a finite matrix for the stochastic operator $\boldsymbol{W}$.

Choosing random fields parallel with or perpendicular to the external field and a quadrupole perturbation they manage to compute Mössbauer spectra reproducing features observed in the experiments.

3.3 *Fluctuations Independent of the Quantum Dynamics*

When the fluctuations are totally independent of the internal dynamics of the atomic system the two operators $\boldsymbol{\Omega}$ and $\boldsymbol{W}$ commute. This is the situation when the fluctuations are due to sources totally outside the system under investigation. One such case is the effective field at the position of the nucleus in magnetic resonance. The crystal environment determines the field and its fluctuations are due to phonons or the effective field of the other, statistically independent spins. Anderson's and Kubo's original work on fluctuations was devoted to this application.

When laser spectroscopy is carried out with a fluctuating light source, e.g., a dye laser, the noise is carried by the light into the interaction region; its properties are fixed at the distant source. Hence they are independent of the state of the atomic system under investigation, and the operator $\boldsymbol{W}$ acts on each density matrix element ρ_{ij} in the same way. This is a model of laser noise which has been applied extensively by Zoller [36] and his collaborators.

When the jump caused by a single perturbation event becomes very small, but their frequency of occurrence increases, we approach the limit of a continuous random function. As for Brownian motion we can describe this case by a differential equation which we can derive from (5) by expanding in the size of the jump in the random parameter ζ. We write

$$\dot{\rho}_{ij}(\zeta,t) = \gamma(\zeta)\left[\sum_{\mu}\Big(\Pi(\zeta|\mu)\,\rho_{ij}(\mu,t) - \rho_{ij}(\zeta,t)\Big)\right]. \tag{48}$$

Then the transition probably is written as

$$\Pi(\zeta|\mu) \equiv \theta(\mu|\zeta-\mu) \tag{49}$$

and we use the expansion

$$\begin{aligned}\sum_{\nu}\theta(\zeta-\nu|\nu)\,\rho_{ij}(\zeta-\nu,t) &= \sum_{\nu} e^{-\nu\frac{\partial}{\partial\zeta}}\,\theta(\zeta|\nu)\,\rho_{ij}(\zeta,t)\\ &= \rho_{ij}(\zeta,t) - \frac{\partial}{\partial\zeta}\left[S_1(\zeta)\,\rho_{ij}(\zeta,t)\right]\\ &\quad + \frac{1}{2}\frac{\partial^2}{\partial\zeta^2}\left[S_2(\zeta)\,\rho_{ij}(\zeta,t)\right],\end{aligned} \tag{50}$$

where

$$S_1(\zeta) = \sum_{\nu}\nu\,\theta(\zeta|\nu)\;, \qquad S_2(\zeta) = \sum_{\nu}\nu^2\,\theta(\zeta|\nu)\;. \tag{51}$$

Inserting this into Eq. (48) we obtain the ordinary Fokker-Planck equation which has been used extensively in connection with noise. The integral or matrix operator of (48) has then been replaced by a differential operator.

When we deal with a fluctuating external field this usually occurs linearly in the Hamiltonian and we can write the quantum mechanical operator of (11) in the form

$$\boldsymbol{\Omega} = \boldsymbol{\Omega}_0 + \hat{\zeta}\boldsymbol{\Omega}_1 ; \tag{52}$$

this is the only explicit occurrence of $\hat{\zeta}$ in the equations.

If we now know the eigenfunctions of the operator $\boldsymbol{W}$ as in Eqs. (9—10) we can use the steady state eigenfunction Φ_0 to define a scalar product. Because Φ_0 is a scalar probability distribution it is a positive quantity and we can define the product

$$(\psi_1 , \psi_2) = \sum_{\mu} \psi_1(\mu)\ \Phi_0(\mu)^{-1}\ \psi_2(\mu) . \tag{53}$$

The equilibrium condition is written

$$\begin{aligned} \gamma(\nu) \sum_{\mu} \Pi(\nu|\mu)\ \Phi_0(\mu) &= \gamma(\nu)\ \Phi_0(\nu) \\ &= \sum_{\mu} \gamma(\mu)\ \Pi(\mu|\nu)\ \Phi_0(\nu), \end{aligned} \tag{54}$$

where we introduced the last step from the state independent version of (7). When detailed balance holds the equality (54) holds for each term separately and we have

$$\gamma(\nu)\ \Pi(\nu|\mu)\ \Phi_0(\mu) = \gamma(\mu)\ \Pi(\mu|\nu)\ \Phi_0(\nu) . \tag{55}$$

By this property the operator $\gamma\Pi$ is Hermitean with respect to the scalar product (53) because

$$\begin{aligned} (\psi_1 , \gamma\Pi\psi_2) &= \sum_{\mu} \psi_1(\mu)\ \Phi_0(\mu)^{-1} \sum_{\nu} \gamma(\mu)\ \Pi(\mu|\nu)\ \psi_2(\nu) \\ &= \sum_{\mu\nu} \psi_1(\mu)\ \gamma(\nu)\ \Pi(\nu|\mu)\ \Phi_0(\nu)^{-1}\ \psi_2(\nu) \\ &= \sum_{\nu} \Big(\sum_{\mu} \gamma(\nu)\ \Pi(\nu|\mu)\ \psi_1(\mu)\ \Phi_0(\nu)^{-1}\ \psi_2(\nu) \Big) \\ &= (\gamma\Pi\psi_1 , \psi_2) . \end{aligned} \tag{56}$$

With the chosen product (53) the problem is hence a Hermitean one and the ordinary conclusions concerning orthogonality and completeness are at our disposal. We expand the density matrix as

$$\rho = \sum_{n=0}^{\infty} \rho_n \Phi_n , \tag{57}$$

where

$$\rho_n = \sum_{\mu} \Phi_n(\mu)\ \Phi_0^{-1}(\mu)\ \rho(\mu) . \tag{58}$$

The operator form of the stochastic parameter $\hat{\zeta}$ in (52) becomes the matrix

$$\zeta_{nn'} = (\Phi_n , \zeta \Phi_{n'})$$

$$= \sum_{\zeta} \Phi_n(\zeta) \Phi_0(\zeta)^{-1} \zeta \Phi_{n'}(\zeta) . \tag{59}$$

With this notation and the equations (9), (12) and (52) we obtain

$$i \dot{\rho}_n = (\boldsymbol{\Lambda}/\hbar)_n + \boldsymbol{\Omega}_0 \rho_n + \boldsymbol{\Omega}_1 \sum_{n'} \zeta_{nn'} \rho_{n'} - i \lambda_n \rho_n . \tag{60}$$

The parameters λ_n appear directly as damping rates in the equation. The stochastic problem is now entirely in the diagolanization of the matrix $\zeta_{nn'}$. The pumping term is given by

$$(\boldsymbol{\Lambda}/\hbar)_n = \sum_{\mu} \Phi_0(\mu)^{-1} \Phi_n(\mu) \boldsymbol{\Lambda}(\mu)/\hbar . \tag{61}$$

When the matrix $\zeta_{nn'}$ can be handled, the calculation has been reduced to the quantum mechanical problem concerned with the operators $\boldsymbol{\Omega}_0$ and $\boldsymbol{\Omega}_1$.

The smallest eigenvalue is $\lambda_0 = 0$ which gives the steady state. The next smallest one $\lambda_1 > 0$ characterizes the long time behaviour and the approach to equilibrium if it is well separated from the others $\lambda_1 \ll \lambda_2, \lambda_3, \ldots$ The corresponding eigenfunction Φ_1 describes the shape of the deviations from equilibrium for large times. If its shape is known we know something about the dynamic long time behaviour.

4. Some Mathematical Models

4.1 *The Finite State Space Model*

In the original Anderson model the stochastic parameter ζ was assumed to take only a finite number of values, $\zeta \in \{\zeta_\mu\}$; correspondingly there exists only a finite number of stochastic states $\{\Phi_\mu\}$. If, in addition, the space of the operator $\boldsymbol{\Omega}(\zeta)$ is of finite dimensionality, $\{\phi_i\}$, we write the relation (44) in the form

$$\sum_{\mu' i'} \langle \phi_i \Phi_\mu | [s + i\Omega(\zeta_{\mu'}) - \boldsymbol{W}] | \phi_{i'} \Phi_{\mu'} \rangle$$
$$\times \langle \phi_{i'} \Phi_{\mu'} | \tilde{\rho}(s) | \phi_j \Phi_\nu \rangle = \langle \phi_i \Phi_\mu | \rho(t_0) | \phi_j \Phi_\nu \rangle . \tag{62}$$

This equation is solved by the inversion of a finite matrix and the averages (16 — 17) or the spectrum (21) can easily be evaluated. Because the operator $\boldsymbol{W}$ is a finite matrix also time dependent properties can be evaluated.

Anderson introduced this type of process in 1954 to describe motional narrowing in spin systems due to a perturbing field. To illustrate the ideas he considered the two-state model with a dipole oscillating at either $+\omega_0$ or $-\omega_0$. With the density matrix element ρ_{21} denoting the oscillating dipole we find the equations

$$\dot{\rho}_{21}(+) = -i\omega_0 \rho_{21}(+) + \Pi_{+-}\rho_{21}(-) - \Pi_{++}\rho_{21}(+) \tag{63}$$

$$\dot{\rho}_{21}(-) = i\omega_0 \rho_{21}(-) + \Pi_{-+}\rho_{21}(+) - \Pi_{--}\rho_{21}(-) . \tag{64}$$

The condition (7), or detailed balance requires that

$$\Pi_{+-} = \Pi_{--}, \qquad \Pi_{++} = \Pi_{-+}. \tag{65}$$

To simplify things further we set $\Pi_{--} = \Pi_{++} = \Pi$.

We assume the initial state to be the linear superposition

$$|\phi_0\rangle = \frac{1}{\sqrt{2}}(|1\rangle + |2\rangle), \tag{66}$$

which maximizes the expectation values of the dipole moment

$$\mu = \begin{bmatrix} 0 & \mu_0 \\ \mu_0 & 0 \end{bmatrix}. \tag{67}$$

Introducing this into the expression (21) for the spectrum we have

$$\begin{aligned} S(\omega) &= \frac{1}{\pi}\,\mathrm{Re}\left\{ Tr \left[\begin{bmatrix} 0 & \mu_0 \\ \mu_0 & 0 \end{bmatrix} \begin{bmatrix} -i(\omega-\omega_0)+\Pi & -\Pi \\ -\Pi & -i(\omega+\omega_0)+\Pi \end{bmatrix}^{-1} \begin{bmatrix} \frac{1}{2} & \frac{1}{2} \\ \frac{1}{2} & \frac{1}{2} \end{bmatrix} \begin{bmatrix} 0 & \mu_0 \\ \mu_0 & 0 \end{bmatrix} \right] \right\} \\ &= \frac{\mu_0^2}{\pi} \frac{2\,\Pi\,\omega_0^2}{(\omega-\omega_0)^2(\omega+\omega_0)^2 + 4\omega^2\Pi^2} . \end{aligned} \tag{68}$$

The maxima in the spectrum are found at

$$\omega = \pm\,\mathrm{Re}\left[\omega_0^2 - 2\,\Pi^2\right]^{\frac{1}{2}}, \tag{69}$$

from which we can see that when Π grows the two resonances at $\pm\omega_0$ come together and merge for

$$\Pi = \frac{\omega_0}{\sqrt{2}} . \tag{70}$$

This is an example where the increase in the rate of the fluctuations decreases the effective line width, i.e., motional narrowing. For small values of Π we can write near the maxima $\omega^2 \approx \omega_0^2$

$$\begin{aligned} &(\omega-\omega_0)^2(\omega+\omega_0)^2 + 4\omega^2\Pi^2 \\ &\qquad \approx \left[(\omega-\omega_0)^2 + \Pi^2\right]\left[(\omega+\omega_0)^2 + \Pi^2\right], \end{aligned} \tag{71}$$

which shows that to lowest order in Π the poles of (68) are shifted by the imaginary quantity $i\Pi$; **the resonances acquire damping.**

The finite state model was further developed by Kubo and his collaborators [37]. It has been applied to Mössbauer spectroscopy by Blume and his collaborators [32-34]. In spite of the fact that the random numbers ζ_μ can be complex and hence simulate both amplitude and phase fluctuations, it has not been used for laser noise calculations.

4.2 *The Strong Perturbation Model*

With the scalar product (53) we can write the operator $\boldsymbol{W}$ as

$$W(\mu|\mu') = \sum_{n=0}^{\infty} \Phi_n(\mu) \frac{-\lambda_n}{\Phi_0(\mu')} \Phi_n(\mu') \tag{72}$$

because its action on any eigenfunction gives the right result

$$\sum_{\mu'} W(\mu|\mu') \, \Phi_m(\mu') = -\lambda_m \Phi_m(\mu) , \tag{73}$$

as follows from the orthonormality of the eigenfunctions with respect to the scalar product defined.

When the range of the variable μ is continuous, it is useful to approximate (72) by a finite number of terms. Then the operator $\boldsymbol{W}$ becomes an integral operator with a separable kernel which can always be inverted by solving a finite set of coupled equations.

The master equation form of the stochastic operator, Eq. (5), separates the rate-in and rate-out terms, which applied to Eq. (72) gives

$$W(\mu|\mu') = \sum_{n=0}^{\infty} \Phi_n(\mu) \frac{\nu - \lambda_n}{\Phi_0(\mu')} \Phi_n(\mu') - \nu \sum_{n=0}^{\infty} \Phi_n(\mu) \, \Phi_n(\mu')^{-1} \, \Phi_n(\mu') . \tag{74}$$

The last sum gives the unit operator due to the completeness of the eigenfunctions Φ_n, and the first term may be truncated to a finite number of terms. In the most usual treatments only the first term, the steady state result, is retained

$$W(\mu|\mu') \approx \nu[\Phi_0(\mu) - \delta(\mu|\mu')] \tag{75}$$

The interpretation is that there are perturbations occurring at the rate γ, and each one is strong enough to force any arbitrary assembly into the equilibrium distribution Φ_0 after only one collision.

This limit is exactly the opposite one to the diffusive motion case described by Fokker-Planck equations. It is called a 'kangaroo process' by van Kampen [22]. In atomic collision theory the corresponding strong collision model has been used extensively by Rautian and Sobel'man [38]. In laser linewidth calculations it has been applied by Burshtein and Zusman to phase fluctuation problems [39].

In collision theory there exists a generalization of the strong collision model called the Keilson-Storer kernel [40]. This has been applied to laser spectroscopy with collisional perturbations by Berman [41] and others.

In optical spectroscopy the laser noise may be due to external uncorrelated jerks, which shift the frequency of the light by an unpredictable amount. If the long time average of the frequency destribution is known, we can identify this with Φ_0 and use the strong collision model. Hence this model can also be called 'the jerky laboratory limit'. In this case we often know very little about how to model the modes with finite eigenvalues.

The kernel (75) has got the one single eigenfunction Φ_0 with eigenvalue zero

$$\sum_{\mu'} W(\mu|\mu')\ \Phi_0(\mu') = 0 . \tag{76}$$

With the right normalization of Φ_0 this function gives the correct steady state distribution. Thus we expect it to give the properties of the steady state correctly. It has the advantage to be easily solved for most properties. There are, however, no reasons to believe that it will give the correct dynamic behaviour, because of the absence of the small eigenvalues providing the long-time approach to equilibrium. It is thus doubtful whether the model can be used to calculate the spectrum of a system in the presence of noise.

4.3 *The Fokker-Planck Model*

When the random variable ζ is regarded as smoothly varying we often apply a Fokker-Planck description. It is then common to employ a damped drift model, i.e., the Ornstein-Uhlenbeck process. The equation of motion is given by

$$\ddot{\zeta} = -\gamma\dot{\zeta} + F(t) , \tag{77}$$

where F is a delta-correlated random force. The corresponding Fokker-Planck equation contains the stochastic operator

$$W = \gamma \frac{\partial}{\partial\zeta}\left[\zeta + \gamma D \frac{\partial}{\partial\zeta}\right] , \tag{78}$$

where we have written $\gamma^2 D$ in front of the diffusion term.

The normalized steady state solution of (78) is

$$\Phi_0(\zeta) = \frac{1}{\sqrt{2\pi\gamma D}}\ e^{-\zeta^2/2\gamma D} , \tag{79}$$

and writing

$$\Phi_n(\zeta) = \Phi_0(\zeta)\, v_n(\zeta) \tag{80}$$

we obtain for v_n the eigenvalue equation

$$\gamma^2 D v_n'' + \gamma\zeta v_n' + \lambda_n v_n = 0 . \tag{81}$$

Introducing the variable

$$x = \frac{\zeta}{\sqrt{2\gamma D}} \tag{82}$$

we find the equation

$$\frac{d^2 v_n}{dx^2} - 2x\frac{dv_n}{dx} + 2\frac{\lambda_n}{\gamma} v_n = 0 , \tag{83}$$

which is solved by the Hermite polynomials when the eigenvalue is

$$\lambda_n = n\gamma . \tag{84}$$

Using the scalar product (53) we find the orthonormality condition

$$\int \Phi_0(\zeta)\, v_n(\zeta)\, v_{n'}(\zeta)\, d\zeta = \delta_{nn'} \tag{85}$$

if we adopt the normalization

$$\Phi_n(\zeta) = \frac{1}{\sqrt{2\pi\gamma D 2^n n!}}\, e^{-\zeta^2/2\gamma D}\, H_n\left(\frac{\zeta}{\sqrt{2\gamma D}}\right) . \tag{86}$$

The functions differ from the harmonic-oscillator wave functions because of the non-Hermiticity of the original Fokker-Planck equation.

The completeness properties of the Hermitean polynomials are well known and hence any operator can be expanded in the functions (86). The use of this method is simplified because the random variable $\hat{\zeta}$ has a simple matrix representation in terms of the basis (86). We find

$$\begin{aligned} \zeta_{nn'} = (\Phi_n, \zeta\,\Phi_{n'}) &= \int \Phi_n(\zeta)\ \zeta\ \Phi_0(\zeta)^{-1}\, \Phi_{n'}(\zeta)\, d\zeta \\ &= \sqrt{\gamma D}\left[\sqrt{n+1}\ \delta_{n',n+1} + \sqrt{n}\ \delta_{n',n-1}\right] . \end{aligned} \tag{87}$$

In field fluctuation models where the stochastic parameter occurs linearly as in (52) this leads to a three-term recurrence relation which suggests solutions in terms of continued fractions. For strong signal treatments of laser noise this implies matrix continued fractions, which have been extensively utilized by Zoller and his collaborators[42]. The relaxation rates $\lambda_n = \gamma n$ form a linear ladder which scales with the single parameters γ. This makes the model especially valuable when one wants to consider the limits of very slow or very fast relaxation processes, see the discussion in Stenholm[19] Part 5.

The Fokker-Planck equation has been extensively utilized to describe noise due to quantum fluctuations; see Louisell[43]. Except near the threshold of laser operation the fluctuations are usually not dominated by quantum noise but by classical, externally imposed perturbations. Be these thermal or not, they can usually be taken to have the Gaussian distribution (79) and then the Fokker-Planck model provides a good approach because of the familiar analytic properties of the eigenfunctions. It can be used to describe Gaussian fluctuations in one real parameter, in lasers often the phase, but it can be generalized to more components as we will see in the next section.

4.4 *The Chaotic Light*

When the electromagnetic field amplitude ε is composed of very many statistically independent contributions we expect the probability distribution of the complex field to be Gaussian of the type

$$P(\varepsilon) = \frac{1}{\pi\,\overline{|\varepsilon|^2}}\, e^{-\varepsilon\varepsilon^*/\overline{|\varepsilon|^2}} . \tag{88}$$

This is a consequence of the central limit theorem; for a specific derivation see Zoller *et al.*[44]. Such a light with both a random phase and a random amplitude is called chaotic. The name has got nothing to do with the recent use of chaotic as related to the behaviour of nonlinear differential evolution equations.

If we introduce the noise correlation time γ^{-1} and the diffusion coefficient D by setting

$$D = \overline{|\varepsilon|^2}\,\gamma^{-1} , \tag{89}$$

we find that (88) is the steady state solution of the following Fokker-Planck operator

$$L = \gamma\left[\frac{\partial}{\partial\varepsilon}\,\varepsilon + \frac{\partial}{\partial\varepsilon^*}\,\varepsilon^* + 2\gamma D\,\frac{\partial^2}{\partial\varepsilon\,\partial\varepsilon^*}\right] . \tag{90}$$

Such a formulation can be derived from the quantized theory of the laser, see e.g. Ref. 43, but it may be used always when a Gaussian distribution of the form (88) is an appropriate description of the light.

By writing $\varepsilon = x + iy$ it is possible to rewrite the operators in (90) as the sum of two uncoupled operators in the orthogonal directions of the complex plane, see, e.g., Ref. 19, but it is more convenient to use the representation

$$\varepsilon = \sqrt{I}\,e^{i\phi} , \tag{91}$$

where I is the field intensity. Transforming the derivatives with respect to ε and ε^* in (90) to derivatives with respect to I and ϕ we find

$$L = 2\gamma\left[\frac{\partial}{\partial I}\,(I - I_0) + I_0\,\frac{\partial^2}{\partial I^2}\,I + \frac{I_0}{4I}\,\frac{\partial^2}{\partial\phi^2}\right] , \tag{92}$$

where the intensity parameter I_0 is given by

$$I_0 = \gamma D . \tag{93}$$

This equation has been treated by Zoller[45]. In steady state we assume no ϕ dependence and write for $P_0(I)$

$$(I - I_0)\,P_0 + I_0\,\frac{\partial}{\partial I}\,I P_0 = 0 , \tag{94}$$

giving

$$P_0 = \frac{1}{2\pi I_0}\,e^{-I/I_0} \tag{95}$$

in accordance with (88), when we notice the difference in the normalization integral.

If we now introduce the scaled variable $x = I/I_0$ and assume the solution in the form

$$L\Phi_n(x) = -\,2\gamma\lambda_n\,\Phi_n(x) \tag{96}$$

we find that

$$\Phi_n(x) = e^{i\alpha\phi}\,P_n^{\alpha}(x) , \tag{97}$$

where

$$\left[x\frac{d^2}{dx^2}+(1+x)\frac{d}{dx}+\left(1-\frac{\alpha^2}{4x}\right)\right]P_n^\alpha(x)=-(n+\tfrac{1}{2}\alpha)\,P_n^\alpha(x)\,, \tag{98}$$

and P_n^α is expressed by the generalized Laguerre polynominal as

$$P_n^\alpha(x)=C_n^\alpha\,e^{-x}\,x^{\alpha/2}\,L_n^{(\alpha)}(x)\,. \tag{99}$$

For $\alpha=n=0$ we regain the equilibrium result (95) and the corresponding eigenvalue

$$\lambda_n^{(\alpha)}=(n+\tfrac{1}{2}\alpha) \tag{100}$$

is indeed zero. In Eq. (97) we need both positive and negative integers α, but for simplicity let us, in the following, write the equations for positive α's only.

The functions (97) are orthogonal for different α's because of the integral over $\phi\in(0,2\pi)$. To determine the normalization coefficient C_n^α we again use the scalar product (53) and get

$$C_n^\alpha\,C_{n'}^\alpha\;2\pi\int_0^\infty e^{-x}\,x^\alpha\,L_n^{(\alpha)}(x)\,L_{n'}^{(\alpha)}(x)\,dx\int_0^{2\Pi}d\phi=\delta_{nn'}\,. \tag{101}$$

But the integral gives $[(\alpha+n)!/n!]$ and hence

$$C_n^\alpha=\frac{1}{2\pi}\sqrt{\frac{n!}{(\alpha+n)!}}\,. \tag{102}$$

The functions (99) with (102) form a complete basis set for the variables ϕ and $I=I_0x$ and we can expand everything again as we did before.

To perform calculations with chaotic light we need to evaluate matrix elements of the field ε. It is of interest to do this for a higher power of ε because some characteristic features of chaotic light emerge only in multi-photon processes. We consequently write

$$\begin{aligned}\langle n\alpha|\varepsilon^\beta|n'\alpha'\rangle &= I_0^{\beta/2}\int d\phi\int dx\;e^{i(\alpha'+\beta-\alpha)\phi}\\ &\quad\times C_{n'}^{\alpha'}C_n^\alpha\;2\pi\,e^{-x}\,x^{\frac{1}{2}(\alpha+\alpha')}\,x^{\beta/2}\,L_{n'}^{(\alpha')}(x)\,L_n^{(\alpha)}(x)\\ &= I_0^{\beta/2}\left[\frac{n!\qquad\qquad n'!}{(n+\alpha)!\;\;(n'+\alpha+\beta)!}\right]^{\frac{1}{2}}\delta_{\alpha',\alpha-\beta}\\ &\quad\times\int_0^\infty e^{-x}\,x^\alpha\,L_{n'}^{(\alpha-\beta)}(x)\,L_n^{(\alpha)}(x)\,dx.\end{aligned} \tag{103}$$

In order to evaluate this integral we use the relation

$$L_n^{(\alpha-1)}(x)=L_n^{(\alpha)}(x)-L_{n-1}^{(\alpha)}(x)\,, \tag{104}$$

which we write symbolically as

$$e^{-\partial/\partial\alpha}=1-e^{-\partial/\partial n}\,. \tag{105}$$

This allows us to write

$$\begin{aligned} L_{n'}^{(\alpha-\beta)} &= e^{-\beta \frac{\partial}{\partial\alpha}} L_{n'}^{(\alpha)} \\ &= \left[1 - e^{-\partial/\partial n}\right] L_{n'}^{(\alpha)} \\ &= \sum_{\mu=0}^{\beta} \binom{\beta}{\mu} (-1)^{\mu} L_{n'-\mu}^{(\alpha)} \quad . \end{aligned} \tag{106}$$

The integral in Eq. (103) is expanded as

$$\begin{aligned} &\int_0^{\infty} e^{-x} x^{\alpha} L_{n'}^{(\alpha-\beta)}(x)\, L_n^{(\alpha)}(x)\, dx = \\ &= \sum_{\mu=0}^{\beta} \binom{\beta}{\mu} (-1)^{\mu} \int_0^{\infty} e^{-x} x^{\alpha} L_{n'-\mu}^{(\alpha)}(x)\, L_n^{(\alpha)}(x)\, dx \\ &= \sum_{\mu=0}^{\beta} \binom{\beta}{\mu} (-1)^{\mu} \frac{(n+\alpha)!}{n!} \delta_{n,n'-\mu} \end{aligned} \tag{107}$$

Finally the matrix element (103) becomes

$$\langle n\alpha | \varepsilon^{\beta} | n'\alpha' \rangle =$$

$$I_0^{\beta/2} \delta_{\alpha',\alpha-\beta} \sum_{\mu=0}^{\beta} \binom{\beta}{\mu} (-1)^{\mu} \delta_{n,n'-\mu} \left[\frac{(n+\alpha)!}{n!} \frac{(n+\mu)!}{(n+\mu+\alpha-\beta)!}\right]^{\frac{1}{2}} \tag{108}$$

For the single photon transition where $\beta = 1$ we obtain

$$\langle \varepsilon \rangle \propto I_0^{\frac{1}{2}} \left[\sqrt{n+\alpha}\; \delta_{n,n'} - \sqrt{n+1}\; \delta_{n,n'-1} \right] , \tag{109}$$

as given in Ref. 45, Eq. (18). For larger values of $\beta > 1$ the β-photon transition rate comes from the lowest term with $\mu = 0$ or

$$\langle \varepsilon^{\beta} \rangle \propto I_0^{\beta/2} \frac{(n+\alpha)!}{n!} \delta_{n,n'-\beta} \tag{110}$$

The transition rate will contain the square of the prefactor. Because the observables averaged over the noise parameters always must include the term with $n' = 0$ and $\alpha' = 0$ we find that to lowest order in I_0^{β} we couple only to $n = 0$ and $\alpha = \beta$. Then the rate from (110) becomes

$$|\langle \varepsilon^{\beta} \rangle|^2 \propto I_0^{\beta}\, \beta! \; . \tag{111}$$

The combinatorical factor $\beta!$ indicates the chaotic enhancement factor; it gives the number of ways we can absorb the β photons from the light. If we had a fully coherent state, the perturbation theory result (111) would have the same factor I_0^{β}, but lack the factor $\beta!$ Further expressions derived under the influence of chaotic light are given by Zoller *et al.*[44-47] Here we have only shown the general structure of an expansion in terms of stochastic eigenfunctions for chaotic light. We have written out the equations explicitly for the case when all α:s are non-negative; the corresponding cases for negative values have to be worked out separately. In particular we must have the eigenvalues (100) in the form

$$\lambda_n^{(\alpha)} = (n + \tfrac{1}{2}|\alpha|) , \tag{112}$$

in order to ensure that they all give decay except $\lambda_0^{(0)}$. For further details we refer to the references.

5. Conclusions and Outlook

We have added conventional stochastic parameters to quantum mechanical systems and found that the calculational techniques developed for classical stochastic variables are applicable to quantum systems, too. As an example we have taken the induced dipole moment in a two-level atom

$$\rho_{21} \propto \exp\left[-i(\omega_{21} - \Omega)t + i\phi(t)\right] , \tag{113}$$

where $\omega_{21} - \Omega$ is the detuning of the atomic levels from the laser frequency Ω; $\phi(t)$ is a random phase.

The physical origins of the random phase can be different. If we associate it with the level spacing ω_{21}, it is a modulation due to random encounters with perturbing atoms; if, on the other hand, we associate it with a modulation of the impinging light frequency, it represents laser noise. Mathematically the two problems are identical and equal to the phase fluctuation problem as originally stated in connection with magnetic resonance.

The general quantum mechanical problem is a direct generalization of the form (113) because in the time evolution operator the random parameters appear in an exponential. Hence we can regard all our discussions here as straightforward extensions of the techniques used in connection with (113). Only the complications due to noncommutativity are added.

To use an explicit time dependence in the stochastic parameter leads to cumbersome calculations. Only in simple cases and perturbation theory is it possible to proceed in this way. It is more advantageous to replace the stochastic function by an abstract stochastic variable which is equipped with its own dynamics. For Markovian problems this is given in terms of a master equation or its equivalent Fokker-Planck equation. The situation is familiar from field theory. In a semiclassical picture we have to cope with fields that depend explicitly on time. When they are quantized they acquire their own Hamiltonian dynamics, and many problems become easier to treat theoretically even when no specific quantum effects are expected. Maybe we, one day, can consistently see the quantization of fields as merely a clever way to treat the fluctuations in the field parameters.

All considerations in this essay regard the random parameters as functions of one real variable, the time $t \in R^{(1)}$. We can, however, generalize the concept of a random function to be defined on a d-dimensional real space $R^{(d)}$. Such structures are called random fields [48] and pose tremendous mathematical problems. They have recently acquired importance in physics as models of disordered solids. [49]

When the field is a scalar we have a random potential; the solution of an electronic Schrödinger equation in this potential leads to the problem originally posed by Anderson [50] in 1958. In one dimension we know that the introduction of disorder localizes the wave functions; for $d \geqslant 2$ the model is still being worked on. The one-dimensional problem is similar to the stochastic cases treated here and can be handled by Markovian techniques. [51] The initial value problem is, however, replaced by a boundary value problem. When the random field is a vector, we approach the problem of a disordered spin system. Much work has recently been devoted to this field [52]. In one dimension the problem is too trivial, but the 2- and 3-dimensional situation, the so-called spin-glass, still poses a challenge to solid state theorists. Thus we see that stochastic models have steadily increased in importance, even if they have been with us for a very long time.

References

1. Ter Haar, D., In *Fluctuations, Relaxation and Resonance in Magnetic Systems*, (ed. D. ter Haar), Oliver & Boyd, Edinburgh (1962).
2. Wangsness, R.K. and Bloch, F., Phys. Rev., **89**, 728 (1953),
3. Redfield, A.G., IBM J. Res. Develop., **1**, 19 (1957).
4. Argyres, P.N. and Kelley, P.L., Phys. Rev., **134**, A98 (1964).
5. Stenholm, S. and ter Haar, D., Physica, **32**, 1361 (1966).
6. Ter Haar, D., Phys., **34**, 1164 (1966).
7. Zwanzig, R., J. Chem. Phys., **33**, 1338 (1960).
8. Fano, U., Phys. Rev., **131**, 259 (1963).
9. Argyres, P.N., In *Statistical Physics and Solid State Physics*, Boulder Summer Institute for Theoretical Physics, Vol.VIII A, (ed. W.E. Brittain), University of Colorado Press, Boulder, (1966).
10. Jancel, R., *Foundations of Classical and Quantum Statistical Mechanics*, Pergamon Press, Oxford (1969).
11. Grabert, H., *Projection Operator Techniques in Nonequilibrium Statistical Mechanics*, Springer Tracts in Modern Physics, Vol. 95, Springer-Verlag, Heidelberg (1982).
12. Lax, M., In *Brandeis Summer Institute Lectures 1966*, Vol. II, (eds. M. Chretien, E.P. Gross and S. Deser), Gordon and Breach, New York (1968).
13. Lax, M., Phys. Rev., **145**, 110 (1966).
14. Haken, H., *Handbuch der Physik*, Band XXV/2c Light and Matter, (ed. L. Genzel), Springer, Berlin (1970).
15. Eberly, J.H., Phys. Rev. Letters, **37**, 1387 (1976).
16. Burshtein, A.I., Soviet Physics JETP, **21**, 567 (1965).
17. Salomaa, R., J. Phys., **B11**, 3745 (1978).
18. Mostowski, J. and Stenholm, S., Physica Scripta, **26**, 211 (1982).

19. Stenholm, S., *Foundations of Laser Spectroscopy*, John Wiley and Sons, New York (1983).
20. Neumann, von J., *Mathematica Foundations of Quantum Mechanics*, Princeton University Press, Princeton, N.J. (1955).
21. Jammer, M., *The Philosophy of Quantum Mechanics*, John Wiley and Sons, New York (1974).
22. Kampen, van N.G., *Stochastic Processes in Physics and Chemistry*, North Holland Publ. Co., Amsterdam (1981).
23. Keldysh, L.V., Soviet Physics JETP, **20**, 1018 (1965).
24. Klein, M.J., Phys. Rev., **87**, 111 (1952).
25. Kubo, R., J. Phys. Soc. Japan, **12**, 570 (1957).
26. Forster, D., *Hydrodynamic Fluctuations, Broken Symmetry, and Correlation Functions*, W.A. Benjamin, Inc., Reading, Mass., (1975).
27. Alekseev, V.A., Andreeva, T.L. and Sobel'man, I.I., Soviet Physics JETP, **35**, 325 (1972).
28. Berman, P.R., Phys. Rev., **A5**, 927 (1972) and *ibid* **A6**, 2157 (1972).
29. Berman, P.R., Appl. Phys., **6**, 283 (1975).
30. Berman, P.R., Physics Reports, **C43**, 101 (1978).
31. Stenholm, S., Phys. Rev., **A2**, 2089 (1970).
32. Clauser, M.J. and Blume, M., Phys. Rev., **B3**, 583 (1971).
33. Blume, M., Phys. Rev., **174**, 351 (1968).
34. Blume, M. and Tjon, J.A., Phys. Rev., **165**, 446 & 456 (1968).
35. Anderson, P.W., J. Phys. Soc. Japan, **9**, 316 (1954).
36. Zoller, P., In *Laser Physics*, Proceedings of the Second New Zealand Summer School in Laser Physics, (eds. D.F. Walls and J.D. Harvey), Academic Press, New York, (1980).
37. Kubo, R., In *Fluctuations, Relaxation and Resonance in Magnetic Systems* , (ed. D. ter Haar), Oliver and Boyd, Edinburgh (1962).
38. Raution, S.G. and Sobel'man, I.I., Soviet Physics USPEKHI, **9**, 701 (1967).
39. Zusman, L.D. and Burshtein, A.I., Soviet Physics JETP, **34**, 520 (1972).
40. Keilson, J. and Storer, J.E., Q. Appl. Math., **10**, 243 (1952).
41. Berman, P.R., Phys. Rev., **A9**, 2170 (1974).
42. Dixit, S.N., Zoller, P. and Lambropoulos, P., Phys. Rev., **A21**, 1289 (1980).
43. Louisell, W.H., *Quantum Statistical Properties of Radiation*, John Wiley and Sons, New York, (1973).
44. Zoller, P., Alber, G. and Salvador, R., Phys. Rev., **A24**, 398 (1981).
45. Zoller, P., Phys. Rev., **A19**, 1151 (1979).
46. Zoller, P., Phys. Rev., **A20**, 1019 (1979).
47. Zoller, P., Phys. Rev., **A20**, 2420 (1979).
48. Adler, R.J., *The Geometry of Random Fields*, John Wiley and Sons. New York, (1981).
49. Ziman, J.M., *Models of Disorder*, Cambridge University Press, Cambridge, (1979).
50. Anderson, P.W., Phys. Rev., **109**, 1492 (1958).
51. Halperin, B.I., Physica Fennica, **8**, 215 (1973).
52. Binder, K., In *Fundamental Problems in Statistical Mechanics V*, p.21, (ed. E.G.D. Cohen) North-Holland Publ. Co., Amsterdam, (1980).

Correlation Effects in Atomic Diffusion†

R. A. Tahir-Kheli
Department of Physics, Temple University, Philadelphia, Pennsylvania 19122, U.S.A.

ABSTRACT

Classical rate equations for vacancy assisted particle motion on regular lattices are solved to determine the diffusional response for long wavelengths and small frequencies. The treatment is exact at both ends of the concentration range and for finite hopping rates also provides an excellent interpolation scheme for all concentrations. The diffusion correlation factors are evaluated for systems which admit a wide variety of particle hops.

1. Introduction

Vacancy assisted atomic diffusion in solids has long been known to exhibit correlations[1] which are neglected by random walk descriptions of the phenomenon. In the limit of vanishing vacancy concentration ν, i.e., for a single vacancy in a macroscopic lattice, exact results for the self diffusion correlation factor in a single component alloy have been available for several lattices for more than a quarter century.[2,3] More recently, the methodology for obtaining these single vacancy limit results has been further developed[4-6] so that exact results can be obtained even when the tracer hopping characteristics are distinct from those of the background atoms. (Note, this is true only when the macroscopic lattice has a *single* tracer **atom.**)

In the opposite limit of large vacancy concentration, namely for $(N-1)$ vacancies in a lattice of N sites, the treatment of the problem (which now consists of only a single atom) is even simpler. Here, in addition to the

† **Supported by the United States National Science Foundation.**

diffusion coefficient, the complete frequency and wave-vector dependent response $S(\boldsymbol{K},\omega)$ can also be exactly calculated.[7]

However, once the system departs from these highly idealized limits, the solution of the problem becomes far from simple. This is true not only for $S(\boldsymbol{K},\omega)$ but also for the simpler quantity embodied in the tracer diffusion correlation factor, f^{tr}. The latter involves knowledge of $S(\boldsymbol{K},\omega)$ only in the limit $K\to 0$, $\omega\to 0$.

Several years ago Fedders and Sankey[8,9] (FS) undertook the task of calculating $S(\boldsymbol{K},\omega)$ throughout the concentration range. They presented a scheme which summed dominant subsets of an infinite series of diagrams. For uniform lattices, their calculations referred to the case where the tracer is identical to the background hopping atoms, i.e., to a single component alloy with arbitrary number of vacancies.

Recently Nakazato and Kitahara[10] have presented a brief account of a treatment employing a projecion operator technique. They found the Laplace transform of the response in the form

$$S(\boldsymbol{K},u) = \left[u + \sum_{\mu,\nu} D_{\mu,\nu}(u)\, k_\mu k_\nu + \dots\right]^{-1} \tag{1.1}$$

where $u=-i\omega$. For small wavevectors this defines the first set of terms in a convergent expansion and using the limit $u\to 0$ they gave an expression for the tracer diffusion correlation factor valid for a simple (hyper) cubic lattice of d-dimensions. While the NK prediction for the diffusion coefficient was noted to be exact in both the concentration limits, i.e., $v=0$ and $v=1$, no rigorous statements concerning either its rate of change in these limits or its accuracy as a function of the particle concentration and the ratio of the rates of hopping of the background particles and the tracer atom were provided.

Very recently Tahir-Kheli and Elliott[11] (TKE) have presented an equations of motion method to calculate $S(\boldsymbol{K},\omega)$ at general K and ω. Using the assumption that the third order fluctuations may be ignored, TKE solved the integral equation involving the second order terms (for both hypercubic and closed packed lattices). The neglect of third order fluctuations in the TKE theory was shown to be justified at both $v=0$ and $v=1$, with the additional statement that the initial rate of change of the diffusion correlation factor with respect to the particle concentration was also given exactly by such a procedure. For general $\boldsymbol{K}$ and ω, the predictions of the TKE theory were found to be in excellent agreement with those of a very careful Monte Carlo work[12] available for the face centred cubic lattice (fcc) where the tracer hopping rate, J^0, coincided with that of the background particles, J.

In the simple hyper cubic lattice, the TKE result for the diffusion coefficient has been found to be identicial [13] to the NK result. In addition for the case $J^0 = J$, the TKE expressions for the diffusion coefficient also appear to reduce to those given in the FS work [8,9].

Independently of TKE, and soon thereafter, Gunther and Gralla [14] presented "an exact solution to diffusion via the vacancy mechanism". On closer examination this work is found to provide an exact solution only for the $v \to 0$ limit. This, of course, is no mean achievement [15] since such a solution for the response is given for general frequency ω (or equivalently, for general time t). Nevertheless, TKE [11] contains an equivalent exact solution for the full response $S(\boldsymbol{K},\omega)$ which for the $v \to 0$ limit holds true even for the rather anomalous case of one dimension [16,17].

Another feature of TKE[11] worth noting is its exactness to the leading order in the background particle concentrations, i.e., to order $(1-v)$. Unlike the $v \to 0$ limit, for which at least the $K \to 0$, $\omega \to 0$ limit has been known for a long time [2-4] (namely the diffusion coefficient), for the case $(1-v) \ll 1$, the corresponding result is of more recent vintage [18]. (Note that a mathematically analogous quantity to the diffusion coefficient, namely the long wavelength spin wave stiffness coefficient, was computed for a simple lattice with small concentration $c = (1-v)$ of non-magnetic impurities by Harris *et al.*[19] in 1974. From this expression the tracer diffusion coefficient for a completely stationary background of $Nc \ll N$ particles can be obtained. In contrast, the recent treatment [18] applies to a kinetic background and is valid for general lattices). Recently, an exact solution for the small concentration limit, $c \ll 1$, has also been given for general $\boldsymbol{K}$ and ω.[20] This work is dependent upon the existence of a Taylor-expansion of the mass operator (for the Green's function that leads to the general incoherent diffusional response). While such an expansion properly exists in $d \geqslant 2$ it completely breaks down in one dimension [16].

For general concentrations, the neglect of the third order fluctuations in the TKE theory is not fully justified. However, using heuristic arguments supplemented by detailed numerical computations, TKE have concluded that in three dimensions, as long as the background particle hopping rate, J, is faster than a number approximately equal to J^0/z (z is the coordination number of the lattice), their theory provides an adequate description for the entire response, $S(\boldsymbol{K},\omega)$, throughout the concentration range. This is also true in d-dimensions with $d \geqslant 2$.[17] Conversely, when the background is slow, i.e., $J < (J^0/z)$, the TKE theory becomes inadequate at small frequencies, i.e. for $\omega < (J^0/z)$. However, for larger frequencies, $\omega \sim zJ^0$, the TKE theory remains valid even for the case of a much slower background. Because the behaviour at small frequencies and long wavelengths pertains to diffusion, the TKE predictions for the diffusion correlation factor are also inadequate when J falls below about J^0/z. In particular, for the worst case, i.e., $J \to 0$, the TKE results are quite unsatisfactory for vacancy concentrations of around 50% or smaller for the fcc lattice.[13] In lower coordination systems, the corresponding figure for the vacancy concentration below which the TKE results for the $J \to 0$ case would become poor is anticipated to be even higher, being about 75% for the quadratic lattice.

In a very recent work, Tahir-Kheli[21] has addressed the problem relating to the above mentioned inadequacies of the original TKE theory. The primary source of the difficulty which causes the quality of the TKE results for the diffusion correlation parameter to deteriorate when the hopping rate of the background atom falls below J^0/z, was identified and the problem was corrected by an improved treatment which embodied a recurrent renormalization of the tracer hopping rate as it diffuses through the kinetic sea of the background particles. It was found that the additional self-consistent dressing of the tracer propagator becomes progressively more important as J decreases below J^0/z towards zero. In particular, for $J \to 0$ the corrections to the propagator given by TKE are most noticeable when the diffusion is substantially slowed. Of course, this occurs as the concentration of the immobile atoms increases towards the percolation threshold.

Our objective here is two-fold: One, it is to extend the fully renormalized treatment[21] to anisotropic lattices. (We note that appropriate anisotropies in the hopping rates can be used to model a change in the system dimensionality.[17]) Another, equally important, objective is to extend the self-consistent dressing procedure of the mass operator to treat longer range hops. As the simplest example of this type of extension, we treat in detail the case of second neighbour hops in two dimensional (quadratic) as well as three dimensional (simple, body-centred and face-centred cubic) lattices.

2. Tracer Diffusion in Isotropic Lattices with Nearest Neighbour Hops

Let us consider first a single distinct tracer atom with a single hopping rate J^0 in a host system consisting of cN atoms, with hopping rate J, distributed randomly over a regular d-dimensional (dD) lattice of N sites. (Here $d \geqslant 2$.) The lattice may be simple hypercubic or close-packed. The vacant sites are to be treated as though they were occupied by *double occupancy avoiding entities* called 'vacancies' whose concentration v is therefore equal to $1-c$.

As usual,[4-14] instantaneous but stochastic hopping of atoms to unoccupied sites is assumed. Forbidding multiple atomic occupancy of any site, the following rate equation is used:

$$\frac{d\sigma_i}{dt} = -\sum_j J^{\sigma}_{ij}\,[\sigma_i V_j - \sigma_j V_i]\,. \tag{2.1}$$

Here $\sigma_i = p_i$ or n_i according as to whether the site i is currently occupied by the tracer or one of the host atoms, respectively. Similarly, V_i refers to the vacancy occupation of a site i.

The hopping integrals, J^{σ}_{ij}, are assumed to be isotropic and their spatial range is restricted to nearest neighbour separation, i.e.,

$$\begin{aligned} J^{\sigma}_{ij} &= J\,, \quad \text{if } i \text{ and } j \text{ are neighbours} \\ &= 0\,, \quad \text{otherwise}\,, \end{aligned} \tag{2.2}$$

and $J^{\sigma} = J^0$ or J depending on whether σ_i refers to p_i or n_i.

Once the tracer and the background particle occupancy variables have been introduced, the vacancy occupancy is completely specified, i.e.,

$$V_i = 1 - p_i - n_i\,. \tag{2.3}$$

In the TKE[11] formulation the tracer occupancy retarded Green's function,

$$G_{gg'}(t) = -2\pi i\Theta(t)\,\langle p_g(t)\,p_{g'}(0)\rangle\,, \tag{2.4}$$

and the frequency and the reciprocal lattice Fourier transforms

$$\begin{aligned} G_{gg'} &\equiv \langle\langle p_g\,;p_{g'}\rangle\rangle \\ &= (2\pi)^{-1}\int_{-\infty}^{+\infty} G_{gg'}(t)\exp(i\omega t)\,d\omega\,, \end{aligned} \tag{2.5}$$

are analyzed. [In this section, without any loss of generality, the length of the elementary cube edge has been chosen to be unity. This, in turn, also fixes the scale for the inverse lattice wave-vectors.]

As demonstrated by TKE, the mass operator of G_K, i.e.,

$$G_K = [\omega + i\textstyle\sum(K,\omega)]^{-1}, \tag{2.6}$$

$$\textstyle\sum(K,\omega) = v J^0\left[\omega_K - \sum_{\delta}[1-\exp(i\boldsymbol{K}\cdot\delta)]\;\rho(\boldsymbol{\delta})\right] \tag{2.7}$$

involves the second order propagator, $\rho(\boldsymbol{\delta})$, which refers to a composite pair consisting of a tracer and a neighbouring background particle fluctuation field (henceforth to be called the composite pair), diffusing with relative momentum K, i.e.,

$$\rho(\boldsymbol{\delta}) = (1/N)\sum_{K'} G^{(2)}_{K-K',K'}\exp(-i\boldsymbol{K}'\cdot\boldsymbol{\delta})\,/G_K\,. \tag{2.8}$$

Here $\boldsymbol{\delta}$ is a nearest neighbour vector and

$$G^{(2)}_{\ell j;g'} \equiv \langle\!\langle p_\ell u_j ; p_{g'} \rangle\!\rangle$$

$$= (1/N)^2 \sum_{K_1\ K_2} G^{(2)}_{K_1,K_2} \exp\left[i\boldsymbol{K}_1\cdot(\ell-\boldsymbol{g}') + i\boldsymbol{K}_2\cdot(j-g')\right], \qquad (2.9)$$

$$u_j = n_j - \langle n_j \rangle = n_j - c\,. \qquad (2.10)$$

The remaining notation is as follows:

$$\omega_K = z\left[1-\gamma_K\right],$$

$$\gamma_K = \sum_\delta \cos(\boldsymbol{K}\cdot\boldsymbol{\delta})/z\,, \qquad (2.11)$$

where z is the coordination number of the lattice.

Within the simplest mean-field approximation, (MFA), the second order propagator, $\rho(\boldsymbol{\delta})$, is neglected and thus

$$\Sigma(\boldsymbol{K},\omega) \underset{\text{MFA}}{\sim} (vJ^0)\,\omega_K\,. \qquad (2.12)$$

For long wavelengths, the tracer motion is easily identified with random walk diffusion since $\omega_K \sim K^2$. Thus the uncorrelated diffusion coefficient D_0 of the tracer is simply equal to vJ^0.

$$D \underset{\text{MFA}}{\sim} vJ^0 = D_0 \text{ (Random Walk Approximation)}\,. \qquad (2.13)$$

If correlations were to be taken into account, the diffusion coefficient of the tracer would, in general, be reduced, i.e.,

$$D = \lim_{(K^2/\omega)\to 0} \left[\Sigma(\boldsymbol{K},\omega)/K^2\right] \equiv D_0 f^{tr}, \qquad (2.14)$$
$$= \lim_{\omega\to 0}$$

and the diffusion correlation parameter $f^{tr} \leqslant 1$ would depend on both the lattice structure as well as the details of the host alloy.

In order to take account of the correlations, TKE analyzed the equation of motion of the second order propagator $G^{(2)}$. Such an equation in turn involves $G^{(3)}$,

$$G^{(3)} \underset{1\neq 2\neq 3}{=\!=} \langle\!\langle p_1 u_2 u_3 ; p_{g'} \rangle\!\rangle\,, \qquad (2.15)$$

which refers to diffusion of larger composites. Because in $d \geqslant 2$, these terms are strictly of order $v^2 c^2 G_K$, their neglect relative to the $G^{(2)}$ terms is justified at both the concentration limits, namely $v \ll 1$ and $c = 1 - v \ll 1$.

Ignoring third order fluctuations, TKE were able to solve the remaining set of equations in a compact form, i.e.,

$$\rho(\boldsymbol{\delta}) = \Phi(\boldsymbol{\delta}) + \sum_{\delta'} \rho(\boldsymbol{\delta}') \left\{\alpha\left[P(\boldsymbol{\delta}-\boldsymbol{\delta}') - P(\boldsymbol{\delta})\right] + \beta\left[P(\boldsymbol{\delta}+\boldsymbol{\delta}') - P(\boldsymbol{\delta}-\boldsymbol{\delta}')\right]\right\} \tag{2.16a}$$

$$\Phi(\boldsymbol{\delta}) = (i\beta/N) \sum_{K'} \exp(i\boldsymbol{K}'\cdot\boldsymbol{\delta})[\omega_{K+K'} - \omega_{K'}]/E(K'), \tag{2.16b}$$

$$E(\boldsymbol{K}') = \omega + i\,[J\omega_{K'} + (vJ^0)\,\omega_{K+K'}], \tag{2.16c}$$

$$P(\boldsymbol{r}) = (1/N) \sum_{K'} \exp(i\boldsymbol{K}'\cdot\boldsymbol{r})/E(\boldsymbol{K}'), \tag{2.16d}$$

$$\alpha = J + vJ^0, \quad \beta = cJ^0. \tag{2.16e}$$

The $\boldsymbol{\delta}'$ sum on the right-hand-side of equation (2.16a) is taken over z-nearest neighbour vectors. This makes equation (2.16a) equivalent to a simple set of z simultaneous linear equations in the z unknowns $\rho(\boldsymbol{\delta}')$.

The following aspects of these simultaneous linear equations are of special interest here. First, the coefficients involve inverse lattice sums, $P(\boldsymbol{r})$ and $\Phi(\boldsymbol{\delta})$, both of which have the same denominator, $E(K')$. Second, these sums can all be carried out analytically in the diffusive regime of interest: that is, when $\frac{K^2}{\omega} \to 0$, $\omega \to 0$. Third, and most important, in this regime all the coefficients depend uniquely on two specific combinations of the rate parameters, namely, α and β, and also $E(K')$ can be replaced by $i\alpha\omega_{K'}$.

The physical roles played by the parameters α and β are quite distinct. Whereas α provides the signature of the composite pair diffusion [e.g., in the denominators on the right-hand-side of (2.16a)], β plays no such corresponding role. In this connection we also note that the signature of the diffusing tracer (without an accompanying background field) is provided by the rate factor (vJ^0).

In the TKE formulation, all correlations altering the tracer diffusion rate within the composite pair were totally ignored. This is clearly justified for $d \geqslant 2$ when either $c \to 0$, or $v \to 0$, because such effects are vanishingly small in these limits. On the other hand, for finite c (or finite v), the effect of correlations on the tracer diffusion both *within* and *without* the composite pair must be taken into account. Moreover, if possible, this should be done self-consistently.

The self-renormalized procedure, therefore, is the following. Calculate f^{tr} as in TKE. Introduce this f^{tr} into the diffusion coefficient of the composite pair by replacing α by its correlated value α' such that

$$\alpha' = J + v f^{\mathrm{tr}} J^0. \tag{2.17}$$

Repeat the cycle until no further changes in f^{tr} occur.

The TKE expression for f^{tr} as evaluated from equations (2.7), (2.14) and (2.16a) – (2.16e) without renormalizing the composite pair diffusion was as follows:

$$f^{tr} \underset{\text{TKE}}{=\!=} \left[1 - \frac{2\beta\overline{\cos\theta}}{\alpha(1+\overline{\cos\theta})}\right]^{-1} . \tag{2.18}$$

Therefore, a fully iterated self-consistent treatment of tracer diffusion both *outside* and *inside* the composite pair is contained in the following:

$$f^{tr} \underset{\text{(RENORMALIZED)}}{=\!=} \left[1 - \frac{2\beta\overline{\cos\theta}}{\alpha'(1+\overline{\cos\theta})}\right]^{-1} . \tag{2.19}$$

The nature of the approximation involved in (2.18) is quite different from that inherent in (2.19). Equation (2.19) assumes the 3rd and higher order propagators have much the *same* effect on the tracer diffusion *within* the composite pair as the 2nd order propagators have on the tracer diffusion of the tracer by itself. In contrast, equation (2.18) ignores all contributions from third and higher order propagators, especially on the diffusing composite pair,

[Note, the quantity $\overline{\cos\theta}$ mentioned in equations (2.18) and (2.19) is the well-known average cosine of the angle between the directions of successive vacancy-tracer interchanges ;

$$\begin{aligned}\overline{\cos\theta} &= -0.209842, \text{ sc}\\ &= -0.157948, \text{ bcc}\\ &= -0.122680, \text{ fcc}\\ &= -0.363380, \text{ quadratic lattice.]}\end{aligned}$$

Clearly for $v \to 0$ the replacement of α with α' does not affect the results. Away from this limit, we write schematically

$$f^{tr} \sim f^{tr}_{\text{TKE}} + \Delta , \tag{2.20}$$

and hence

$$\Delta \sim -2cv\left(f^{tr} f^{tr}_{\text{TKE}}\right)^{-1} \frac{\overline{\cos\theta}}{(1+\overline{\cos\theta})} \frac{(f^{tr}-1)}{(v+\eta)(vf^{tr}+\eta)} \tag{2.21}$$

where

$$\eta = J/J^0 . \tag{2.22}$$

Therefore, when c is small or when η is large, the new formalation reduces to TKE. Any differences, Δ, between the two are important only when η is small compared to unity. Moreover, this is especially true for that range of vacancy concentration where $f^{tr} \ll 1$. As is well known, the latter occurs in the 'forbidden' range of concentrations [22] (where for $\eta \to 0$, $f^{tr} \to 0$).

The new equation for f^{tr} given in (2.19) has a simple solution. This is especially the case when the background is immobile, i.e., $J = 0$,

$$f^{tr} = 1 + 2c\,\overline{\cos}\,\theta/[v(1+\overline{\cos}\,\theta)] \,. \tag{2.23}$$

This gives an acceptable (that is, 'positive') value for f^{tr} until the vacancy concentration falls to the percolation threshold, v_p,

$$v_p = 1 - c_p = 1 - \frac{(1+\overline{\cos}\,\theta)}{(1-\overline{\cos}\,\theta)} \equiv 1 - f^0 \,, \tag{2.24}$$

where $f^{tr} \to 0$. Below v_p, the tracer cannot diffuse at all, hence the 'forbidden' vacancy concentration region, i.e., $v_p < v \leqslant 0$.

It is interesting to compare both the 'bare' and the 'dressed' diffusion coefficients given in equations (2.18) and (2.19) to corresponding results available in the literature. As mentioned earlier for the case $J = J^0$, the earlier work of Fedders and Sankey[8,9] contains equivalent numerical results to the 'bare' diffusion coefficient given in equation (2.18). Similarly, Nakazato and Kitahara's results are also identical to the undressed result (2.18) for the cases of a square and the simple cubic lattices to which they apply. None of these authors, however, include any self-consistent renormalization that is caused by the correlated diffusion of the tracer within the composite pair (described above) and hence their results do not correspond to the dressed diffusion coefficient given in equation (2.19).

In a similar context, it is interesting to also make contact with another series of calculations[23-25] which have dealt with the computation of f^{tr} in the $v \ll 1$ regime. In these works, the leading correction to the $v = 0$ result is estimated. Writing, with Ishioka and Koiwa[25], for $J = J^0$

$$f^{tr} = f^0\,(1 + \beta_0 v) \tag{2.25}$$

the coefficient β_0 is easily determined from equations (2.18) and (2.19). These give, respectively,

$$\beta_0 = -4\,\overline{\cos}\,\theta/(1-\overline{\cos}\,\theta) \equiv \beta_{00} \tag{2.26a}$$

and

$$\beta_0 = \beta_{00} + (1-f^0)\,f^0 \tag{2.26b}$$

where f^0 is as defined in equation (2.24).

For the cubic lattices (sc, bcc and fcc, respectively) the coefficient β_0 in the TKE (unrenormalized) approximation, i.e., equation (2.18), is 0.6938, 0.5456 and 0.4371, respectively. The corresponding estimates given by Ishioka and Koiwa are 0.7123, 0.5787 and 0.5205, respectively. In comparison, our renormalized theory, Eq. (2.20), gives slightly larger numbers, i.e., 0.9203, 0.7440 and 0.6079.

It should be emphasized that none of the three sets of results for β_0 given above can claim to be exact. Nevertheless, chances are that the renormalized theory results would be somewhat closer to the mark (especially when

$J/J^0 < 1/z$).

Let us briefly examine next the opposite concentration limit with only a dilute background, namely $c = 1-v \ll 1$. Here both the original TKE result as well as its renormalized version lead to identical results to the leading two orders, i.e.

$$f^{tr} = 1 - \beta_1 c - 0(c^2) , \tag{2.27a}$$

$$\beta_1 = -\left(\frac{2J^0}{J+J^0}\right)\frac{\overline{\cos\theta}}{1+\overline{\cos\theta}} . \tag{2.27b}$$

The above result is exact to the linear order in c for all ratios of J/J^0. For the particular case $J = J^0$ the works of McKee[26] and Wolf[27] also yield values of β_1. For this case our results for β_1 are[18] $= 0.2655695, 0.187574$ and 0.139835 for the sc, bcc and fcc lattices, respectively. The corresponding three results obtained from Wolf's work appear to be 0.070, 0.039, and 0.020, respectively, whereas McKee's work yields $\beta_1 = 1/9$ for the fcc lattice.

An analogous physical situation to the above obtains in an isotropic Heisenberg ferromagnet at zero temperature with nearest neighbour exchange coupling J^0 and a small concentration c of the spins replaced by randomly distributed non-magnetic impurities. (Note, the corresponding background hopping rate J has to be considered as being equal to zero for such a system). Writing the spin-wave stiffness as

$$D_{\text{spin-wave}} = J^0 f^{sw}(1-c), \tag{2.28a}$$

we have the exact result

$$f^{sw} = 1 - 2\beta_1 c - 0(c^2) . \tag{2.28b}$$

The coefficient $2\beta_1$ was first calculated by Harris *et al.*[14] (only for a simple cubic lattice) who estimated it to be 0.532. It should be noted that (2.28b) refers only to the limiting case of a long wavelength, small frequency spin-wave. For more general response, the corresponding results are available in TKE[11] (also compare, Ref. 20).

It is possible to test the accuracy of our renormalized theory at least for the face centred cubic lattice. Here Monte Carlo data for f^{tr} are available for the case of an immobile background,[13] i.e., $J = 0$ — which, of course, provides the most stringent test for equation (2.19) or its equivalent (for this case) given in equation (2.23). Fortunately, an additional set of Monte Carlo data referring to the vacancy diffusion for this system is also available. By relating the vacancy diffusion to the tracer diffusion in such a system,[21] we can obtain a comprehensive set of Monte Carlo data with which equation (2.23) may be compared.

Such a comparison is displayed in Fig. 1. Here f^{tr} is plotted as a

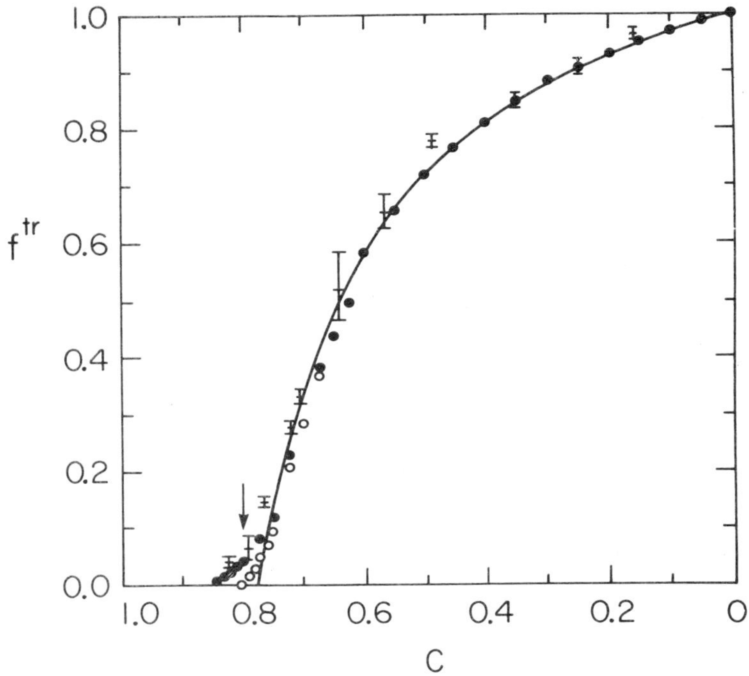

Fig. 1 The tracer diffusion correlation factor, f^{tr}, is given as a function of the background particle concentration c. The lattice is fcc and the background particles, which are assumed to be immobile, are considered to be randomly distributed. The full curve represents equation (2.23) of the text. The data marked ⫱ is taken from Kutner and Kehr's Monte Carlo work (see Ref. 13, Fig. 7). Both the open and the closed circles represent different sets of Monte Carlo results for the *same* quantity (see section 3 of the text) given by Murch and Rothman (see Fig. 1 of Ref. 28).

function of the immobile background particle concentration, c. The continuous curve represents equation (2.23), full and open circles are the Monte Carlo data given in Ref. 28, whereas data points marked as ⫱ are those provided by Kutner and Kehr.[13] The closeness of the fit is a convincing testimony favouring the renormalized approximation. (Note, the TKE unrenormalized theory would yield a poor fit when c is $\geqslant 0.45$).

3. Tracer Diffusion in Anisotropic Lattices

Let us consider next the case of anisotropic diffusion. Such diffusion mechanisms have been treated by Montet [6] who has provided exact results for a variety of lattices. Unfortunately, however, Montet's analysis [6] is restricted to the special case of zero vacancy concentration. Thus, a more complete treatment valid for the whole range of vacancy concentration is wanting.

To provide such a treatment we consider a d-dimensional hypercubic lattice. The tracer hopping rate along direction ν (note, $\nu = 1, \ldots, d$) is assumed to be J_ν^0 whereas the corresponding hopping rate for the background particles, whose concentration is $c = (1-v)$, is J_ν. We note that this system is somewhat more general than that treated by Montet [6], who considers only the case of self-diffusion, i.e., $J_\nu^0 = J_\nu$. The nearest neighbour lattice spacing along direction ν is taken to be a_ν.

The simple rate equation relevant to such a system is

$$\frac{d\sigma_\ell}{dt} = -\sum_{\mathbf{j}} J^\sigma(\ell - j)\left[\sigma_\ell V_{\mathbf{j}} - \sigma_{\mathbf{j}} V_\ell\right]. \tag{3.1}$$

Variables σ_ℓ and $V_{\mathbf{j}}$ are defined as in equation (2.1). The hopping rate $J^\sigma(\boldsymbol{\ell} - \boldsymbol{j})$ is $= J_\nu^\sigma$ if the vector $\boldsymbol{\ell} - \boldsymbol{j}$ is a nearest neighbour vector along the direction ν. [Refer to equation (2.2) for explanation of the superscript on J^σ.]

Working with a generalized version of the notation used in section 2, equations (2.16a) – (2.16e) take the following form:

$$\rho(\boldsymbol{\delta}_\nu) = \Phi(\boldsymbol{\delta}_\nu) + i\sum_{\mu=1}^{\mathbf{d}}\sum_{\delta'_\mu}\rho(\boldsymbol{\delta}'_\mu)\,R(\boldsymbol{\delta}_\nu, \boldsymbol{\delta}'_\mu), \tag{3.2a}$$

$$R(\boldsymbol{\delta}_\nu, \boldsymbol{\delta}'_\mu) = (\alpha_\mu - \beta_\mu)\,P(\boldsymbol{\delta}_\nu - \boldsymbol{\delta}'_\mu) - \alpha_\mu P(\boldsymbol{\delta}_\nu) + \beta_\mu P(\boldsymbol{\delta}_\nu + \boldsymbol{\delta}'_\mu), \tag{3.2b}$$

$$\alpha_\mu = J_\mu + v f_\mu^{\mathbf{tr}} J_\mu^0; \quad \beta_\mu = c J_\mu^0 \tag{3.2c}$$

$$\Phi(\boldsymbol{\delta}_\nu) = \frac{c}{N}\sum_{\mathbf{K}'}\frac{\exp(i\boldsymbol{K}'\cdot\boldsymbol{\delta}_\nu)\left[\Omega_{\mathbf{K}+\mathbf{K}'} - \Omega_{\mathbf{K}'}\right]}{E(K')}, \tag{3.2d}$$

$$P(\boldsymbol{r}) = -\frac{i}{N}\sum_{\mathbf{K}'}\frac{\exp(i\boldsymbol{K}'\cdot\boldsymbol{r})}{E(K')}, \tag{3.2e}$$

$$E(\boldsymbol{K}') = \sum_{\mu=1}^{\mathbf{d}}\sum_{\delta_\mu}\alpha_\mu\left[1 - \exp(i\boldsymbol{K}\cdot\boldsymbol{\delta}_\mu)\right], \tag{3.2f}$$

$$\Omega_{\mathbf{K}'} = \sum_{\mu=1}^{\mathbf{d}}\sum_{\delta_\mu} J_\mu^0\left[1 - \exp(i\boldsymbol{K}\cdot\boldsymbol{\delta}_\mu)\right]. \tag{3.2g}$$

Because of the choice of a hypercubic lattice geometry, there are two possible nearest neighbours, $\boldsymbol{\delta}'_\mu$, along any particular direction μ. [Note that for a more general lattice, equations (3.2a)–(3.2g) apply as given except that the sum over $\boldsymbol{\delta}'_\mu$ must then take into account the relevant number of nearest neighbours across which the hopping rates J^0_μ and J_μ obtain.] Therefore, for lattices under consideration, equations (3.2f) and (3.2g) can be written more simply as follows:

$$E(\boldsymbol{K}') = 2 \sum_{\mu=1}^{d} [1 - \cos(K'_\mu a_\mu)]\alpha_\mu , \tag{3.3a}$$

$$\Omega_{K'} = 2 \sum_{\mu=1}^{d} [1 - \cos(K'_\mu a_\mu)] J^0_\mu . \tag{3.3b}$$

It will be noticed that these equations are already in the renormalized form and thus lead to the result (2.10) for an isotropic system. The point to note here is the presence of the correlation factor, f^{tr}_μ, i.e., equation (3.2c), which indicates that correlations are anisotropic, reflecting the anisotropy in the hopping characteristics of both the tracer and the background particles.

The mass operator for the occupancy Green's function for the anisotropic system [compare equations (2.6) and (2.7)] is readily seen to be the following:

$$\Sigma(\boldsymbol{K},\omega) = \sum_{\mu=1}^{d} \Sigma^{(\mu)}(\boldsymbol{K},\omega) , \tag{3.3a}$$

$$\Sigma^{(\mu)}(\boldsymbol{K},\omega) = v J^0_\mu \left\{2[1-\cos(K_\mu a_\mu)] + \sum_{\delta_\mu} [\exp(i\boldsymbol{K}\cdot\boldsymbol{\delta}_\mu) - 1]\ \rho(\boldsymbol{\delta}_\mu)\right\} \tag{3.3b}$$

For $|\boldsymbol{K}| \to 0$, $\omega \to 0$, the mass operator can be cast into the form:

$$\lim_{(k_\mu a_\mu)^2/\omega \to 0\ ;\ \omega/J^\mu \to 0} \Sigma^{(\mu)}(\boldsymbol{K},\omega) = v J^0_\mu (k_\mu a_\mu)^2 f^{tr}_\mu + 0(k_\mu a_\mu)^4 , \tag{3.4}$$

whereby the quantity of interest, f^{tr}, is given by the following:

$$f^{tr}_\mu = 1 + \frac{2 c J^0_\mu T_\mu}{1 - (2\beta_\mu - \alpha_\mu) T_\mu} , \tag{3.5a}$$

where

$$T_\mu = (1/N) \sum_{K'} [\cos(2 k'_\mu a_\mu) - 1] / E(K') . \tag{3.5b}$$

Equations (3.5a) and (3.5b) represent a set of d self-consistent equations for the desired correlation factors f^{tr}_μ.

In the limit $v = 0$, and $c = 1$ the above equations reduce to the following simple expressions,

$$f^{tr}_\mu = [1 + J_\mu t_\mu] / [1 + (J_\mu - 2J^0_\mu)\, t_\mu] , \tag{3.6a}$$

where

$$t_\mu = \frac{1}{N}\sum_{K'} \frac{[\cos(2k'_\mu a_\mu) - 1]}{2\sum_\mu J_\mu[1-\cos(k_\nu a_\nu)]} \tag{3.6b}$$

Montet's work deals only with the case $J^0 = J$. With this substitution Montet's equations (4.2) are identical to those given above. [Note, however, that the relevant dimensionality is 3 for Montet's equation (4.2).]

For comparison at the opposite end of the concentration, i.e., $c \ll 1$, the only suitable results appear to be the perturbation theory spin-wave dispersion curves given by McGurn and Tahir-Kheli.[29] Setting $v = 1-c$, $J_\mu = 0$, and $J_1^0 = J_2^0 = J_+^0$, $J_3^0 = J_{11}$, the identity of these two sets of results is readily confirmed. As an aside, the relative simplicity of the present treatment as compared to the usual perturbation expansion should be noted.

For general concentration, the set of equations (3.5a) and (3.5b) have to be solved self-consistently. For $d \geqslant 2$ this can be done only numerically. In Fig. 2, a numerical solution for a three dimensional hypercubic lattice is given for two different sets of hopping parameters. Curves 1, 2, 3, display f_1^{tr} f_2^{tr} and f_3^{tr} for such a system with $J_1^0 = J_1 = 1$, $J_2^0 = J_2 = 10$, $J_3^0 = J_3 = 100$ whereas the other set of three curves, marked 4, 5, 6, represent the same parameters for $J_1 = J_2 = J_3 = 0$ and $J_1^0 = J_2^0/10 = J_3^0/100$. Notice that the latter three curves lead to a unique result for the percolation threshold $v_p = 1-c_p$, where $c_p = [1+\overline{\cos\theta}]/[1-\overline{\cos\theta}] = 0.653108$. Because v_p ought to be independent of the actual magnitude of the hopping integrals (as long as they are all positive), a unique result for v_p is therefore a satisfying feature of this treatment.

The pair approximation, which is the basis of the present analysis, takes an exact account of the neighbouring shell, consisting say of z sites in an isotropic lattice, and an approximate account of shells lying further away. This aspect of the theory, offers a convenient characterization of its accuracy in terms of a perturbation expansion in power of z^{-1}.

It is readily seen that for $z \to \infty$, the following asymptotic expansion holds,

$$v_p \sim \frac{2}{z} + \frac{e}{z^2}. \tag{3.7}$$

Accordingly, for finit but large z, the estimate for v_p should be in error by a few times z^{-2}. [An equivalent statement to this is the following: v_p itself is located to a relative accuracy of order $1/z$.]

The percolation considerations referred to above are, however, germane only to the case of a stationary background, i.e., $J_\mu \to 0$, for which the present treatment is at its weakest. It turns out[21] that for a system with an arbitrarily slow background, i.e., $J_\mu/J_\mu^0 > 0$, the theory can be expected to be valid to a relative accuracy of order z^{-3} or better when the system is in the concentration range $a \geqslant c \gtrsim z^{-1}$, and to an accuracy of order z^{-2} when the system falls within either $z^{-1} \leqslant c \leqslant (c_p - z^{-1})$ or $0 \leqslant v < v_p - z^{-1}$. Outside these concentration ranges lies the critical region where for $J_\mu/J_\mu^0 \leqslant z^{-2}$ the relative accuracy of the present results can be expected to be the poorest, being approximately of order z^{-1}. However, for $J_\mu/J_\mu^0 \gtrsim z^{-1}$, improvement to order z^{-2} can again be expected.

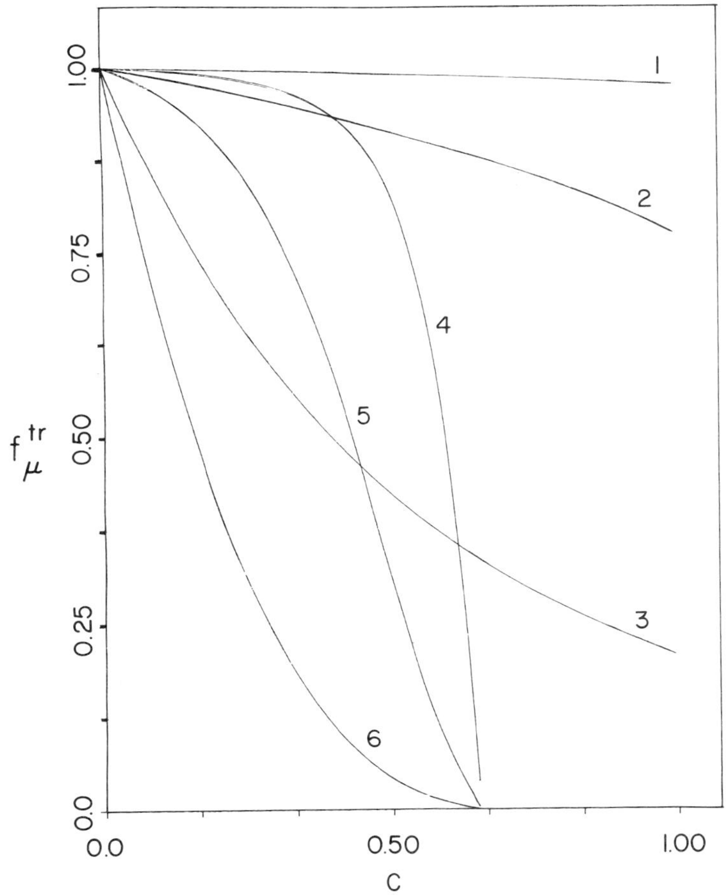

Fig. 2 For a three dimensional anisotropic hypercubic lattice with $J_1^0 = 1$, $J_2^0 = 10$ and $J_3^0 = 100$, the tracer diffusion correlation parameters, f^{tr}, are plotted as a function of the background particle hopping rates J_i and concentration c. Curves 1, 2,3 refer to f_1^{tr}, f_2^{tr} and f_3^{tr}, respectively. Here $J_1 = J_1^0$, $J_2 = J_2^0$ and $J_3 = J_3^0$. Curves 4, 5, 6 again refer to f_1^{tr}, f_2^{tr} and f_3^{tr} with the difference that $J_1 = J_2 = J_3 = 0$. Note, that the latter curves all converge to zero at $c = 0.653108$.

4. Tracer Diffusion in Isotropic Lattices with Second Neighbour Hopping

As discussed in the preceding sections, the solution of rate equations which treat the particle hops to be restricted to neighbouring sites is already hard enough, so that it is not completely tractable (except in certain concentration limits). Nevertheless, because a useful theoretical description has now been developed which treats the entire concentration regime adequately, it is worthwhile examining systems which allow longer range hops.

A possibility of the occurrence of both first and second neighbour hops has been raised by some recent experimental observations of Lottner *et al.*[30]. These authors have attempted to fit the quasi-elastic incoherent neutron scattering data from diffusing hydrogen in *Nb* and other bcc metals with $S(\boldsymbol{K},\omega)$ calculated from the model using only nearest neighbour hops between tetrahedral sites. This process led to an estimated hopping rate J which apparently varies with the momentum transfer involved. On the other hand, satisfactory fits of the data were found possible [31] with a model which also included hops between 'topological' second neighbours. Indeed, the contribution arising from second neighbour jumps appeared to increase with the elevation of system temperature.

We study below the properties of a system containing a single tracer with nearest and next nearest neighbour isotropic hopping rates J_1^0, J_2^0 on a regular lattice with a general concentration c of background atoms which too are allowed to jump across both first and second neighbour separations with hopping rates J_1 and J_2. We shall assume the relevant number of nearest neighbours and next nearest neighbours to be z_1 and z_2, respectively. [If the tracer is identical to the background atoms, then $J_i^0 = J_i$, $i \equiv 1, 2$.]

By following the procedure outlined in section 2, the frequency and the reciprocal lattice Fourier transform of the occupancy Green's function of the tracer can be cast into the following form:

$$[\omega + i v \Sigma(\boldsymbol{K},\omega)]\, G_K = 1 , \qquad (4.1)$$

where

$$\Sigma(\boldsymbol{K},\omega) = \Sigma^{(1)}(\boldsymbol{K},\omega) + \Sigma^{(2)}(\boldsymbol{K},\omega) \qquad (4.2a)$$

and

$$\Sigma^{(i)}(\boldsymbol{K},\omega) = \omega_i^0(\boldsymbol{K}) + J_i^0 \sum_{\delta_i} [\exp(i\boldsymbol{K}\cdot\boldsymbol{\delta}_i) - 1]\, \rho(\boldsymbol{\delta}_i) . \qquad (4.2b)$$

In equation (4.2b) $\sum_{\delta_i}$ implies a sum over the set of z_i i-th nearest neighbour vectors. With this notation, $\rho(\boldsymbol{\delta}_\nu)$ can once again be represented by equations (3.2a) – (3.2e). The definition of $E(K')$ and $\Omega_{K'}$ have, however, to be changes as follows:

$$E(K') = \sum_{i=1}^{2} \left[z_i - \sum_{\delta_i} \exp(i\boldsymbol{K}\cdot\boldsymbol{\delta}_i) \right] \alpha_i \, , \tag{4.3a}$$

$$\Omega_K = \omega_1^0(K) + \omega_2^0(K) \, , \tag{4.3c}$$

$$\omega_i^0(K) = J_i^0 \left[z_i - \sum_{\delta_i} \exp(i\boldsymbol{K}\cdot\boldsymbol{\delta}_i) \right] . \tag{4.3c}$$

The set of equations (3.2a) — (3.2e) and (4.1) — (4.3c) have to be solved self-consistently in the limit

$$(Ka)^2/\omega \longrightarrow 0 \quad \text{and} \quad \omega/[J_1^0 z_1 + J_2^0 z_2] \longrightarrow 0 \, . \tag{4.4}$$

[Here a is a convenient elementary length in the problem, such as the cube edge length for the lattices under consideration.] This solution in turn leads to the quantity of interest f^{tr}. (see Appendix 'A').

The results given for f^{tr} in Appendix 'A' are exact in the limit $v = 0$. Moreover, they are also exact to the linear order in c when $c \ll 1$. In the intermediate concentration regime, the accuracy is a function both of the relative sizes of J_1, J_2, J_1^0 and J_2^0 and the concentration. However, as long as the background is not too slow, i.e., $J_i/J_i^0 \geqslant (z_1 + z_2)^{-2}$, the overall expected error of these results is quite low, i.e., of order one out of $(z_1 + z_2)^{-2}$ or better. Again, problems arise as we go across the percolation threshold and the accuracy is somewhat diminished. However, because $(z_1 + z_2)$ in three dimensions is generally large compared to unity, the results remain satisfactory over the entire range of concentrations.

To give a feel for these results, we have plotted f^{tr} as a function of the background particle concentration c, first, for the case of only nearest neighbour hopping (see Figs. 3) and then for the more general case being considered in this section (see Figs. 4).

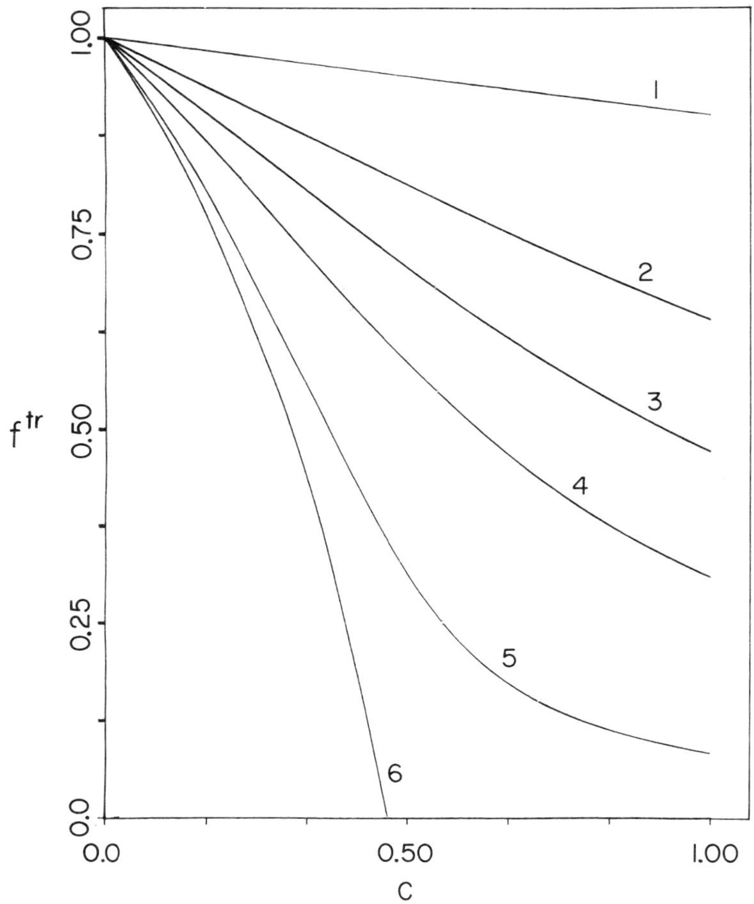

Fig. 3 For a nearest neighbour hopping model, with $J_1^0 = 1$, the tracer diffusion correlation factor f^{tr} is plotted as a function of c. For the curves labelled 1, 2, 3, 4, 5 and 6, the background particle hopping rate, J, is $= 10$, 2, 1, 1/2, 1/10 and 0, respectively. Figures (a), (b), (c) and (d) refer to the quadratic, the simple cubic, the body centred and the face centred cubic lattices, respectively.

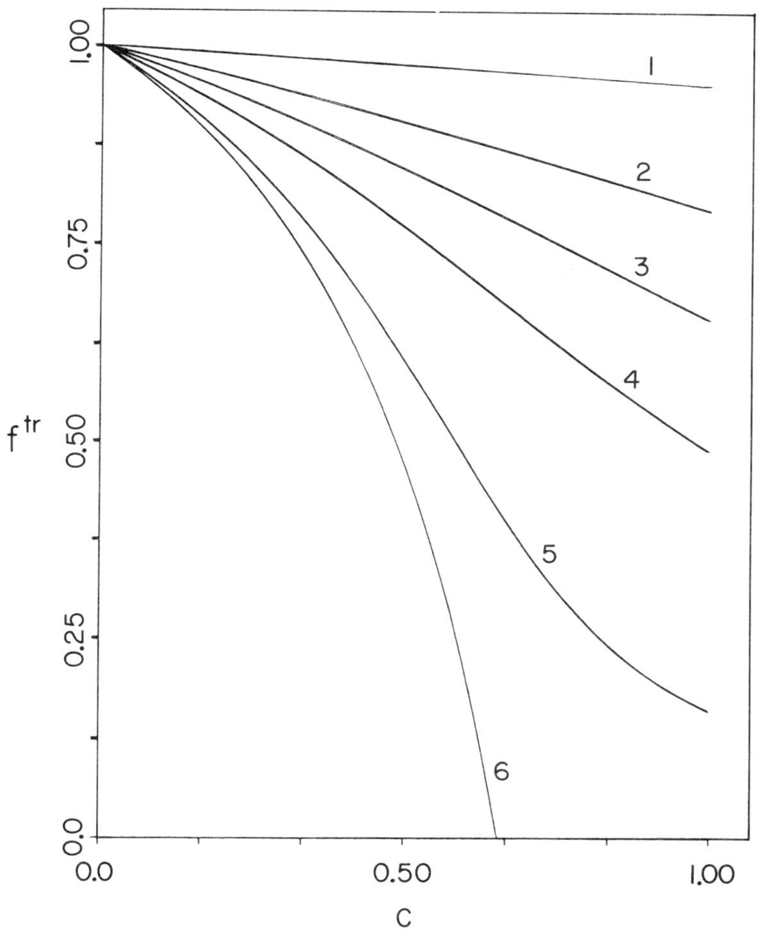

Fig. 3(b) The simple cubic lattice

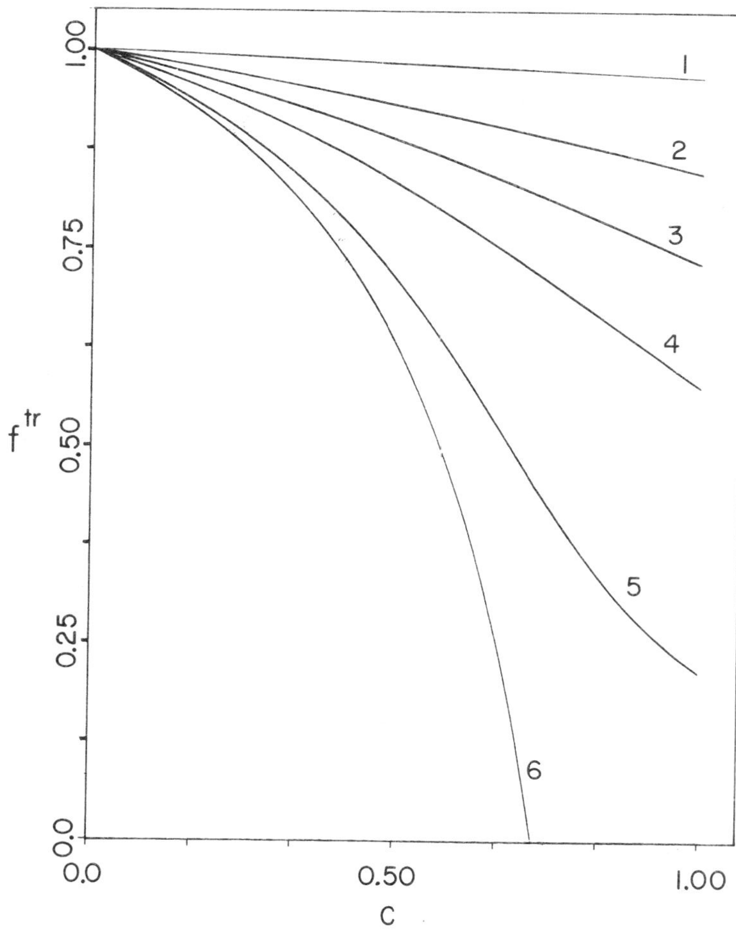

Fig. 3(c) The body centred cubic lattice

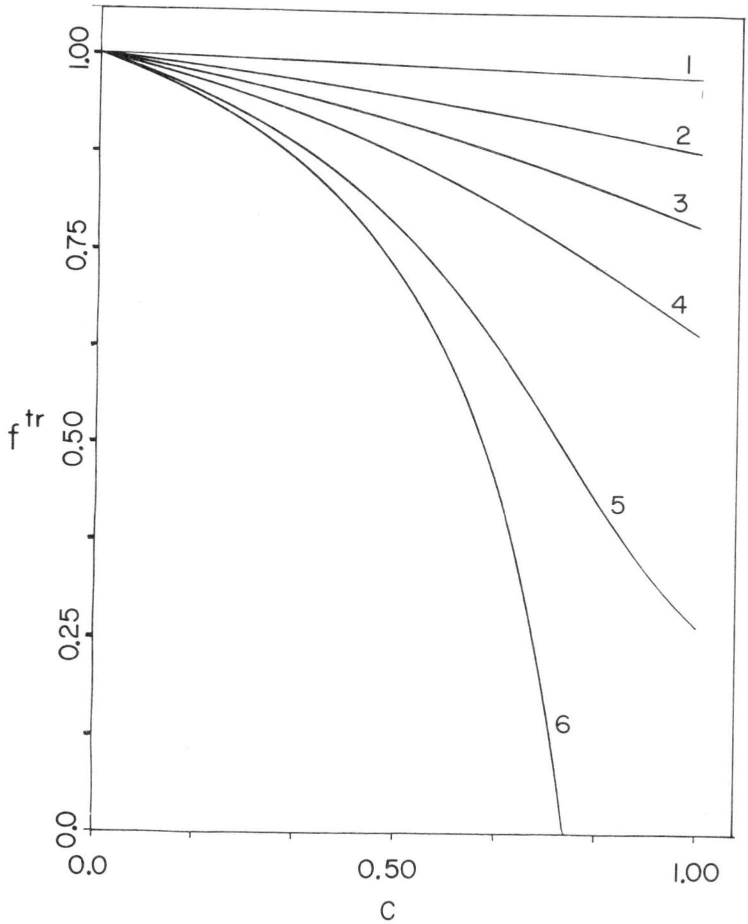

Fig. 3(d) The face centred cubic lattice

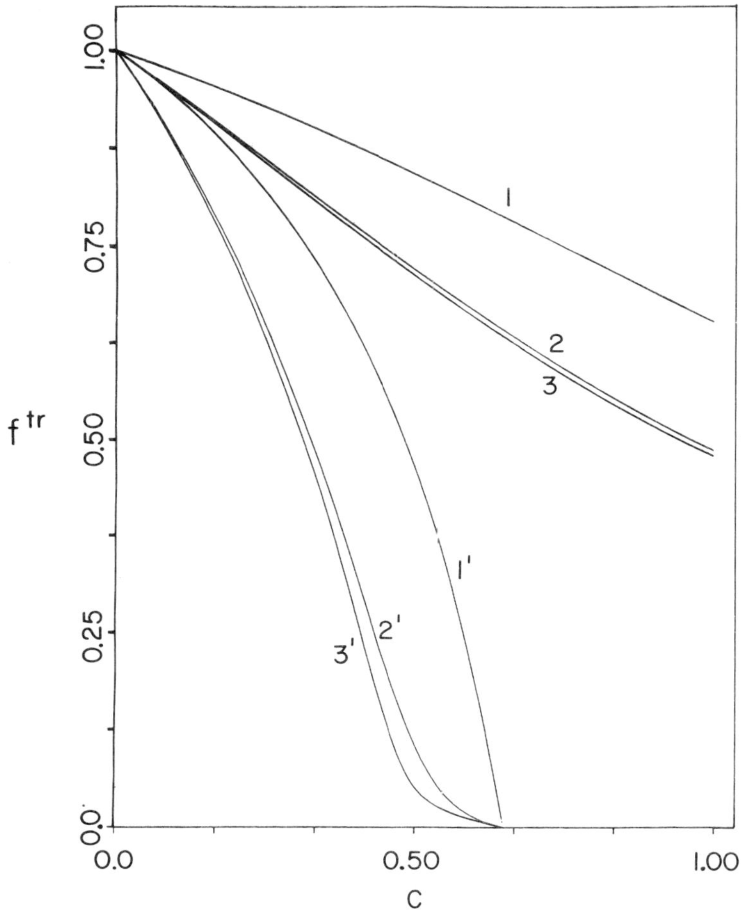

Fig. 4 For a model with both nearest and next nearest neighbour hopping, f^{tr} is plotted as a function of c. Setting $J_1^0 = J_1$ and $J_2^0 = J_2$, and varying the ratio, J_2^0/J_1^0, we get curves 1, 2, 3 which refer to J_2^0/J_1^0 = 1, 100 and 1/100, respectively. Curves, 1′, 2′, 3′, are similar except for the difference that here $J_1 = J_2 = 0$. Figures (a), (b), (c) and (d) refer to the quadratic, the simple cubic, the body centred, and the face centred cubic lattices, respectively.

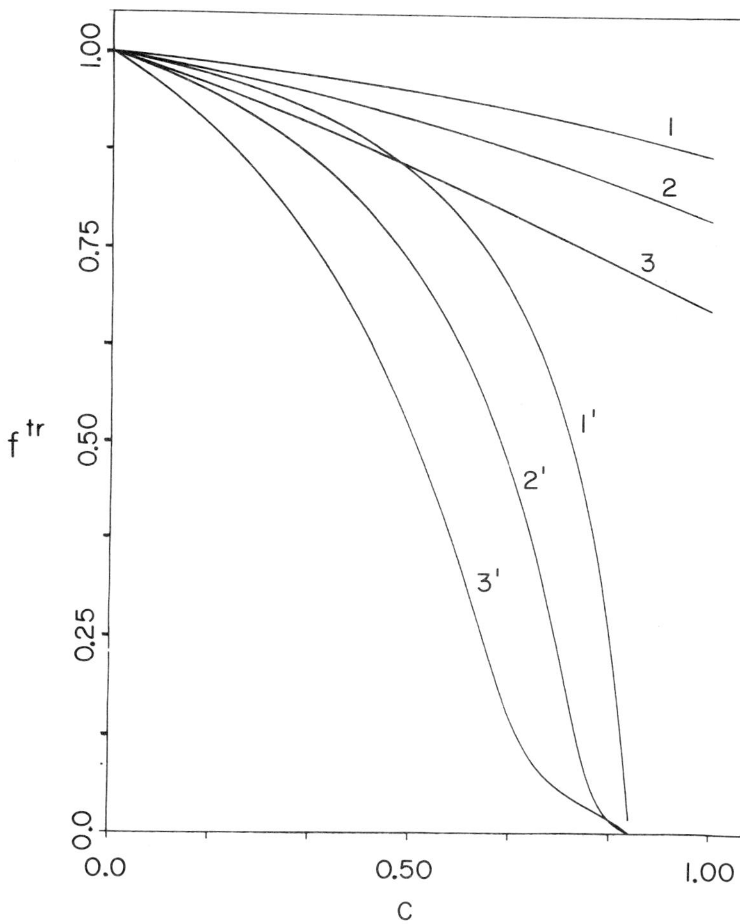

Fig. 4(b) The simple cubic lattice

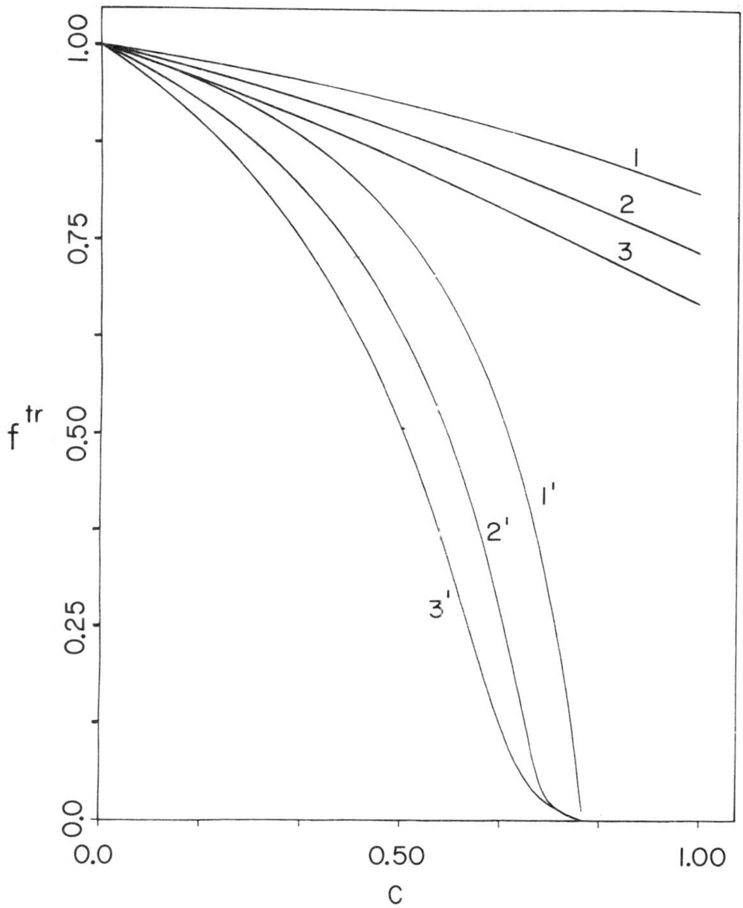

Fig. 4(c) The body centred cubic lattice

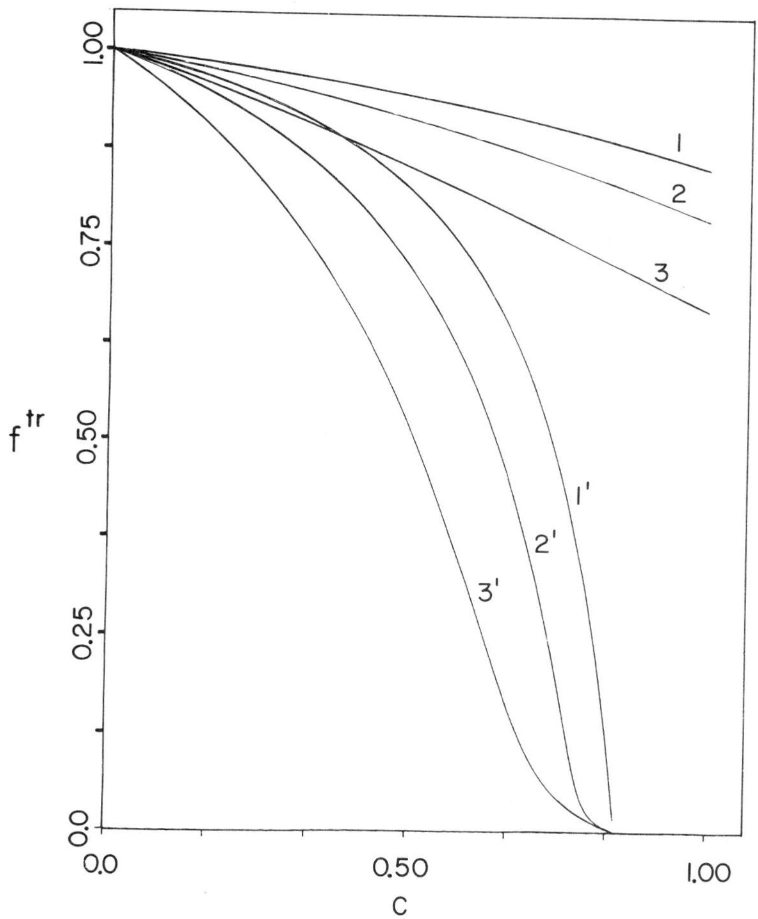

Fig. 4(d) The face centred cubic lattice

5. Concluding Remarks

In the foregoing a procedure for treating vacancy assisted classical particle diffusion in regular lattices over the entire concentration range has been described which makes contact with both the known results for the zero vacancy limit and the perturbatively soluble small concentration limit. Except for a completely empty lattice, first analyzed in detail by Chudley and Elliott[29], the results are strongly affected by correlations caused by the hard-core repulsion that forbids double occupancy.

In actual systems, atoms experience interparticle interactions that extend beyond the range of single site, hard-core repulsion[33]. Such systems have recently been treated within a mean-field like format by Kutner, Binder and Kehr[34,35]. Because of the additional complexity of nearest (and more distant) neighbour interparticle interactions, the most successful methods for studying these systems so far are the Monte Carlo computer simulation procedures[36,37]. However, we have recently begun to apply an extension of the procedures described in the foregoing to treat these systems. The essence of our approximation is to sum out all the static degrees of freedom explicitly associated with three and more particles, thereby reducing the problem to a generalized pair problem. For high temperatures, this procedure is quite successful, reducing at $T \to \infty$ to the treatment given in section 4. However, in the ordered region the success of this approach is somewhat uncertain.

APPENDIX 'A'

The analysis of section 4 is applied separately to the following four lattices:

(i) The quadratic lattice with $z_1 = 4$ and $z_2 = 4$, $|\boldsymbol{\delta}_1| = a$ and $|\boldsymbol{\delta}_2| = \sqrt{2}a$.

(ii) The simple cubic lattice with $z_1 = 6$ and $z_2 = 12$, $|\boldsymbol{\delta}_1| = a$ and $|\boldsymbol{\delta}_2| = \sqrt{2}\,a$.

(iii) The body centred cubic lattice with $z_1 = 8$, $z_2 = 6$, $|\boldsymbol{\delta}_1| = \sqrt{3}\,a/2$, $|\boldsymbol{\delta}_2| = a$.

(iv) The face centred cubic lattice with $z_1 = 12$, $z_2 = 6$, $|\boldsymbol{\delta}_1| = a/\sqrt{2}$, $|\boldsymbol{\delta}_2| = a$.

For these lattices, the tracer diffusion correlation factor $f^{\mathbf{tr}}$ is related to the 'partial correlation factors' $f_1^{\mathbf{tr}}$, $f_2^{\mathbf{tr}}$ as follows:

$$f^{\mathbf{tr}} = (J_1^0 f_1^{\mathbf{tr}} + \varepsilon J_2^0 f_2^{\mathbf{tr}}) \,/\, (J_1^0 + \varepsilon J_2^0)\,, \tag{A.1}$$

where

$$\begin{aligned} \varepsilon &= 2, \text{ quadratic lattice} \\ &= 4, \text{ simple cubic lattice} \\ &= 1, \text{ body and face centred cubic lattices.} \end{aligned} \tag{A.2}$$

The quantities $f^{\mathbf{tr}}$ are not directly observable. Rather, they are an arte-fact of the present formulation which gives

$$\begin{aligned} \Sigma^{(1)}(K\to 0\,,\,\omega\to 0) &= J_1^0\,(Ka)^2 f_1^{\mathbf{tr}} \\ \Sigma^{(2)}(K\to 0\,,\,\omega\to 0) &= J_2^0\,(Ka)^2 f_2^{\mathbf{tr}}\,\varepsilon\,. \end{aligned} \tag{A.3}$$

Although each lattice has to be analyzed separately, it is convenient to re-cast the relevant expressions in a uniform format. At the very least this procedure results in a considerable saving of space. Accordingly we write

$$f_1^{\mathbf{tr}} = 1 + 2c\,S_1\,, \quad f_2^{\mathbf{tr}} = 1 + 2wc\,S_2\,, \tag{A.4}$$

where c is, as usual, the particle concentration in the background dynamic stream, w is a parameter which is unity for the square, the simple cubic and the face centred cubic lattices but is $= \frac{1}{2}$ for the body centred cubic lattice. The parameters S_1 and S_2 can be represented in the following compact form:

$$\begin{aligned} S_1 &= [\psi_1(1-\sigma_2 B_2) + \sigma_2 A_2 \psi_2]/S \\ S_2 &= [\psi_1 \sigma_1 B_1 + (1-\sigma_1 A_1)\,\psi_2]/S \\ S &= (1-\sigma_1 A_1)(1-\sigma_2 B_2) - \sigma_1 \sigma_2 B_1 A_2\,. \end{aligned} \tag{A.5}$$

Here the quantities σ_i are independent of the lattice and are related to the atomic hopping rates as well as the partial diffusion correlation parameters, i.e.,

$$\sigma_i = 2cJ_i^0 - (J_i + v f_i^{tr} J_i^0) \quad , \quad i \equiv 1,2 \, . \tag{A.6}$$

Other parameters in (A.5) are lattice dependent

$$\begin{aligned} \psi_1 &= J_1^0 A_1 + x J_2^0 A_2 \, , \\ \psi_2 &= J_1^0 B_1 + x J_2^0 B_2 \, , \end{aligned} \tag{A.7}$$

where $x = 1$, for the simple cubic and the quadratic lattices and $x = 2$ for the body and the face centred cubic lattices.

The remaining quantities that appear in (A.5), i.e. A_1, A_2, B_1, and B_2, involve reciprocal lattice sums over the entire Brillouin zone and are, in a sense, generalized forms of the $\overline{\cos\theta}$ referred to in the text.

It is convenient to introduce the notation

$$\alpha_i = J_i + v J_i^0 f_i^{tr} \, , \tag{A.8}$$

$$c_{n\zeta} = \cos(\lambda_\zeta na/2) \quad , \quad \zeta = x, y, z \, . \tag{A.9}$$

In (A.9) n is an integer and λ_ζ is the ζ-th Cartesian component of the d-dimensional ($d = 2$, for the quadratic lattice and 3 for the cubic ones) reciprocal lattice vector $\boldsymbol{\lambda}$. Using this notation we can represent the quantities A_i, B_i, etc., as follows:

(i) *The Quadratic Lattice*

For brevity, the $\boldsymbol{\lambda}$-summation over the two dimensional Brillouin zone is assumed on the right-hand-side of the equation whenever the equal sign is replaced by an arrow.

$$\begin{aligned} A_1 &\longrightarrow (c_{4x} - 1)/\mathcal{D} \\ B_1 &\longrightarrow (c_{4x} - 1)\, c_{2y}/\mathcal{D} \\ A_2 &= 2B_1 \\ B_2 &\longrightarrow (c_{4x} c_{4y} - 1)/\mathcal{D} \\ \mathcal{D} &= 4\alpha_1 [1 - (c_{2x} + c_{2y})/2] + 4\alpha_2 [1 - c_{2x} c_{2y}] \, . \end{aligned} \tag{A.10}$$

(ii) *The Simple Cubic Lattice*

In all the following equations, whenever the equal sign is replaced by an arrow, the summand on the right-hand-side is assumed to be summed over the entire three dimensional Brillouin zone.

$$\begin{aligned} A_1 &\longrightarrow (c_{4x} - 1)/\mathcal{D} \\ B_1 &\longrightarrow (c_{4x} - 1)\, c_{2y}/\mathcal{D} \\ A_2 &= 4B_1 \\ B_2 &\longrightarrow [c_{4x} c_{4y} - 1 + 2(c_{4x} - 1)\, c_{2y} c_{2z}]/\mathcal{D} \end{aligned}$$

$$\mathcal{D} = 6\alpha_1[1-(c_{2x}+c_{2y}+c_{2z})/3] + 12\,\alpha_1\,[1-(c_{2x}\,c_{2y}+c_{2y}\,c_{2z}+c_{2z}\,c_{2x})/3] \qquad \text{(A.11)}$$

(iii) *The Body Centred Cubic Lattice*

$$A_1 \longrightarrow [c_{2x}\,c_{2y}\,c_{2z}-1+(c_{2x}-1)\,c_{2y}]/\mathcal{D}$$

$$B_1 \longrightarrow 8\,(c_{2x}-1)\,c_x\,c_y\,c_z/\mathcal{D}$$

$$A_2 = B_1/4$$

$$B_2 \longrightarrow (c_{4x}-1)/\mathcal{D}$$

$$\mathcal{D} = 8\,\alpha_1\,(1-c_x\,c_y\,c_z)+6\,\alpha_2[1-(c_{2x}+c_{2y}+c_{2z})/3] \qquad \text{(A.12)}$$

(iv) *The Face Centred Cubic Lattice*

$$A_1 \longrightarrow [(c_{2x}\,c_{2y}-1)+2\,(c_{2x}-1)\,c_y\,c_z]/\mathcal{D}$$

$$B_1 \longrightarrow 8\,(c_{2x}-1)\,c_x\,c_y/\mathcal{D}$$

$$A_2 = B_1/4$$

$$B_2 \longrightarrow (c_{4x}-1)/\mathcal{D}$$

$$\mathcal{D} = 12\,\alpha_1\,[1-(c_x\,c_y+c_y\,c_z+c_z\,c_x)/3] + 6\alpha_2[1-(c_{2x}+c_{2y}+c_{2z})/3]\,. \qquad \text{(A.13)}$$

To conclude, a few words about the above results are in order. First, it should be mentioned that although the algebra leading to these results is straightforward, it can be a little tedious. Comparatively speaking, however, this algebra is much simpler, almost by two orders of magnitude that is, than that needed for diagonalizing an 18×18 matrix. (The latter would be necessary for treating the small v or the small c limits in sc and fcc lattices with second neighbour interactions.) Secondly, we note that there are some simple internal checks on these results which are very helpful in ascertaining their correctness. For example, when the second neighbour hops are disallowed, i.e., $J_2^0 = J_2 = 0$, we must retrieve the results of section 2. Similarly, if the first neighbour hops are forbidden, i.e., $J_1^0 = J_1 = 0$, the sc lattice result should lead to that given in section 2 for an fcc lattice (with 'a' replaced by '2a'). Similarly, the second neighbours of the bcc and the fcc form a simple cubic lattice. Hence the appropriate check. Finally, when the background is stationary, i.e., $J_1 = J_2 = 0$, for each of the lattices as c is increased from zero towards unity, a unique value of c is found for which both f_1^{tr} and f_2^{tr} simultaneously (and smoothly) go to zero. This feature of the results is quite sensitive to the correctness of the final expressions given above.

References

1. Bardeen, J. and Herring, C., In *Imperfections in Nearly Perfect Crystals*, (ed. W. Shockley), Wiley, New York, (1952).
2. LeClaire, A.D. and Lidiard, A.B., Phil. Mag., **1**, 518 (1956).
3. Compaan, K. and Haven, Y., Trans. Faraday Soc., **52**, 786 (1956); *ibid*, **54**, 1498 (1958).
4. Manning, J.R., *Diffusion Kinetics for Atoms in Crystals*, Van Nostrand, Princeton, New Jersey, (1968).
5. Howard, R.E., Phys. Rev., **144**, 650 (1966).
6. Montet, G.L., Phys. Rev., **B7**, 650 (1973).
7. Chudley, C.J. and Elliott, R.J., Proc. Phys. Soc. London, **77**, 33 (1961).
8. Fedders, P.A. and Sankey, O.F., Phys. Rev., **B15**, 3580 (1977); *ibid*, **B18**, 5938 (1978).
9. Sankey, O.F. and Fedders, P.A., Phys. Rev., **B15**, 3586 (1977); Phys. Rev., **B20**, 39 (1979); *ibid*, Phys. Rev., **B22**, 5135 (1980).
10. Nakazato, K. and Kithara, K., Prog. Theoret. Phys., **64**, 2261 (1980).
11. Tahir-Kheli, R.A. and Elliott, R.J., Phys. Rev., **B27**, 844 (1983).
12. Kehr, K.W., Kutner, R. and Binder, K., Phys. Rev., **B23**, 4931 (1981).
13. Kutner, R. and Kehr, K.W. (to be published).
14. Gunther, L. and Gralla, B., J. Phys., **C16**, 1863 (1983).
15. Compare, for example, Tahir-Kheli, R.A. and Elliott, R.J., J. Phys., **C15**, L445 (1982), where the general response $S(\boldsymbol{K},\omega)$ was first calculated exactly for the $v \to 0$ limit. However, unlike Ref. 11, Ref. 15 is restricted to the tracer being identical to the background particles.
16. Tahir-Kheli, R.A. and El-Meshed, N., Phys. Rev., **B27**, 7759 (1983).
17. Tahir-Kheli, R.A., Phys. Rev., **B27**, 6072 (1983).
18. Tahir-Kheli, R.A., Elliott, R.J. and Kaski, K., Phys. Letters, **88A**, 473 (1982)
19. Harris, A.B., Leath, P.L., Nickel, B.G. and Eilliott, R.J., J. Phys., **C7**, 1693 (1974).
20. Tahir-Kheli, R.A. and Elliott, R.J., J. Phys., **C15**, 5751 (1982).
21. Tahir-Kheli, R.A., Phys. Rev., **B28**, 3049 (1983).
22. Manning, J.R., Phys. Rev., **B4**, 1111 (1971).
23. Benoist, P., Bocquet, J.L. and Lafore, P., Acta Metall., **25**, 265 (1977).
24. Koiwa, M., J. Phys. Soc. Japan, **45**, 781 & 1327 (1978).
25. Ishioka, S. and Koiwa, M., Phil. Mag., **A41**, 385 (1980).
26. McKee, R.A., Phys. Rev., **B21**, 4269 (1980).
27. Wolf, D., J. Phys. Chem. Solids, **41**, 1053 (1980).
28. Murch, G.E. and Rothman, S.J., Phil. Mag., **A43**, 229 (1981).
29. McGurn, A.R. and Tahir-Kheli, R., Phys. Rev., **B17**, 2316 (1978)
30. Lottner, V., Haus, J.W., Heim, A. and Kehr, K.W., J. Phys. Chem. Solids, **40**, 557 (1972).
31. Kehr, K.W., *Theory of Diffusion in Hydrogen Metals*, In 'Proceedings of the International Symposium on the Electronic Structure and the Properties of Hydrogen in Metals', (ed. P. Jena), Plenum, New York, (1982).
32. Chudley, C.J. and Elliott, R.J., Proc. Phys. Soc., **77**, 33 (1961).
33. Dieterich, W., Fulde, P. and Peschel, I., Advances in Physics, **29**, 527 (1980).
34. Kutner, R., Binder, K. and Kehr, K.W., Phys. Rev., **B26**, 2967 (1982).
35. Kehr, K.W., Kutner, R. and Binder, K., (to be published).
36. For a recent review see, Kehr, K.W. and Binder, K: *Simulation of Diffusion in Lattice Gases and Related Kinetic Phenomena;* (to be published).
37. Sadiq, A. and Binder, K., Surface Science, **128**, 350 (1983).

The Gibbs Paradox

N. G. van Kampen
Institute for Theoretical Physics of the University of Utrecht, The Netherlands

ABSTRACT

The dependence of the entropy on the number of molecules can never be found from studying closed systems. The argument based on counting states of particles, indistinguishable or not, is therefore spurious. On the other hand, interaction with particle reservoirs leads to different results for distinguishable and indistinguishable molecules. The choice between both depends on whether the experimentalist is able and willing to make the distinction, not on the fundamental properties of the molecules. Thus the paradox is resolved by replacing the Platonic idea of entropy with an operational definition. Quantum mechanics has no bearing on the question.

Beste Dik,

It is a long time since we both attended the lectures by Mrs de Haas-Lorentz on thermodynamics. They were excellent from a paedagogical point of view since they forced you to figure out almost everything yourself. That is a much better way of learning physics than the spoonfeeding of predigested formalisms which is nowadays regarded as the highest wisdom in education. I remember how puzzled I was by the sudden appearance of the term $-kN\log N$ in the entropy. After I figured it out I found that there is still much confusion about it in the literature. That is my excuse for bothering you with such a time-worn subject.

1. When a volume V contains a mixture of N_A molecules A and N_B molecules B with the same mass, the entropy is not the same as when it contains $N = N_A + N_B$ molecules of one and the same kind, say A. The difference is for classical ideal gases

$$-kN_A \log N_A - kN_B \log N_B + k(N_A + N_B) \log (N_A + N_B) . \tag{1}$$

The Gibbs paradox consists in the fact that this entropy of mixing remains finite when B becomes very similar to A, and even when it is indistinguishable from A for all practical purposes.[1] This fact is a consequence of the way in which the entropy depends on the number of molecules, and I therefore start out in section 2 by sketching how this dependence is established in thermodynamics.

In *statistical mechanics* this dependence is obtained by inserting a factor $1/N!$ in the partition function. Quantum mechanically this factor enters automatically and in many textbooks that is the way in which it is justified. My point is that this is irrelevant: *even in classical statistical mechanics it can be derived by logic* — rather than by the somewhat mystical arguments of Gibbs[2] and Planck.[3,4] Specifically I take exception to such statements as: "It is not possible to understand classically why we must divide by $N!$ to obtain the correct counting of states",[5] and: "Classical statistics thus leads to a contradiction with experience even in the range in which quantum effects in the proper sense can be completely neglected".[6]

Actually the problem of the $N!$ was completely cleared up by Ehrenfest and Trkal,[7] but their argument is ignored in most of the literature.[8] It may therefore be of some use to take up the matter again starting from scratch, as a service to future textbook writers.

2. First, consider ordinary macroscopic *thermodynamics*, which deals with systems in equilibrium. Our system is a gas contained in a vessel whose volume V can be varied by means of a piston. In addition there is a supply of heat reservoirs, which can be brought in thermal contact with our system. If in equilibrium, it has a pressure P and a temperature T, and these two parameters fully specify the thermodynamic state. All other thermodynamic quantities are functions of them, and any two may serve to specify the state.[9]

The state can be changed by slowly moving the piston and by supplying successive small amounts of heat δQ from the heat reservoirs. The change must be so slow that at all times the gas is in equilibrium, which implies that the change is *reversible*. Such a change can be represented by a curve in the P,T plane, from some initial state 1 to a final state 2. The *first law* states that the integral

$$\int_1^2 (\delta Q - PdV) \tag{2}$$

is independent of the curve connecting 1 and 2 and therefore defines an internal energy $E(P,T)$, up to an additive constant. In the same way the *second law* defines the entropy S by

$$\int_1^2 \frac{\delta Q}{T} = S(P_2,T_2) - S(P_1,T_1). \tag{3}$$

For a classical ideal gas of N monatomic molecules one knows $P = NkT/V$ and $E = \frac{3}{2} NkT$ + constant, and consequently

$$S(P,T) = \tfrac{5}{2} Nk \log T - Nk \log P + C. \tag{4}$$

The purpose of these trite remarks is to emphasize that *the second law defines only entropy differences between states that can be connected by a reversible change*. The value of C is only required to be the same for all states so connected, that is, it must be independent of P and T. Otherwise it may depend on anything, even the price of the apparatus or the number of molecules N. It is merely for convenience that this freedom is restricted by two conventions.

Convention 1. If I have a second system, identical with the first one, I agree to assign to it the same C and therefore the same $S(P,T)$. This is obviously convenient but not logically necessary and by itself contains no information about the physical world.

Convention 2. Two separate systems may be regarded as forming together a single system. I agree to take as the entropy of the combined system the sum of the entropies of the two parts. Of course each part must be in equilibrium but they don't need to be in equilibrium with one another. Hence this convention may serve to define entropy of systems that are not in equilibrium themselves but can be divided into subsystems that are.[10]

3. Consider a set of systems, consisting of identical vessels containing *the same gas* but in different amounts. One is free to represent all possible states obtained in this way in a space with coordinates P,T,N. There is no way, however, to compare entropy values belonging to different N, unless one introduces a new kind of process by which N can be varied in a reversible way. This is done by the familiar device of taking two identical vessels with the same P,T,N and opening up a channel between them. Together with convention 1 this leads to the conclusion that $S(P,T,N)$ must be proportional to N. Hence the formula (4) for the ideal gas now becomes

$$S(P,T,N) = \tfrac{5}{2}\, Nk \log T - Nk \log P + cN \tag{5}$$

where c does not depend on P,T,N, but may depend on anything else, for instance on the kind of gas.

If two vessels contain *different gases* A, B the entropy of the combined system is defined by convention 2. When a channel is opened up an irreversible mixing occurs, so that this device cannot be utilized to define the entropy $S(P,T,N_A,N_B)$ of the resulting mixture. In its stead one now uses the familiar reasoning based on semi-permeable walls and obtains for the ideal gas

$$S(P,T,N_A,N_B) = \tfrac{5}{2}(N_A+N_B)\log T - N_A k \log P_A - N_B k \log P_B + c_A N_A + c_B N_B, \tag{6}$$

where P_A, P_B are the partial pressures

$$P_A = \frac{N_A}{N_A+N_B}\, P \qquad P_B = \frac{N_B}{N_A+N_B}\, P\,. \tag{7}$$

The fact that (6) with (7) does not reduce to (5) when $A = B$ constitutes the Gibbs paradox. The difference is given in (1).

4. The origin of the difference is that two different processes had to be chosen for extending the definition of entropy. They are mutually exclusive: the first one cannot be used for two different gases and the second one does not apply to a single one. But suppose that A and B are so similar that the experimenter has no physical way of distinguishing between them. Then he does not have the semi-permeable walls needed for the second process, but on the other hand the first one will look reversible to him. He will then be led to (5), i.e.,

$$S(P,N_A,N_B) = (N_A+N_B)\, k \left\{\tfrac{5}{2}\log T - \log P + c\right\}.$$

The point is, that *this is perfectly justified* and that he will not be led to any wrong results. If you tell him that 'actually' the entropy increased when he opened the channel he will answer that this is a useless statement since he cannot utilize the entropy increase for running a machine. The entropy increase is no more physical to him than the one that could be manufactured by taking a single gas and mentally tagging the molecules by A or B.

In fact, this still holds when the experimenter would be able to distinquish between A and B, by means of a mass spectrograph for instance, but is not interested in the difference because it is not relevant for his purpose. This is precisely what engineers do when they make tables of the entropy of steam, ignoring the fact that it is actually a mixture of normal and heavy water. Thus, whether such a process is reversible or not depends on how discriminating the observer is. The expression for the entropy (which he constructs by one or the other processes mentioned above) depends on whether or not he is able and willing to distinguish between the molecules A and B.[11] This is a paradox only for those who attach more physical reality to the entropy than is implied by its definition.

5. Now *statistical mechanics*. If our vessel is not in contact with a heat reservoir and has reflecting walls the microscopic motion of all molecules is described in classical mechanics by a Hamilton function $H(q,p)$. The Hamilton equations define a trajectory in phase space, which lies on a hypersurface of constant energy. The trajectory covers that surface evenly: the time spent in any region is proportional to the area of that region.[12] Hence if I observe the system at an arbitrary moment, the probability of finding it in that region is proportional it its area. If I measure one or more physical quantities $A_r(q,p)$ the probability to obtain values between a_r and $a_r + \Delta a_r$ is proportional to the area of the region on the energy surface cut out by $a_r < A_r(q,p) < a_r + \Delta a_r$. This region has been called the Z-star.[13]

This fact makes it possible to establish equilibrium statistical mechanics by means of measures in phase space alone, without taking the explicit equations of motion into account — as was done by Gibbs.[2] Yet the fact itself is based on the existence of these equations, and their properties of conserving energy and measure — as emphasized by the Ehrenfests.[13] The probability distribution in the space of the a_r is determined by that measure and by the nature of the A_r. It is *not* determined by the information or lack of information of the observer, or by a fair representation of his ignorance and other anthropomorphic considerations.[14] No-one seriously believes that statistical mechanics ceases to be valid when he is not looking.[15]

One could add, as an aside, that the energy surface can be partitioned in $N!$ equivalent parts, which differ from one another only by a permutation of the molecules.[16] The trajectory, however, does not recognize this equivalence because it cannot jump from one point to an equivalent one. There can be no good reason for identifying the Z-star with only one of these equivalent parts. In fact, that merely amounts to dividing its measure by $N!$, which in the expression for the probability itself is cancelled by the normalization.

6. Having established the microcanonical ensemble as a description of an *isolated gas* in equilibrium we may now restore the contact with the heat reservoir. Together they constitute one big isolated system to which the

previous considerations apply. The energy E of the gas itself is no longer constant, but is now one of the observables A_r. Hence its probability distribution can be computed by measuring phase space with the familiar result

$$W(E) = \text{const.}\ e^{-\beta E}\ \Omega(E)\ , \tag{8}$$

where $\Omega(E)$ is the area of the energy surface $H(q,p) = E$. Since we know already the distribution within each individual energy surface we may conclude that the probability in the phase space of our gas is

$$W(q,p) = \text{const.}\ e^{-\beta H(q,p)}\ . \tag{9}$$

The parameter β is all that is left from the heat bath and can be identified with $1/kT$ by computing the mean kinetic energy from (9). The constant may still depend on N.

It is not surprising that the canonical distribution can only be derived by allowing the system to interact with a heat reservoir, because it is only correct in that case. This fact is somewhat obscured by the common practice of using the canonical distribution (9) even for isolated systems. That leads to the same values for most quantities since (8) is so sharply peaked that it is practically confined to a single energy surface. However this practice is merely a mathematical trick.

7. Similarly, when we want to regard N as a variable and ask for the probability $W(E,N)$ we have to bring our gas in contact with a particle reservoir. Take therefore a second volume V' with N' molecules of the same gas and open up a channel between both volumes. Together they form a single system with $N+N'=N^*$ molecules and volume $V+V'=V^*$. For convenience we also add a heat reservoir so that we may use (9). The quantities E and N are now observables A_r, whose probability distribution is found by computing phase space volumes.

Let H_N be the Hamilton function for N molecules in V, and similarly $H'_{N'}$. Then the total H^* can be written $H_N + H'_{N^*-N}$ for each separate value of N, provided that the interaction energy in the channel is negligible. To find the probability $W(N,q,p)$ for having N molecules in V at the points $q_1, p_1; \ldots ; q_N, p_N$ in phase space we have to start from the distribution (9) applied to the combined volumes and integrate over all states compatible with the prescribed $N, q_1, \ldots, p_N$. One first has to select N out of the N^* molecules to be located in V and then integrate over all possible locations of the remaining $N^*-N=N'$ molecules in V'. The result is

$$W(N,q,p) = \text{const.}\binom{N^*}{N}\, e^{-\beta H_N(q,p)} \int_{V'} e^{-\beta H'_{N'}(q',p')}\, dq'\, dp'\ .$$

If one now takes the limit $N^* \to \infty$, $V' \to \infty$ with constant density N^*/V' one obtains the grand canonical distribution

$$W(N,q,p) = \text{const.}\ \frac{z^N}{N!}\ e^{-\beta H_N(q,p)}\ , \tag{10}$$

where z is a function of that density and of β. For an ideal gas this can be verified by explicit calculation.

If V contains a mixture of A and B one can introduce two separate reservoirs of molecules A and B respectively, and connect them with V through semi-permeable walls. One then finds in the same way

$$W(N_A, N_B, q_A, p_A, q_B, p_B) = \text{const.}\ \frac{z_A^{N_A}\ z_B^{N_B}}{N_A!\ N_B!}\ e^{-\beta H_{N_A,N_B}(q_A, p_A, q_B, p_B)}\ . \tag{11}$$

The fact that this does not reduce to (10) when $A = B$ is again the Gibbs paradox and the answer is again: The question is not whether they are identical in the eye of God, but merely in the eye of the beholder.

8. The essential point is that (10) follows as a mathematical consequence from the microcanonical ensemble, which itself was based on the dynamics of the system. *The $N!$ arises from the computation of phase space volume according to the original rules without any additional postulate*, either classical or quantummechanical.

Gibbs [2] argued that, since the observer cannot distinguish between different molecules, "it seems in accordance with the spirit of the statistical method" to count all microscopic states that differ only by a permutation as a single one. Actually it is exactly opposite to the basic idea of statistical mechanics, namely that the probability of a macrostate is given by the measure of the Z-star, i.e., the number of corresponding, macroscopically indistinguishable microstates. As mentioned in section 5, it is impossible to justify the $N!$ as long as one restricts oneself to a single closed system. Unfortunately, Planck[3,4] revived the Gibbs mystique and this view seems dominant in the literature.

9. Finally the connection between statistical mechanics and thermodynamics. Since a heat reservoir is indispensable in thermodynamics we have to use a canonical distribution. For simplicity I take the distribution (9) for closed systems, which suffices to make my point clear.

According to our general rules the macroscopic energy E is the average of H, which can be written

$$E = -\frac{\partial}{\partial\beta} \log \int_V e^{-\beta H(q,p)}\, dq\, dp \tag{11}$$

$$= -\frac{\partial}{\partial\beta} \log Z(V,\beta) . \tag{12}$$

Here the partition function Z is defined as a calculational aid. Similarly the pressure is the average force on the piston (per unit area); which gives

$$P = \frac{1}{\beta}\frac{\partial \log Z}{\partial V} . \tag{13}$$

These two equations determine E and P as functions of V,T and contain all the thermodynamicist ever wants to know abcut the system.

One of the properties of the system is that it has an entropy, and this property must be contained in these equations. In fact, (12) and (13) imply

$$d \log Z(V,\beta) = -E d\beta + \beta P dV ,$$

which may be rearranged to give

$$Td\left\{k \log Z + \frac{E}{T}\right\} = dE + PdV .$$

This shows that the Gibbs-Duhem relation is satisfied and that

$$S = k \log Z + E/T + K , \tag{14}$$

where K is an arbitrary quantity independent of E,V, that is, independent of the state of our closed system. A more familiar way of writing (14) is

$$e^{-\beta F} = e^{K} \int_V e^{-\beta H(q,p)} dq^N dp^N .$$

We defined Z by the integral in (11), but we could have inserted an arbitrary factor provided it does not depend on β, V. (In fact, some people add a factor h^{-3N} for cosmetic reasons.) Such a factor need not, however, affect the value of S, owing to the intervention of K.

If our gas is in contact with a particle reservoir as well one can do the same calculation with the grand-canonical distribution (10). The result is [17]

$$e^{-\beta PV} = e^{K} \sum_{N=0}^{\infty} \frac{z^N}{N!} \int_V e^{-\beta H_N(q,p)} dq^N dp^N ,$$

where z is related to the chemical potential by $\mu = T \log z$. The constant K is now independent of β, V, z.

The point I want to make is that the arbitrary factor in the definition of the partition function and the arbitrary constant in the entropy are not linked together, unless one chooses to put $K = 0$ by convention.

10. A few remarks about *quantum mechanics*. The Gibbs paradox seems less virulent, because the distinction between identical particles and non-identical particles is an inherent feature of quantum mechanics itself.[18] It

is possible, however, to take for A and B the same molecule in two different quantum states, which by linear combination can be made as similar as one wishes.[19] Nonetheless our general answer remains valid: if the experimenter chooses to ignore the difference he may use the entropy expression for a single gas, but if he observes the difference he must use the expression for a mixture.[20]

Thermodynamics is affected by quantum mechanics only through the existence of the third law. Nernst found experimentally that the entropy difference (3) vanishes when $T_1 = T_2 \to 0$.[21] In quantum mechanics it is also true that (3) remains finite for $T_1 \to 0$ with fixed T_2. These two facts make it possible to fix the arbitrary constant in the entropy by putting $S = 0$ at $T = 0$, but of course that is another convention.[4] The familiar argument based on the non-degeneracy of the ground state is wrong for several reasons.[22]

In *quantum statistics* the $1/N!$ enters naturally in the partition function owing to the fact that the permutation of identical molecules now really does not give another microscopic state. This factor happens to be the same as the one needed to obtain the correct dependence of S on N. Hence the prevailing view that quantum mechanics resolves the counting problem.[23] Yet this is misleading, because it is still true that *for a closed system* one may incorporate any factor $f(N)$ in the partition function, just as in sections 5 and 6. Moreover this factor is linked to the dependence of S on N by convention only, as shown in section 9. The correct N-dependence can only be determined by means of a process in which N changes reversibly. In principle, therefore, *the Gibbs paradox is no different in quantum mechanics, it is only less manifest.*

References

1. Gibbs, J.W., *On the Equilibrium of Heterogeneous Substances.* Reprinted in *The Scientific Papers of J. Willard Gibbs.* I. Dover, New York, p.167.
2. Gibbs, J.W., *Elementary Principles in Statistical Mechanics.* Yale University Press, New Haven, 1902; Dover, New York, p.187 (1960).
3. Planck, M., *Theorie der Wärmestrahlung* 5th ed., Barth, Leipzig, p.206 (1923); *Theorie der Wärme*, 2nd ed., Hirzel, Leipzig, p.234 ff (1932); See also L. Rosenfeld in: *Max Planck Festschrift 1958*, (eds. B. Kockel *et al.*) Deutscher Verl. Wissens., Berlin, p.203; translated as chapter 17 of: *Selected Papers of Léon Rosenfeld* (eds. R.S. Cohen and J.J. Stachel), Reidel, Dordrecht (1979).
4. Planck, M., Ann. Physik, **66** (4), 365 (1921) = *Physikalische Abhandlungen und Vorträge* II, Vieweg, Braunschweig, p.527 (1958).
5. Huang, K., *Statistical Mechanics*, Wiley, New York, p.154 (1963).
6. Münster, A., *Statistical Thermodynamics*, Vol.I., Springer, Berlin, p.57 (1969).
7. Ehrenfest, P. and Trkal, V., Ann. Physik, **65**, 609 (1921) = P. Ehrenfest, *Collected Scientific Papers*, North-Holland, Amsterdam, p.414 (1959).

8. However: Uhlenbeck, G.E. and Ford, G.W., *Lectures in Statistical Mechanics*, Amer. Mathem. Soc., Providence, R.I., p.31 (1963).
9. Self-gravitating systems, such as stellar clusters, do not have an equilibrium and are therefore outside the realm of thermodynamics. See Thirring, W., Z. Physik, **235**, 339 (1970); Miller, R.H., In *Advances Chemical Physics*, (eds. I. Prigogine and S.A. Rice), Wiley, New York, p.107 (1974). The *entropy of the universe* is therefore not defined. See also Bridgman, P.E., *The Nature of Thermodynamics*, Harvard Univ. Press (1941); Harper, New York, p.148 ff (1961).
10. Kampen, van N.G., Physica, **25**, 1294 (1959).
11. It may be that this is the same as the opinion of P.W. Bridgman (ref. 9) p.168. It is not the same as the interpretation thereof in T.H. Boyer, Amer. J. Phys., **38**, 771 (1970).
12. The ergodic problem is not our subject. We may also ignore that the area should be measured in units $|\text{grad } H|^{-1}$.
13. Ehrenfest, P. and T., In *Encyklopädie der mathematischen Wissenschaften* IV.32, Teubner, Leipzig (1911) = *Conceptual Foundations of the Statistical Approach in Mechanics*, (translated by M.J. Moravcsik), Cornell University Press, Ithaca (1959).
14. Tolman, R.C., *The Principles of Statistical Mechanics*, Clarendon, Oxford, (1938); Jaynes, E.T., Phys. Rev., **106**, 620 and **108**, 171 (1957), and in: *The Maximum Entropy Formalism*, (eds. R.D. Levine and M. Tribus) M.I.T. Press, Cambridge, Mass., (1978).
15. It is legitimate to use the idea of *disorder* to illustrate the fact that the most probable macroscopic states are those that correspond to the least restricted phase space area. It is *not* legitimate to use it as a justification of that fact. As all illustrations, it is dangerous when taken too seriously, and it has actually caused many idle speculations.
16. Planck, M., Sitz. Ber. Preuss. Akad. Wissens. 653 (1916) = *Physikalische Abhändlungen und Vorträge* II (Vieweg, Braunschweig, p.420 (1958).
17. Kramers, H.A., Proc. Roy. Acad. Amsterdam, **41**, 10 (1938) = *Collected Scientific Papers*, North-Holland, Amsterdam, p.738 (1956).
18. See, e.g., Slater, J.C., *Introduction to Chemical Physics*, McGraw-Hill, New York, p.129 (1939); Stern, O., Rev. Mod. Phys., **21**, 534 (1949).
19. Neumann, v. J., *Mathematische Grundlagen der Quantenmechanik*, Springer, Berlin, p.197 (1932).
20. Klein, M.J., Amer. J. Phys., **26**, 80 (1958); Ned. T. Natuurk., **25**, 73 (1959); Lesk, A.M., J. Phys. (London), **A**13, L111 (1980).
21. Hence also the entropy of mixing, in contrast with the high temperature case (1). See P.T. Landsberg and D. Tranah, Amer. J. Phys., **46**, 228 (1978).
22. Kampen, van N.G., Physica, **20**, 603 (1954); Casimir, H.B.G., Z. Phys., **171**, 246 (1963) and in: *Fundamental Problems in Statistical Mechanics* IV. (eds. E.G.D. Cohen and W. Fiszdon), Proc. Intern. Summer School Jadwisin, Ars Polona, Warszawa (1978).
23. Often on the authority of E. Schrödinger, *Statistical Thermodynamics*, University Press Cambridge, p.58 ff. (1946).

Wave Packets and Ghosts

Tzuru Fujiwara* and Harald Wergeland**
*Department of Mathematical Science, U.O.P. 591 Sakai, Japan
**N-2090 Hurdal, Norway

1. Introduction

Subsequent to the famous papers[1] which established Wave Mechanics, Schrödinger adjoined a short communication: *Der stetige Übergang von der Mikro-zur Makromechanik*[2] on the concept of mechanical motion under the undulatory aspect of Quantum theory.

In this he showed that a pulse of the de Broglie waves pertaining to a Harmonic Oscillator will not only follow the classical orbit for all times, it will even remain stable, in contrast to light waves and their inevitable dispersion in an analogous optical signal

The particular superposition of stationary state wave functions for the Harmonic Oscillator: $\Psi_n = \langle x|n\rangle(t)$, selected by Schrödinger:

$$X(t) = \text{const.} \sum_{0}^{\infty} \frac{(A/2)^n}{n!} \Psi_n \tag{1.1}$$

where

$$\Psi_n = \mathcal{H}_n(\xi)\, \frac{e^{-i\omega t(n+\frac{1}{2})}}{\sqrt{n!\, 2^n \sqrt{\pi}}} \quad ; \quad \xi = \sqrt{\frac{m\omega}{\hbar}} \cdot x$$

he could sum by one of Hermite's formulae (*Sur un nouveau developpement en series des fonctions*, C.R. 58 (1908) pp.93, 266) in the closed form:

$$X(t) = \left(\frac{1}{\pi}\right)^{\frac{1}{4}} e^{-\frac{1}{2}(\xi - A\cos\omega t)^2 - iA\sin\omega t(\xi - \frac{A}{2}\cos\omega t)} .$$

Hence the coordinate probability density :

$$|X(t)|^2 = \frac{e^{-(E-A\cos\omega t)^2}}{\sqrt{\pi}}$$

ETP-U

indeed looks like an immutable ball, tracing out the classical amplitude of the oscillator: $\langle x;\ p_0 = 0\rangle(t) = x_0 \cos\omega t$, $x_0 = \sqrt{\hbar/m\omega}\cdot A$.

This may have seemed to promise restoration, upon some deeper level, of the classical ideas about motion, in particular the concept of mechanical orbit, which, until the encounter with quantum phenomena, had not been called in question since the time of Parmenides and Zenon. Anyway, Schrödinger, equally attentive to the schools of Elea and Abdera[3] saw fit to conclude his essay on a mechanistic note:

> "*Es lässt sich mit Bestimmtheit voraussehen, dass man auf ganz ähnliche Weise auch die Wellengruppen konstruieren kann, welche auf hochquantigen Keplerellipsen herumlaufen und das undulationsmechanische Bild des Wasserstoffelektrons sind* .....".

Heisenberg, on the other hand, in expounding the concept of a wave packet[4,5] which came to play such a dominant role in the epistemological discussion, pointed out that wave packets of band width $\Delta\omega = \Delta E/\hbar$ in stationary states: E_n, in general diffuse in the course of times of the order: $h/\left(\frac{\partial^2 E}{\partial n^2}\cdot\Delta n\right)$.

This then explained the infinite life-time of Schrödinger's wave packet ($E_n = (n+\frac{1}{2})\hbar\omega$), and the Harmonic Oscillator as an exception case.

2. A Ghostly Wave Packet

Now, Schrödinger's wave packet (1) is in fact a very special one, even for the Harmonic Oscillator, in as much as its initial form:

$$X(t=0) = \left(\frac{1}{\pi}\right)^{\frac{1}{4}}\cdot e^{-\frac{1}{2}(\xi - A)^2} \tag{2.1}$$

happens to be precisely what is now denoted as a 'Coherent State' (Appendix A.1) in the theory of linear fields.

A slightly more general type:

$$X(t=0) = \left(\frac{\alpha}{\pi}\right)^{\frac{1}{4}}\cdot e^{-\alpha/2(\xi - A)^2} \tag{2.2}$$

which is no longer 'coherent' when $\alpha \neq 1$, leads to (Appendix A.2)

$$X(t) = \left(\frac{\alpha}{\pi}\right)^{\frac{1}{4}} \frac{e^{\frac{(\xi + i\alpha A\sin\omega t)^2}{2\sin^2\omega t(\alpha - i\,\mathrm{ctg}\,\omega t)} + \frac{i}{2}\mathrm{ctg}\,\omega t\cdot\xi^2 - \frac{\alpha}{2}A^2}}{(2\pi i)^{\frac{1}{2}}\ (\alpha\sin\omega t - i\cos\omega t)^{\frac{1}{2}}} \tag{2.3}$$

and the probability density:

$$|X(t)|^2 = \left(\frac{\alpha/\pi}{\cos^2\omega t + \alpha^2\sin^2\omega t}\right)^{\frac{1}{2}} e^{-\alpha\frac{(\xi - A\cos\omega t)^2}{\cos^2\omega t + \alpha^2\sin^2\omega t}} \tag{2.4}$$

from which the subtle intuition ($\alpha = 1$!) so characteristic to Schrödinger's mathematical devices can again be seen.

As before $\langle x\rangle(t)$ traces out the classical orbit, but in contrast to

Schrödinger's wave packet — which always preserves a dignified countenance: $\langle x|X(t)|\rangle$, this one is periodically sucking and puffing its cheeks in and out, jester fashion:

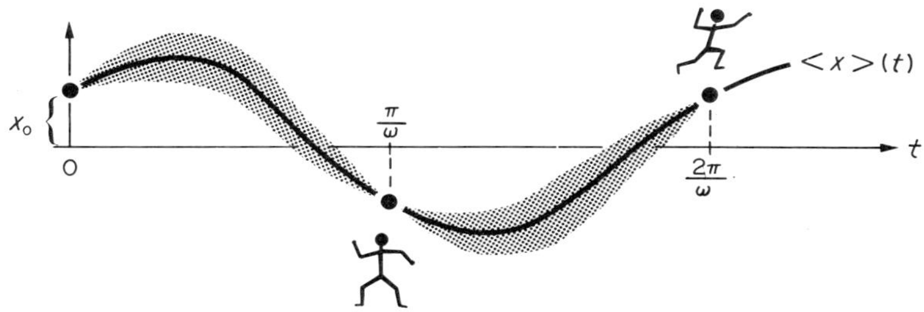

$$\langle \Delta x^2 \rangle(t) = \frac{\hbar}{m\omega}\cdot\frac{1}{\alpha}\left[\cos^2\omega t + \alpha^2\sin^2\omega t\right]$$

$$= \frac{\hbar}{m\omega}\begin{cases}\alpha\,, & t = (n+\frac{1}{2})\cdot\frac{\pi}{\omega}\\ 1/\alpha\,, & \text{"} = n\cdot\text{"}\end{cases}\quad n = 0,1,2,\ldots \tag{2.5}$$

The most amusing feature of this 'jester' is of course the limit $\alpha\to\infty$:

$$|\langle x|X(t)\rangle|^2 \longrightarrow \begin{cases}\delta(x-x_0)\,, & t = 0\,,\frac{2\pi}{\omega}\,,\ldots\\ \delta(x+x_0)\,, & t = \frac{\pi}{\omega}\,,\frac{3\pi}{\omega}\,,\ldots\end{cases} \tag{2.6}$$

$$x_0 = \sqrt{\frac{\hbar}{m\omega}}\cdot A$$

whence the wave packet reappears at the classical turning points: as a 'ghost' of δ-function shape, once every half period, in the mean time disappearing from this world, i.e., $X(t) = 0$ *a.e.*

3. Resurrection of Wave Packets

That Poincaré Cycles may occur also in the evolution of quantum mechanical states, was shown on intuitive grounds already by N.S. Krylov (1932) (private communication from Mark Kac) and more exactly proved by Bocchiery and Loinger (1954). Accordingly we must expect, that for suitably confined systems, even wave packets of δ-function shape, will sometimes recur, not only for the Harmonic Oscillator.

Furthermore, a special class of quantum systems is distinguished by having exactly periodic recurrences — namely those with energy eigenvalues proportional to any subset of the natural integers. Next to the Harmonic Oscillator, the simplest example is of course the free particle on a line:

$x \subset [0, l]$ with

$$E_n = \frac{\pi^2 \hbar^2}{2ml^2} \cdot n^2 ; \qquad n = 1, 2, 3, \ldots \tag{3.1}$$

Pauli [6], Born [7] and others have certainly pointed out how wave packets of this system will diffuse more or less rapidly towards uniform density: $|\langle x|X\rangle|^2 = 1/l$.

Yet, this holds for not too long times. According to Eq. (3.1), *any* wave packet :

$$X(t) = \sum_n C_n \sin(n\pi x/l)\, e^{-iE_n t/\hbar}$$

must be completely restored at epochs: $(4ml^2/\pi\hbar)$ times an integer.

4. The Free Particle on a Line

The special status of the Harmonic Oscillator, embodied in such properties as the relation of its energy eigenvalues to those of the general number operator, $a^\dagger a$, and the equivalence of the oscillator packet constructed by Schrödinger [1] with a right eigenvector of the annihilation operator:

$$a = \left(\frac{m\omega}{2\hbar}\right)^{\frac{1}{2}} \cdot [q + i(p/m\omega)] \tag{4.1}$$

or 'coherent state' — is certainly unique among dynamical systems. Likewise, the concomitant stability of Schrödinger's wave packet is undoubtedly special, as explained by Heisenberg l.c. (ref.5, p.9 ff).

Yet, the possibility of ghostly, recurring wave packets exemplified in Eq. (2.6), is not limited to the Harmonic Oscillator. Here only the free particle on a line will be considered, since this model was so much used in the profound discussion between Einstein [8] and Born l.c. [7]: "*Continuity, Determinism and Reality*".

In order to avoid some phase factors which are not needed in this context, use is made of the relation: [9]

$$|\langle q'|\Psi(t)\rangle|^2 = \langle \Psi(0)|\delta[Q-q']|\Psi(t)\rangle \tag{4.2}$$

where Q is the position operator in the Heisenberg Picture.

For the Harmonic Oscillator we can write the Heisenberg position operator:

$$Q(t) = e^{it\mathcal{H}/\hbar} \cdot q \cdot e^{-it\mathcal{H}/\hbar}$$

in the form

$$Q = q \cdot f(t) + p \cdot g(t) \tag{4.3}$$

where $f = \cos\omega t$, $g = \frac{\sin\omega t}{m\omega}$ and q, p are the corresponding Schrödinger operators.

Writing in Fujiwara's Eq. (4.2)

$$\delta(Q-x) = \frac{1}{2\pi}\int_{-\infty}^{+\infty} ds\; e^{is(qf+pg-x)}$$
$$e^{is(qf+pg)} = e^{\frac{1}{2}s^2 fg\hbar}\cdot e^{isqf}\cdot e^{ispg} \tag{4.4}$$

and choosing the initial wave packet :

$$|\langle x|X(0)\rangle| = \left(\frac{\beta}{\pi}\right)^{\frac{1}{4}}\cdot e^{-\beta/2(x-x_0)^2+i(p_0x/\hbar)} \tag{4.5}$$

a little more general than in (2.2), one finds:

$$|\langle x|X(t)\rangle|^2 = \int \frac{ds}{2\pi}\, e^{\frac{\hbar}{2}s^2fg}\left(\frac{\beta}{\pi}\right)^{\frac{1}{4}}\int dx\langle X(0)|x\rangle\langle x+sg\hbar\,|X(0)\rangle e^{isxf}$$
$$= \left(\frac{\beta/\pi}{f^2+\beta^2g^2\hbar^2}\right)^{\frac{1}{2}} e^{-\beta(x-x_0f-p_0g)^2/(f^2+\beta^2g^2\hbar^2)} \tag{4.6}$$
$$= \left(\frac{\beta/\pi}{\cos^2\omega t+\alpha^2\sin^2\omega t}\right)^{\frac{1}{2}} e^{-\beta\frac{[x-x_0\cos\omega t-(p_0/m\omega)\sin\omega t]^2}{\cos^2\omega t+\alpha^2\sin^2\omega t}}$$

where $\alpha = (\hbar/m\omega)\beta$.

$$\langle x\rangle(t) = x_0\cos\omega t + (p_0/m\omega)\sin\omega t$$
$$\Delta x^2(t) = \frac{1}{2\beta}\left[\cos^2\omega t+\alpha^2\sin^2\omega t\right] \tag{4.7}$$
$$\langle p\rangle(t) = p_0\cos\omega t - m\omega x_0\sin\omega t$$

in agreement with, and generalization of Eq. (2.4).

Turning now to the free particle on the line: $x \subset \left[-\frac{l}{2}, +\frac{l}{2}\right]$ a classical orbit starting at the position x_0 with velocity v between reflecting walls at $x=\pm l$, can be depicted as below

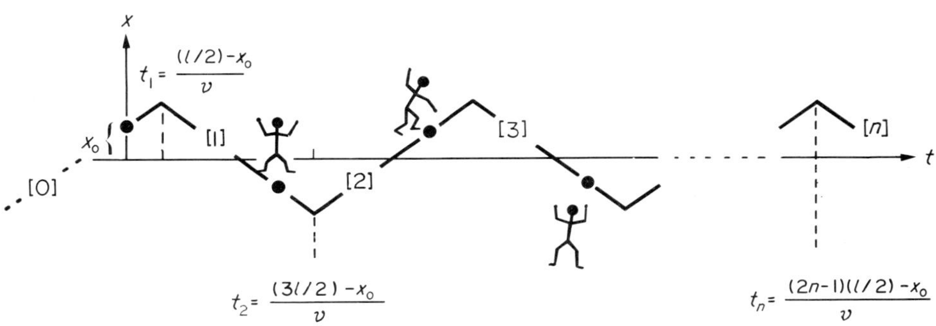

$$
\begin{array}{llll}
[0] & t \subset [0\ , t_1] & : & x(t) = x_0 + vt \\
[1] & t \subset [t_1\ , t_2] & : & x(t) = \frac{l}{2} - v(t-t_1) = l - x_0 - vt \\
[2] & t \subset [t_2\ , t_3] & : & x(t) = -\frac{l}{2} + v(t-t_2) = -2l + x_0 + vt \\
[3] & t \subset [t_3\ , t_4] & : & x(t) = +\frac{l}{2} - v(t-t_3) = 3l - x_0 - vt \\
\vdots & & & \vdots \\
[n] & t \subset [t_n\ , t_{n+1}] & : & x(t) = (-1)^n (x_0 + vt - nl)
\end{array}
$$

where n = the smallest integer $> v(t-t_1)/l$.

The saw tooth functions $[n]$ can be variously comprised by Fourier series, but there seems to be no special point in doing so, since we shall only use the classical picture to obviate scme tedious limit operations implied in the time translation:

$$x(t) = e^{t\{\mathcal{H},\}}\ x(0) \tag{4.8}$$

with a discontinuous Hamiltonian :

$$\mathcal{H} = (p^2/2m) + V(x)\ ; \qquad V = \begin{cases} 0\ , & |x| < l/2 \\ \infty, & |x| > l/2 \end{cases} .$$

Likewise, the Heisenberg coordinate operator:

$$Q(t) = e^{(it/\hbar)[\mathcal{H},\,]}\ q \tag{4.9}$$

can be written in the form of Eq. (4.3) with

$$f(t) = (-1)^n\ ,\ mg(t) = (-1)^n (t - n\ \frac{T}{2})\ ,\ T = \frac{2l}{v}$$

$$t \subset \left[t_1 + (n-1)\frac{T}{2},\quad t_1 + (n+1)\ \frac{T}{2}\right] \tag{4.10}$$

since the transformation (4.8) and (4.9) — with Poisson bracket and commutator as Lie-product respectively — are isomorphic.

To be sure, the wave function (4.5) does not quite satisfy the boundary conditions: $\langle x = \pm\frac{l}{2}|X\rangle = 0$ of the present mcdel, but inside the walls, i.e., $|x_0| < l/2$ and when β is sufficiently large — the error is negligible

Here we are only interested in the limit: $\beta\to\infty$ of

$$|\langle x|X(t)\rangle|^2 = \left(\frac{\beta/\pi}{1 + (\beta\hbar/m)^2 (t - m\ \frac{T}{2})^2}\right)^{\frac{1}{2}} e^{-\beta \frac{[x \mp x_0 \mp (p_0/m)(t - n\frac{T}{n})]^2}{1 + (\beta\hbar/m)^2\ (t - n\frac{T}{2})^2}}$$

$$\lim_{\beta\to\infty}\ |\langle x\,|X(t = \frac{n}{2}\ T)\rangle|^2 = \begin{cases} \delta(x - x_0)\ , & n = 2, 4, 6, \ldots \\ \delta(x + x_0)\ , & n = 1, 3, \ldots \end{cases} \tag{4.11}$$

Whence, starting with $|\langle x|X(0)\rangle|^2 = \delta$-function, the apparition shows up, once in every half period, at $x = \pm x_0$.

In this case however, the wave function can of course not disappear completely in the mean time. The proof that — apart from the recurrence — the coordinate probability tends to uniform density: $1/l$, requires a somewhat longer calculation which we shall here forego.

APPENDIX 'A.1'

THE COHERENT STATES

In his lectures to the Seminar of Theoretical Physics in Trondheim (1956), H. Koppe[10,11] gave a very complete account of these states, as the field eigenstates envisaged in Heisenberg's "Wellenbild" – complementary to the "Partikelbild" of latter, *loc cit.* ref. 5, p.26 ff. Their formal origin was traced back to F. London's representation[12] of the commutation relations

$$aa^{\dagger} - a^{\dagger}a = 1 \tag{A1.1}$$

on the basis of which Koppe expounded their Inner Product, Resolution of Unity, overcompleteness and cetera.

The denotation 'Coherent' was introduced by R.J. Glauber[13] who discovered the great importance of these states for Quantum Optics.

The most complete mathematical exposition is probably the one given by V. Bargmann[14-16] who also independently discovered this alternative representation of the non-canonical commutation relation (A1.1).

Schrödingers wave packet Eq. (2.1) is evidently an eigenstate corresponding to the eigenvalue: $a' = A/\sqrt{2}$, of the annihilation operator a, Eq. (4.1) and hence 'coherent'.

APPENDIX 'A.2'

TEMPORAL EVOLUTION

The Harmonic Oscillator and the particle in a constant field of force, are the only cases, apart from the free particle, in which the phase, S, of Feynman's Propagator:

$$K(x,t,x_0,0) = Ne^{iS/\hbar}$$

is everywhere equal to Hamilton's Principal Function or extremum of the Action functional:

$$S[x(t);\ (x_0,x,t)] = \int_0^t \mathcal{L}(\dot{x},x)\,dt\ ;\ \text{Extr.} = S(x_0,x,t)$$

realised by the classical orbit: $x(t)$, from x_0 to x.

This follows from the Schrödinger equation:

$$i\hbar\dot{K} = -\frac{\hbar^2}{2m}K'' + V(x)\cdot K$$

or, separating real and imaginary parts,

$$\dot{S} + \frac{1}{2m}\left(S'\right)^2 + V - \frac{\hbar^2}{2m}\frac{N''}{N} = 0$$

$$\frac{\dot{N}}{N} + \frac{1}{m}\frac{N'}{N}\cdot S' + \frac{1}{2m}S'' = 0\,.$$

Stipulating that the normalization factor, N, be a function of time only:

$$\frac{\dot{N}}{N} + \frac{1}{2m}S'' = 0\,,$$

the phase, S, must obey the Hamilton Jacobi Equation.

This is only possible in the three cases mentioned above, for which S is accordingly equal to the classical Action.

For the Harmonic Oscillator and the particle under constant acceleration: a, we have

$$\mathcal{L} = \begin{cases} \frac{m}{2}\left(\dot{x}^2 - \omega^2x^2\right) \\ \frac{m}{2}\left(\dot{x}^2 + 2ax\right) \end{cases}$$

$$S = \begin{cases} \frac{m\omega}{2\sin\omega t}\left[(x^2+x_0^2)\cos\omega t - 2xx_0\right] \\ \frac{m}{2}\left[\left(\frac{x-x_0}{t}\right)^2 + a\,(x+x_0)t - \frac{a^2}{12}\cdot t^3\right] \end{cases}$$

and

$$N = \begin{cases} \left(\frac{m\omega}{2\pi i\hbar\sin\omega t}\right)^{\frac{1}{2}} \\ \left(\frac{m}{2\pi i\hbar t}\right)^{\frac{1}{2}} \end{cases}$$

respectively.

Hence the functional integrals

$$\langle x | X(t) \rangle = N \int d[x(t)]\ e^{(i/\hbar)\, S[x(t)]}\ |X(0)\rangle$$

condense into simple quadratures, leading to the equations (2.3) and (4.11)

APPENDIX 'A.3'

THE FREE PARTICLE ON A LINE

In contradistinction to the Harmonic Oscillator: $x \subset [-\infty, +\infty]$ the wave packet must now vanish on the boundaries of a finite domain: $x \subset \left[-\frac{l}{2}, +\frac{l}{2}\right]$. This is perhaps most naturally satisfied by a wave packet of theta-function shape

$$|X(0)\rangle = C \sum_{1}^{\infty} \sin n\xi \cdot \sin n\eta \cdot e^{-n^2/2\gamma} \tag{A3.1}$$

where

$$\xi = \frac{\pi}{l}\left(x - \frac{l}{2}\right), \quad \eta = \frac{\pi}{l}\left(x_0 - \frac{l}{2}\right),$$

γ = positive real.

Then $\gamma \to \infty$ entails $\langle x|X\rangle \longrightarrow \delta(x - x_0)$. Since here, however, we rather want the absolute square: $|\langle x|X\rangle|^2$ to be a δ-function we have first to inspect

1° *The Normalization*

$$C^{-2} = \int_{-l/2}^{+l/2} dx\, |\langle x|X\rangle|^2 = \frac{l}{2} \sum_{1}^{\infty} \sin^2\eta \cdot e^{-n^2/\gamma}$$

$$= \frac{l}{2} \cdot \frac{1}{2} \sum_{-\infty}^{+\infty} \left(1 - e^{2in\eta}\right) \cdot e^{-n^2/\gamma}$$

$$= \frac{l}{4} \sum_{-\infty}^{+\infty} [f_1(n) - f_2(n)] \, .$$

By the Poisson summation formula:

$$\sum_{-\infty}^{+\infty} f(n) = \sum_{-\infty}^{+\infty} \int_{-\infty}^{+\infty} dx\, f(x)\, e^{2\pi i n x}$$

$$C^{-2} = \frac{l}{4} \sqrt{\pi\gamma} \sum_{-\infty,\, +\infty} \left\{ e^{-\gamma\pi^2 n^2} - e^{-\gamma \frac{\pi^2}{l^2}\left[x_0 + (n - \frac{1}{2})l\right]^2} \right\}$$

which for $\gamma \gg 1$ and $|x_0| < \frac{l}{2}$ (that is: Initial position well inside the classical turning points) leads to

$$C^2 = \frac{4}{l\sqrt{\pi\gamma}} \quad + \text{ negligible terms.} \tag{A3.2}$$

2° *The Initial State*

Writing Eq. (A3.1) in the form

$$\langle x|X(0)\rangle = C\cdot\tfrac{1}{2}\sum_{1}^{\infty}\left(\cos n\bar{\xi}-\cos n\overset{+}{\xi}\right)e^{-n^2/2\gamma}$$

$$= C\cdot\tfrac{1}{4}\sum_{-\infty}^{\infty}\left(e^{in\bar{\xi}}-e^{in\overset{+}{\xi}}\right)e^{-n^2/2\gamma}$$

where

$$\overset{\pm}{\xi} = \begin{cases}\dfrac{\pi}{l}\,(x-x_0+l)\\[2mm] \dfrac{\pi}{l}\,(x-x_0)\end{cases}$$

we have

$$|\langle x|X(0)\rangle|^2 = \frac{C^2}{16}\sum_{-\infty}^{+\infty}\left[e^{i(m+n)\bar{\xi}}+e^{i(m+n)\overset{+}{\xi}}+2e^{in\bar{\xi}+im\overset{+}{\xi}}\right]e^{-(m^2+n^2)/2\gamma}$$

$$= \frac{C^2}{16}\sum\sum e^{-(M^2+N^2)/4\gamma}\left[e^{iM\bar{\xi}}+e^{iM\overset{+}{\xi}}-2e^{iM\xi}\cdot e^{iN\eta}\right]$$

$$= \frac{C^2}{4}\pi\gamma\sum_{-\infty}^{+\infty}e^{-4\pi^2\gamma^2M^2}\sum_{-\infty}^{+\infty}\left\{e^{-\gamma\frac{\pi^2}{l^2}(x-x_0-2Nl)^2}+\right.$$

$$\left.+\,e^{-\gamma\frac{\pi^2}{l^2}(x-x_0+2Nl)^2}-2e^{-\gamma\frac{\pi^2}{l^2}\left[\left(x-\frac{l}{2}+2Nl\right)^2+\left(x_0+\frac{l}{2}+2Ml\right)^2\right]}\right\}.$$

For large γ and $|x_0|<l/2$ only the terms $M=0$, $N=0$ prevail. Insertion of the normalization factor (12) shows that

$$|\langle x|X(0)\rangle|^2 = \frac{\pi\gamma}{l^2}\cdot e^{-\frac{\pi^2}{l^2}\cdot\gamma(x-x_0)^2} \quad + \text{ negligible terms,}$$

or in the limit $\gamma\to\infty$: $\delta(x-x_0)$; q.e.d.

References

1. Schrödinger, E., Ann. d. Phys., **79**, 361, 489, 734 (1926).
2. Schrödinger, E., Naturwissensch., **14**, 622 (1926).
3. Schrödinger, E., *Nature and the Greeks*, p.30 ff, Cambridge University Press (1954).
4. Heisenberg, W., Z. f. Phys., **43**, 184 and **46**, 172 (1927).
5. Heisenberg, W., *Phys. Prinzipien d. Quantentheorie*, pp.9, 26, 28, Chicago Lectures (1929).
6. Pauli, W., *Wellenmechanik*, ETH-Lectures (1956).
7. Born, M., Dan. Fys. Medd., **30** (2), (1955).
8. Einstein, A., *Scientific Papers presented to Max Born*, p.33, Oliver & Boyd, London (1953).
9. Fujiwara, I., *et al.*, Progress of Theor. Phys., **64**, 715 (1980).
10. Koppe, H., Ark. Fys. Sem. Trondheim, No.3, (1959).
11. Koppe, H.,K. Norske, Vid. S. Forh., **35**, 25 (1962).
12. London, F., Z. f. Phys., **40**, 193 (1927).
13. Glauber, R.J., Phys. Rev., **131**, 2766 (1963).
14. Bargmann, V., Comm. P. & Appl. Math., **14**, 187 (1961).
15. Bargmann, V., Proc. Nat. Acad., **48**, 199 (1962).
16. Bargmann, V., Rep. Math. Phys., **2**, 211 (1971).

About the Authors

JAMES BINNEY

James Binney was born on April 12, 1950 in London. After graduating from Cambridge in 1971 he spent a year at the Albert Ludwigs Universitaet in Freiburg, West Germany before going on to write a doctoral thesis in Oxford on the formation of galaxies.

Most of his research has been concerned with the structure, dynamics and origins of galaxies. He has paid particular attention to what can be learned about these questions from the shapes of galaxies and rich clusters of galaxies.

He has been a Fellow of Magdalen College Oxford, a Lindemann Fellow and a Visiting Assistant Professor at Princeton University, and a Fairchild Distinguished Scholar at the California Institute of Technology. He is currently a Lecturer in Theoretical Physics at Oxford and a Tutorial Fellow of Merton College, Oxford.

BRIAN GEORGE SPENCER DOMAN

Brian Doman was born on 2nd August, 1936 in England. He went to Balliol College, Oxford, in 1955 with a Domus Open Scholarship, and graduated in 1958 with a 1st class Honours degree in Physics. From then until 1961, he was a research student working with D. ter Haar in Theoretical Solid State Physics. Then he became an assistant Lecturer in Mathematics at Westfield College, London. In 1964 he received his D. Phil. and took up a Lectureship in the Department of Applied Mathematics in Liverpool, where he has been ever since.

His work has covered a number of different areas of Theoretical Physics. In the theory of the Ising Model, he has recently investigated the random one dimensional system; earlier investigations have included other one, two and three dimensional systems and the use of Green functions and Statistical Mechanics to derive certain Thermodynamic Identities. He has examined a number of properties of electrons in magnetic fields in two and three dimensions, including localization. He has worked on problems related to the Kondo effect, and recently on magnetic field effects on charge density waves.

He is a member of the Institute of Physics and of the European Physical Society.

TZURU FUJIWARA

Tzuru Fujiwara was born in Osaka on 15th January, 1923. He was awarded his Dr.Phil. at Kyoto University in 1950. From 1950 to 1965 he held various teaching positions in Japan, Scandinavia and Spain, and in 1965 he was made a professor of the University of Osaka Prefecture, Sakai, a post which he still holds. In 1982 he was made a Foreign Member of the Royal Norwegian Academy of Science and Letters.

VITALY LAZAREVIC GINZBURG

Vitaly Ginzburg, born in Moscow on the 4th October, 1916, is currently Head of the I.S. Tamm Department of Theoretical Physics at the P.N. Lebedev Institute of Physics of the Academy of Sciences of the USSR.

Between 1940 and 1942 he carried out research for his Dr. Sci Doctorate at the same P.N. Lebedev Institute and later, from 1944 to 1971, he held the post of Vice Chairman of the Department there. From 1945 he has also worked in the Faculty of Radiophysics of Gorky University and in 1968 he took the Chair of Physics and Astrophysics at the Moscow Institute of Physics and Technology.

In the years of his doctorate (Ph. Dr. 1938-1940 in Moscow Univeristy) he analyzed problems in the Cerenkov effect, elaborating a quantum treatment of the phenomenon and studying its possible application in crystals and for the generation of micro-waves. During the same period he began a programme of research in the field of electromagnetic waves propagation and radio-astronomy.

In the 1950's, collaborating with V.V. Zeleznyikov, he established the bases on which he was to elaborate a theory of radio-emission in the sun. With L.D. Landau, in 1950, he constructed a phenomenological theory of super-conductivity and the Ginzburg-Landau equations were later to enable the development of the theory of superconducting alloys. In 1968, together with others, he received the Lenin Prize as a recognition of the scientific value of this complex of studies.

Another field that has been analyzed by Ginzburg, beginning in 1958, is that of the theory of excitons and of crystal optics. A book on this topic was published by Ginzburg and V.M. Agranovic in 1965 (second edition in 1979). Also worthy of mention is the monograph of 1963 (Pergamon Press, 1964) in which Ginzburg and S.I. Syrovatsky treated the origin of cosmic rays based on ideas first put forward by Ginzburg himself in several papers, and some other books (published by Pergamon Press) between 1951 and 1953.

Ginzburg has been a full member of the Academy of Sciences of the USSR and the International Academy of Sciences since 1966. He is also a member of the Royal Astronomical Society, of the American Academy of Arts and Sciences, and of some other Scientific bodies.

ALEX HEWSON

Alex Hewson was born on 21st July, 1938 in the United Kingdom. He received his BA degree in Physics at the University of Oxford in 1960 and his D.Phil. in Theoretical Solid State Physics, also at Oxford, in 1964. Since 1963 he has been Lecturer in the Department of Mathematics at Imperial College London.

His research interests have been in the application of many-body techniques in Solid State Physics. These have concerned magnetism in insulating and itinerant systems, magnetic impurities, and more recently, rotational tunnelling of molecular groups and intermediate valency in rare earth systems.

FREDERICK K. LAMB

Frederick Lamb was born 30th June, 1945, in Manhattan, Kansas. He received a B.S. in Physics, with Honors, from the California Institute of Technology in 1967. He attended the University of Oxford on a Marshall Scholarship and received a D.Phil. in Theoretical Physics in 1970. He won a Prize Fellowship at Magdalen College, Oxford, in 1970 and during the period 1970-72 spent time there and at the University of Illinois, where he was a Research Associate. In 1972 he joined the faculty of the University of Illinois, becoming Professor of Physics in 1978. In 1980 he was also made Professor of Astronomy.

His research interests include the quantum mechanics of radiation and matter and the astrophysics of magnetic white dwarfs, neutron stars, pulsars, X- and gamma-ray burst sources, and active galaxies and quasars. In 1974 he was awarded an Alfred P. Sloan Foundation Fellowship to pursue these interests. In recent years he has also become actively involved in questions of arms control and international security. He is currently a member of the Executive Committee of the High Energy Astrophysics Division of the American Astronomical Society.

ANTHONY J. LEGGETT

Tony Leggett was born in London on 26th March, 1938. At School and, initially, at Oxford University he studied classics and obtained a BA in Greats in 1959. Thereafter he switched to physics, taking the BA in 1961 and completing his D.Phil. under the supervision of Dr Dirk ter Haar in 1964. From 1967 to 1982 he worked at the University of Sussex; he is currently Macarthur Professor at the University of Illinois.

His main fields of interest are quantum liquids, especially the superfluid phase of ^{3}He discovered in 1972, superconductivity and the foundations of quantum theory. He is especially interested in the possibility of using the coherence properties and low dissipation of superconductors and superfluids to test the extrapolation of quantum mechanics to the macroscopic level.

CARLOS ALBERTO SILVA LIMA

Carlos Lima was born on 20th December, 1940 in Brazil. He received the M.Sc. degree in theoretical nuclear physics at the University of Brasilia, Brazil, in 1968. He earned the Ph.D. degree in experimental high energy nuclear physics from the University of California, Berkeley, USA, in 1972, working on antiproton physics in the E.Segre-O. Chamberlain Bubble Chamber Group, as a participating guest at the Lawrence Berkeley Laboratory. Moving back to Brazil he joined the staff of the Institute of Science of the University of Brasilia where he later was appointed Head of the Physics Dept. in 1974. He continued some work on particle physics phenomenology and then became increasingly more evolved with the physics of intense field effects upon matter. He joined the staff of the Institute of Physics of the University of Campinas, Brazil in 1977, where he is currently co-leader of the Laser Development and Applications Group at the Department of Quantum Electronics. He works on both experimental and theoretical physics of atomic, molecular and condensed matter systems under strong laser fields and in photoacoustics. He is presently one of the Directors of the Brazilian Physical Society.

MALCOLM E. LINES

Malcolm Lines was born on 26th April, 1936 in Banbury, England. He was educated at Magdalen College School, Brackley and was awarded an Open Exhibition to Magdalen College, Oxford, in 1954. After serving two years in the RAF he went up to Oxford in 1956, receiving a First Class Honours Degree in Natural Science in 1959, and a D.Phil. (in theoretical physics) in 1962. In 1961 he was elected to a Fellowship by Examination at Magdalen, a position he held until 1966 when he emigrated from the U.K. to take up an appointment at Bell Laboratories, New Jersey, USA where he is still involved in active research.

His early work was in crystal field theory and magnetic cooperative phenomena, particularly involving antiferromagnetism and the influences of low magnetic dimensionality. From 1969 he pursued lattice dynamical calculations in formulating soft mode theories of ferroelectricity and pyroelectricity, work which culminated in the publication of a book *Principles and*

Applications of Ferroelectrics and Related Materials, published by Oxford University Press in 1977. More recently his interests have shifted towards research on the structure of glasses, including work on computer simulation, amorphous magnetism and ferro-electricity, and fibre optics.

LUIZ C.M. MIRANDA

Luiz Miranda was born on 10th July, 1943 in Rio de Janeiro, Brazil. He received his D.Phil. in astrophysics at Oxford University in 1972, under Dr D. ter Haar's supervision, and was a John Simon Guggenheim Fellow in 1975. He is currently Professor of Physics, on leave from the University of Brasilia, and Technical Director of the Institute of Advanced Studies of the Aerospace Technical Centre at São José dos Campos, Brazil.

Professor Miranda has carried out research in a variety of areas which includes plasma astrophysics, atomic and molecular physics, and solid state physics. Since 1975 his primary activity has been in the fields of laser-matter interaction and photoacoustic spectroscopy. He has served as member of the editorial boards of some international and national scientific journals and is currently the leader of the laser-plasma programme of the Aerospace Technical Centre.

WILLIAM E. PARRY

Bill Parry was born in England in 1936. He was one of ter Haar's first Oxford undergraduate pupils, and later worked under his supervision for a doctorate. He was made a Fellow of Oriel College, Oxford, and a University Lecturer in 1961. His research interests are mainly in many-body physics.

CHRISTOPHER PETHICK

Christopher Pethick was educated at Magdalen College, Oxford and he received a BA in physics in 1962. His graduate work, done under the supervision of Dirk ter Haar, was on problems in liquid ^{4}He. In 1965 he was elected to a Fellowship by Examination at Magdalen College, Oxford, and during the years 1966-1969 he was on leave at the University of Illinois. After a year spent in Oxford and Copenhagen he rejoined the Illinois faculty, and since 1975 he has also held a position at Nordita, Copenhagen. His scientific interests are in theoretical condensed matter physics and astrophysics.

ANDREW J.R. PRENTICE

Andrew Prentice was born in Melbourne in 1943, and educated at Melbourne Grammar School and the University of Melbourne. He graduated for a B.Sc. in 1965 and an M.Sc. in theoretical plasma physics in 1967. He travelled to Oxford University in 1967 to take up a Royal Commission for the Exhibition of an 1851 research scholarship, and commenced D,Phil. studies in theoretical astrophysics under the supervision of Dr D. ter Haar. He was appointed to the first Junior Research Fellowship of Wolfson College, Oxford, 1967 – 1970. Following the completion of his D.Phil. in 1970, he spent eighteen months as a Research Scientist in theoretical physics at Carnegie-Mellon University; Pittsburgh, prior to his Lectureship appointed in applied mathematics at Monash University, Victoria in 1972. Currently he is a Senior Lecturer there.

His research interests have centred on the development of a modern Laplacian theory for the origin of the solar system. He has also made contributions to understanding the origin and composition of the satellite systems of Jupiter and Saturn, as well as to the structure and evolution of the sun, in connection with the solar neutrino problem. He was appointed to a NASA /National Research Council Senior Research Associate, at the Jet Propulsion Laboratory, Pasadena, at the time of the Voyager 1 and 2 spacecraft encounters with Saturn in 1980 and 1981. He was a council member of the Astronomical Society of Australia, 1979 – 1982.

SOURIRAJA RAMADURAI

Souriraja Ramadurai is with the Theoretical Astrophysics Group of Tata Institute of Fundamental Research, Bombay, India. He was born on 26th May, 1942 in Tamilnadu, India. After leaving National College, Tiruchirapalli India, he joined the Atomic Energy Training School, Bombay, as a Physics Trainee in 1962. After the successful completion of the training, he joined the Cosmic Ray Group of Tata Institute of Fundamental Research, Bombay, India. He received his Ph.D. Degree from Bombay University in 1971 doing research on the Cosmic Rays in Interplanetary and Interstellar Space and in Source Regions. He joined the Theoretical Astrophysics Group in 1973. He spent a year as a Leverhulme Fellow of the Royal Astronomical Society at the Institute of Astronomy, University of Cambridge, U.K. during 1975 – 1976. He joined Dirk ter Haar in the Department of Theoretical Physics, Oxford University for a year during 1976 – 1977 after spending a summer in the Department of Applied Mathematics, University College, Cardiff. He was a Visiting Associate Professor in the Department of Astronomy, University of Pennsylvania, Philadelphia, USA during 1981.

Although he started his research career as an experimental cosmic ray physicist, he is currently interested in the theoretical investigations. He has contributed to models of propagation and acceleration of cosmic rays. He is interested and has contributed to the stellar evolution and nucleosynthesis. The neutral current modifications to the neutrino energy loss rates as well as its effects on stellar evolution has attracted his attention. He is investigating the isotopic anomalies in the solar system as well as in some stars. He is currently engaged in the study of the Chemical Evolution of Galaxies.

STIG T. STENHOLM

Stig Stenholm was born on the 26th February, 1939. He graduated from Helsinki University of Technology and Helsinki University, Helsinki, Finland, in 1964, and received the D.Phil. degree from Oxford University, England, in 1967.

He worked at Yale University, New Haven, CT, from 1967 to 1968 and subsequently at the Helsinki University of Technology, the University of Helsinki, and the Academy of Finland. In 1974 he was with IBM, San Jose, CA. From 1974 to 1980 he was an Associate Professor of Physics at the University of Helsinki. Presently, he is the Scientific Director at the Research Institute for Theoretical Physics as the University of Helsinki. He has spent a sabbatical leave at the Optical Sciences Center, University of Arizona, AZ during 1982 – 1983.

His main research interest is in theoretical laser physics, nonlinear spectroscopy and the theory of interaction of matter and radiation fields. In addition he is interested in the theory of condensed matter, disordered systems and basic questions in statistical physics. At the Research Institute he is a Group Leader responsible for work in Solid State Physics including all low energy activities.

RAZA ALI TAHIR-KHELI

Raza Tahir-Kheli was born on 1st May, 1936 in The North West Frontier Province of British India (now Pakistan). He read Physics at Oriel College, Oxford (1955 – 1958). After completing his D.Phil. in Spring 1962, he worked for the Pakistan Atomic Energy Commission and The University of Pennsylvania. He joined Temple University in 1966 and has been a Professor of Physics since 1971.

His research interests have been in the field of magnetism and alloys.

N.G. van KAMPEN

N.G. van Kampen was born in 1921 and studied physics, with some obvious interruptions, at Leiden and Groningen. He received his Ph.D. in 1952 with H.A. Kramers and stayed some time at the Niels Bohr Institute in Copenhagen, the Institute for Advanced Study at Princeton, N.J. and the Columbia University in New York. From 1958 he is Professor of Theoretical Physics at the University of Utrecht. He has worked in optics, scattering theory, plasma physics, statistical mechanics and stochastic processes. He is a member of the Royal Academy at Amsterdam and is concerned about the present demise of scientific research at the Dutch universities.

HARALD NICOLAI WERGELAND

Harald Wergeland was born on the 14th March, 1912. He graduated with a Diploma of Physical Chemistry from the Norwegian Institute of Technology, Trondheim in 1936, and received the degree of Dr.Phil. in Theoretical Physics from the University of Oslo in 1946. From that time he has been Professor at the University of Trondheim.

Index